T0173885

Pharmaceutical Isothermal Calorimetry

Pharmaceutical Isothermal Calorimetry

Simon Gaisford
The School of Pharmacy
University of London, U.K.

Michael A. A. O'Neill
Department of Pharmacy and Pharmacology
University of Bath, U.K.

CRC Press
Taylor & Francis Group
Boca Raton London New York

CRC Press is an imprint of the
Taylor & Francis Group, an **informa** business

CRC Press
Taylor & Francis Group
6000 Broken Sound Parkway NW, Suite 300
Boca Raton, FL 33487-2742

First issued in paperback 2019

ISBN-13: 978-0-8493-3155-8 (hbk)
ISBN-13: 978-0-367-39020-4 (pbk)

Library of Congress Cataloging-in-Publication Data

Pharmaceutical stress testing : predicting drug degradation / Steven W. Baertschi, editor.
    p. ; cm.
  Includes bibliographical references and index.
  ISBN-13: 978-0-8247-4021-4 (alk. paper)
  ISBN-10: 0-8247-4021-1 (alk. paper)
  1. Drug stability--Testing. I. Baertschi, Steven W.
  [DNLM: 1. Drug Stability. 2. Pharmaceutical Preparations--analysis. 3. Technology, Pharmaceutical--methods. QV 754 P5359 2005]

  RS424.P42 2005
  615'.19--dc22
                                                2005047546

Visit the Taylor & Francis Web site at
http://www.taylorandfrancis.com

and the CRC Press Web site at
http://www.crcpress.com

Dedication

This text is dedicated to the memory of two men who played important roles in instilling the virtues of calorimetry into the authors.

First to Richard Lipscombe, who worked tirelessly in Professor Beezer's laboratory ensuring the calorimeters (and students) were working at maximum efficiency; Richard's legacy, as any student who worked with him will testify, was to inculcate the values of experiment design, instrumental set-up and sample handling into all those around him. The advice he gave us is as relevant today as it was then and underpins Chapter 2.

Secondly, to Tom Hofelich who spent many happy years working for Dow Chemicals. Although the authors only met Tom on a few occasions, mostly at international conferences, his infectious humour and good nature shone through. One memory that stands out is his response to the question that was asked of him, "Why did you present your data in Calories when the SI unit for heat is Joules?" when presenting at a conference in London. Tom replied, "When they rename the Calorimetry Conference the Joulerimetry Conference I will present my data in Joules!" A priceless comment from a priceless man who will be sadly missed, not just by the authors but by all those in the field of calorimetry.

S. Gaisford and MAA O'Neill
May 2006

Preface

Almost all reactions take place with a change in heat content or enthalpy and the quantitative study of such heat changes in chemical reactions has been pursued for many years. For years thermochemical studies were concentrated in University research departments. Now, however, these studies have increasingly been seen by industry, particularly the pharmaceutical industry, as having importance in the discovery, development and characterization of their products. This stems from the universality of the application of the micro-calorimetric technique: it is nondestructive, noninvasive and invariant to the physical form of the sample. Until recently the utility of this method was limited by a need to have relatively rapid (minutes rather than hours or days) reactions for study—industry, for example, is concerned with recording product stability data that are valid for years. New data analysis techniques allow for the determination of both thermodynamic and kinetic parameters that has opened these new areas for study.

Commercial calorimeters have been available for some years and by now most major pharmaceutical companies have purchased instruments. There has not been a concomitant increase in the number of trained calorimetrists. An earlier book (Thermometric Titrimetry by Tyrrell and Beezer) noted that "Widespread acceptance of any instrumental technique of analysis is always delayed until a rapid, reliable instrument is commercially available..." Thus it is not the instruments that present the problem but rather the absence of a hand-book on the principles of the method, the data analysis methods that have been developed, and a guide and review of the procedures that have been proposed to exploit such instruments. This book, therefore, fills a significant void—it offers the possibility of learning of the practice of calorimetry (both instrument and experimental design and data analysis methods) through a carefully developed text. Indeed, given the increasing industrial use of calorimetry it is important that the topic is treated in a text designed carefully to guide both experienced and potential (undergraduates and postgraduates) practitioners through modern applications. This is particularly aimed at preformulation and formulation issues and issues encountered during the design and development of novel drugs and delivery systems.

The authors are experts in the subject matter and deal with its complexities with a thoughtful and considered development. Their experience and sensitivity is shown in the style and manner with which they discuss practical problems; choice of instrument, experimental design and the appropriate data analysis method. It is apparent from the material presented that their own experiences of learning about the practice of calorimetry has informed their text and this ensures that any good scientist, whether experienced or not, will be able to use this book as a source for advice and guidance.

A.E. Beezer
April 2006

Contents

1

Principles of Calorimetry

INTRODUCTION

It is an inevitable consequence of the laws of nature that if a material can change from its current state to a more stable state, then it will do so. This may involve an individual material changing its physical or chemical form, or it may be the result of an interaction between two or more materials. Diamond, for example, is the metastable form of carbon and over time it will convert to graphite. The only consideration to be taken into account in determining if such change is of importance is the length of time over which it will occur. In the case of the transition of diamond to graphite the process occurs over millions of years and it may hence be ignored. For other materials, the rate of change may be more of an issue and chemical or compositional alterations may be needed to ensure stability over an acceptable period.

For manufactured materials, it is generally the case that the initial state is the one with the desired properties, and over time changes in the material will lead to deterioration in its properties and hence, by definition, a loss of quality. The measurement and, consequently, quantification of change is therefore of critical importance if use-by dates are to be specified. Using the product before the stated use-by date ensures that its performance will be within the manufacturer's limits. Of course, use-by dates are more important for some products than others; knowing the lifetime of a pacemaker battery is likely to be more important than knowing that of a radio battery, for instance. Similarly, an emulsion paint that has phase separated can be reconstituted by mixing, but it is unlikely that a patient would do the same with a pharmaceutical cream.

As intimated earlier, knowledge of change (or stability) is especially important for pharmaceuticals. Two principal reasons for this are that (*i*) it is often the case that the bioavailability of a drug depends to a large degree on

1

its physical form (whether it is formulated as a particular polymorph or in an amorphous form, for instance), and (*ii*) a drug may degrade to an inactive or, worse, toxic compound. The manufacturing specifications for pharmaceuticals are thus tight, but most products require a use-by date that is sufficient to cover the time taken for manufacture, distribution, and storage. Typically, a pharmaceutical may be expected to degrade by less than 2% over a two-year period.

The analytical challenge, therefore, is to be able to measure change in a given product (and from an industrial view, to measure change as rapidly and cheaply as possible, using a minimum of sample and operator time), using a *stability assay*. In the specific case of pharmaceuticals, it is desirable to be able to measure small changes (both chemical and physical) over a short experimental time in order to predict long-term stability. There are many analytical tools available to meet this challenge (such as spectroscopy and chromatography, for instance), but most of these require that the sample being studied possess a particular feature to enable successful analysis. In a spectroscopic analysis, the study molecule must possess a suitable chromophore and, in a chromatographic analysis, the study molecule (or the other components) must interact to some extent with the column material. In the best case, this results in a specific assay being developed for any particular sample and, in the worst case, renders some samples unsuitable for analysis. Nevertheless, it is the case that the vast majority of pharmaceutical stability assays are conducted using high-performance liquid chromatography (HPLC).

A further problem with conventional stability assays is that most products are formulated to be stable and hence will not degrade to any considerable extent during the measurement period. In other words, if an assay cannot detect any change in a product, a decision must be made as to whether that is because the product has not actually changed or because any change was below the detection limit of the measuring technique. To ameliorate this problem, stability assays are often conducted under stress conditions. Usually this would involve an increase in both temperature and/or relative humidity (RH). The resulting increase in degradation rate allows a rate constant (k) to be determined at each temperature, although study periods may still extend for weeks or months. The data are then plotted in accordance with the Arrhenius relationship (ln k vs. $1/T$) and the predicted rate constant under storage conditions is obtained.

Necessarily, it is assumed that the analysis results in a linear relationship and that any reaction processes occurring under stress conditions are the same as those that would occur under storage conditions. There are many reasons why this may not be so: different reaction pathways may predominate at higher temperatures, the sample may change its crystal state or go through a glass transition, or water may be involved in degradation. If there is any doubt about the extrapolation, then long-term storage studies (holding the sample directly under expected storage conditions) are conducted to confirm the findings.

HPLC analysis of an active pharmaceutical ingredient (API) cannot, of course, detect if a solid drug has changed polymorph, because dissolution of the sample before analysis removes any solid-state history, nor will it be of use if it is a change in the properties of an excipient, or an interaction between excipients, that causes the product to fail to meet its specification.

Calorimetry, the measurement of heat, offers an alternative approach for the measurement of change. It is an extremely useful technique, because when change occurs, it invariably occurs with a change in heat content. In other words, *heat is a universal accompaniment to chemical and physical change.* The result is that calorimetry can detect, and potentially quantify, the changes in a wide range of materials. The only properties required of the sample are that the process being followed results in a detectable quantity of heat and that the sample (or at least a representative part of it) fits within the calorimetric vessel. Isothermal calorimetry (IC), the measurement of heat as a function of time, in particular, is suited to the measurement of pharmaceutical samples, because it is sensitive enough to allow the analysis of samples non-destructively (i.e., it does not cause any extra degradation other than that which would have occurred upon storage), directly under storage conditions. This means that there is no need for elevated temperature stability studies (and hence no requirement for extrapolation of data) and samples can be recovered and used in other studies (particularly important in the early stages of the drug discovery process, when samples may exist only in milligram quantities). Calorimetry is also invariant to the physical form of a sample, meaning that complex heterogeneous systems are open to investigation. Furthermore, because both physical and chemical processes occur with a change in heat content, the technique is not limited in its detection ability to chemical degradation, in the way HPLC is. This unique combination of qualities makes IC ideally suited to the characterization of pharmaceutical materials.

Careful experimental design allows the investigation of virtually any system, and recent advances in data analysis and interpretation methodologies have resulted in the increasing application of the technique to stability testing of pharmaceuticals. Indeed, careful data analysis can result in a description of the reaction process in terms of both thermodynamics and kinetics, the only technique for which such a complete analysis is possible. It is the purpose of this book to explain, in the context of pharmaceutical development and formulation, the basic principles of IC and good experimental practice, to review the latest developments in experimental techniques and data interpretation and to explore some of the future challenges and applications of the technique. The first half of the book (Chaps. 1–3) deals with the fundamentals of calorimeter operation, good experimental design and data analysis, while the remainder of the text focuses on applications to specific areas; these chapters progress from issues in preformulation, through formulation, to final product characterization and quality assurance, and have been organized to match the development process of a pharmaceutical.

BRIEF HISTORY OF CALORIMETRY

Calorimetry [from calor (Latin), heat; metry (Greek), measurement] can trace its origins back to Black's ice calorimeter, although the first practical calorimeter design is usually regarded to be that of Lavoisier and Laplace in 1780 (1). In these simple designs, a heat-radiating body is surrounded by ice. The heat released causes some of the surrounding ice to melt, whereupon the water is routed out of the calorimeter into a collecting vessel. By knowing the quantity of the water produced and the latent enthalpy of fusion of water, the total heat output for the process, Q, can be determined (kJ kg^{-1}). This is a thermodynamic term, although it is not a very useful measure of the energy of a reaction. This is because the change in enthalpy determined cannot be compared directly with the change in enthalpy for a different reaction, unless the molecular weights of the reagents in both reactions are the same. It is more useful to determine the change in enthalpy in units of kJ mol^{-1}, although this requires that the number of moles of material available for reaction is known.

In one of their first experiments, Lavoisier and Laplace (1) measured the heat of combustion of carbon, finding that "one ounce of carbon in burning melts six pounds and two ounces of ice." This allowed a value for the heat of combustion of -413.6 kJ mol^{-1} to be calculated. When compared with the most accurate current value of -393.5 kJ mol^{-1}, it can be seen that the ice calorimeter was remarkably accurate. In a later experiment, a guinea pig was placed in the sample cell. By comparing the heat evolved from the guinea pig with the amount of oxygen consumed, they concluded (1)

> ... respiration is thus a combustion, to be sure very slow, but otherwise perfectly similar to that of carbon; it takes place in the interior of the lungs, without emitting visible light, because the matter of the fire on becoming free is soon adsorbed by the humidity of these organs. The heat developed in the combustion is transferred to the blood which traverses the lungs, and from this is spread throughout all the animal system.

This conclusion is remarkably astute, given the limited number of data on which it is based, and shows the power of the calorimetric technique. It is also somewhat ironic that the heat-flow recorded from the guinea pig experiment reflects the most complex case that could be encountered, arising from the heat-flows of many thousands of simultaneous metabolic reactions. More than 200 years later, it is the lack of a general method capable of analyzing data derived from such complex reactions that is limiting the widespread use of IC.

It is also possible to gain kinetic information about the sample if the number of water drops falling per unit time is measured. This is because the rate of droplet production is quantitatively proportional to the power output of the sample. Thus, even from this simple experiment, it should be clear that calorimetric measurements allow both thermodynamic and kinetic analysis of a sample.

Modern calorimeters do not differ greatly in principle from these early designs; they simply measure power directly with greater accuracy and are capable of measuring endothermic as well as exothermic events (an ice calorimeter can only measure exothermic events). Measurement of the heat output from a reaction (q) gives thermodynamic information, whereas measurement of heat output as a function of time (power, dq/dt) conveys kinetic information.

Until about the late 1950s, most researchers had no option but to use homemade instruments; even so, careful design and construction resulted in instruments that were capable of recording data very accurately. In a review of the development of calorimeters and calorimetrists over the latter half of the 20th century, Wadsö (2) notes the following opening paragraph from Chapter 9 of the second volume of IUPAC's (International Union of Pure and Applied Chemistry) monograph, *Experimental Thermochemistry* (3):

> The design and construction of a suitable calorimeter is one of the first problems facing the experimental thermochemist planning to measure directly the heat of a chemical reaction. During the past 30 years [i.e. between ~1930–1960] over 300 papers on reaction calorimetry have been published, and more than 200 different reaction calorimeters have been described. This evident need for variety in calorimeter design reflects the very variegated nature of the chemical reactions that have been thermochemically studied.

Thus, the expectation of any user of calorimetry during that period was that the first task to be accomplished when undertaking research was the construction of a bespoke calorimeter. However, advances in data acquisition hardware and electronics led to the first commercially-available instruments becoming available in the 1960s and today there are a number of manufacturers that produce highly sensitive and well-developed calorimeters that allow the investigation of a wide range of samples. (The properties of a range of these instruments are to be found later in this chapter, and a list of manufacturers is provided at the end.) Indeed, the availability and ease of use of these instruments have led to the widespread use of calorimetry for the routine characterization of the physical and chemical properties of many industrial materials.

Wadsö (2), in the same review, also notes that the introduction of commercial microcalorimeters during the 1960s marked a paradigm shift in the use and applications of the technique, with an increasing number of (predominantly) biologists using the instruments as "process monitors" (in a similar manner to that of Lavoisier and Laplace); in other words, the focus of calorimetry moved away from classically designed quantitative thermodynamic measurements on pure materials toward qualitative measurements of complex systems. It is still the case that many isothermal calorimeters are used as process monitors, but new methods are being developed that allow quantitative data analysis rather than qualitative analysis. A discussion of the application of these methods is a central theme of this text.

All calorimeters measure the heat changes that occur in a sample over time and can be classified into two main types, depending upon the temperature control they maintain over the course of an experiment. In differential scanning calorimetry (DSC), the sample is subjected to a preprogrammed temperature change over time (this can be linear or modulated by some mathematical function; see section "Differential Scanning Calorimetry"), whereas in IC the sample is held at a constant temperature. In practice, this means that DSC is limited to studying thermally driven events, usually phase transitions such as melting or recrystallization, whereas IC is used to monitor longer-term events, such as chemical degradation, aging, recrystallization, or the formation of hydrates/solvates. Wadsö (4) has reviewed some of the applications of IC at near ambient temperatures.

Even though it might be expected from the above discussion that IC would be more suited to the study of pharmaceuticals, it is the case that DSC is much more commonly used in the pharmaceutical arena; indeed, the point has been reached where DSC is regarded as a standard analytical tool, in the way HPLC is, and is treated as a "black-box," simply requiring an operator (not necessarily a calorimetrist) who knows how to load samples. IC, conversely, is perceived to be a research tool that requires careful sample preparation, which produces difficult-to-interpret data and that requires the services of an experienced operator. In part, this is correct; proper use of IC does require a knowledgeable and careful operator, but this is also true for DSC. Additionally, recent developments in data analysis routines are making IC much more accessible to nonspecialists; it should also be remembered that the operating principles of IC are much simpler than those of DSC.

The focus of this book is on the potential applications and benefits of using IC in pharmaceutical development, and the reader is referred to other texts for detailed discussions of the application and use of DSC (a list of information sources is appended at the end of this chapter). However, this chapter discusses the principles of both IC and (briefly) DSC, so the reader can draw a comparison, and recognize the important differences between the techniques, and visualize the types of applications for each. It is appropriate first to discuss the thermodynamic basis for both instruments, and explain the benefits of using heat as a marker for change, so that these sections can be fully appreciated.

HEAT AS AN INDICATOR OF CHANGE

All analytical instruments measure the change in some property of the sample under investigation. As stated earlier, the property change being measured in calorimetry is heat (q, given the SI unit of Joules, J). (Note that the use of calories, Cal, is still prevalent but that all heat data in this text conform to SI nomenclature; to convert from J to Cal multiply by 4.184.) As, necessarily, calorimetric measurements must be made as a function of time, calorimeters record an output of power (dq/dt, given the SI unit of Watts, W; note that 1 W is equivalent to 1 Js^{-1}). Thus, the raw output from a calorimetric experiment is a plot of power versus time (often called a thermogram). Integration of the data results in the total quantity of heat released (Q).

The primary reason that calorimetry has so many potential applications is that, as noted earlier, *heat is a universal indicator of chemical or physical change.* This means that potentially all processes are open to calorimetric investigation, because there are very few events that occur without a change in heat content.[a] This is a unique advantage of calorimetry compared with other analytical techniques. There is no need for a sample to possess a specific functional group, as there is in a spectroscopic or chromatographic analysis, for example, and nor is there a need to design a specific assay for a particular compound. A sample can simply be placed in a calorimetric ampoule and be monitored (indeed, the only requirement of sample is that it, or at least a representative fraction of it, will fit into an ampoule).

Moreover, in IC, the sample is studied nondestructively (i.e., the calorimeter does not cause any additional degradation to that which would have occurred upon storage during the same time period). This is a massive advantage for the study of pharmaceutical samples, where only a few milligrams of material may be available or the material may be very expensive. In a DSC experiment, the sample is often destroyed or is returned in a different physical state (a different polymorph for instance, reflecting the different cooling rates exerted upon the sample in the calorimeter compared with during its manufacture).

A further benefit of using changes in heat to monitor samples is that both chemical and physical processes can be observed; this is in contrast to other analytical tools, such as HPLC or UV spectroscopy, where solid samples need to be dissolved prior to analysis. As well as potentially increasing the rate of degradation prior to quantification through hydrolysis, dissolving samples clearly removes the solid-state history of the sample, and information on the number of polymorphs or percent amorphous content will be lost.

However, the universal nature of heat (as eloquently phrased, heat does not come in different colors) is also the technique's biggest drawback. It is often the case that calorimetric data are complex in form and derive from several simultaneous (including both physical and chemical) processes. Furthermore, calorimetric data are extremely (and perhaps uniquely) susceptible to systematic errors introduced by the accidental measurement of one or more of a range of

---

[a]Some reactions may occur at constant enthalpy. For example, in the Joule–Thomson experiment, a gas is allowed to flow from a region of high pressure to a region of low pressure through a porous plug. The expansion of the gas can be shown to occur at constant enthalpy (5). The Joule–Thomson coefficient is defined as the change of temperature with pressure at constant enthalpy. As a gas undergoes a pressure change through the plug, it changes temperature. Most gases, at room temperature, are cooled by the Joule–Thomson effect.

processes (such as solvent evaporation and/or condensation, erosion, side reactions, and so on) that may occur concurrently with the study process(es). Care must therefore be taken to ensure that erroneous or unexpected powers have not been accidentally introduced as a corollary of poor experimental design or execution. This is the main reason why calorimetric data need to be analyzed assiduously, often in combination with other complementary data.

Some of these issues are discussed in Chapter 3, which describes methods by which calorimetric data may be analyzed and interpreted; however, as heat is a thermodynamic term, a brief description of thermodynamics is given below to aid the reader through the later chapters.

INTRODUCTION TO THERMODYNAMICS

Thermodynamics is a general term, which means the study and quantification of transformations of energy. The part of the universe being studied is referred to as the *system*, around which are its *surroundings*, where experimental measurements are made. The system and the surroundings are separated by a *boundary*; it is the description of the boundary that characterizes the system and the surroundings. If matter and/or energy can be transferred through the boundary, the system is described as *open*. If matter cannot be transferred through the boundary but energy can, the system is *closed*. If neither matter nor energy can be transferred through the boundary, the system is *isolated*.

Work, energy, and heat are the basic concepts of thermodynamics. Energy is defined as the capacity of a system to do work. (Usually, work has been done *by* a system if a weight has been raised in the surroundings, and work has been done *on* the system if a weight has been lowered in the surroundings.) When the energy of a system changes as a result of a temperature difference between it and its surroundings, the energy change occurs by a transfer of heat. If a system loses heat, it is said to have undergone an *exothermic* process (and, hence, the change in heat is given a negative sign), and if the system gains heat it is said to have undergone an *endothermic* process.

The total energy content of a system is known as its internal energy (U). It is impossible to quantify in absolute terms the internal energy of a system, so thermodynamics deals only with changes in internal energy (ΔU). Hence, for a given process:

$$\Delta U = U_f - U_i \tag{1}$$

where U_f and U_i are the internal energies of the system in its final and initial states, respectively.

At this point it is important to note that internal energy is both a state property and an extensive property. A state property is one that depends only on the current state of a system, not on how the system arrived at that state (other state properties including, for instance, temperature, pressure, and density). Thus, a 1-L sample of water at 80°C and atmospheric pressure has the same internal energy regardless of whether it was heated to 80°C from room temperature or cooled from boiling. An extensive property is one that is dependent upon the size (extent) of the system (and includes properties such as mass and volume). Considering water (at 80°C and atmospheric pressure) again, a 2-L sample will contain twice the internal energy as a 1-L sample. Temperature, by contrast, is an *intensive* property (i.e., it is not dependent on the extent of the system), so while the internal energies of the 1-L and 2-L water samples differ, they are at the same temperature. Other intensive properties include density, pressure, and all normalized quantities (such as molar terms). Energy, heat, and work are all measured in Joules (J; $1\,J = 1\,kg\,m^2\,s^{-2}$).

There are two further properties of internal energy that are of importance; the first is that heat and work are equivalent ways of changing the internal energy of a system. The second is that the internal energy of an isolated system is constant. Together, these properties form the basis for the first law of thermodynamics.

First Law of Thermodynamics

One way of stating the first law is "the internal energy of a system is constant unless it is altered by doing work or by heating." Mathematically, this is formulated as:

$$\Delta U = q + w \tag{2}$$

where q represents the transfer of heat to or from, and w represents the work done on or by the system. Conventionally, both heat loss and work done by the system are given negative signs, because the first law is usually written from the perspective of the system under study, not the surroundings.

It is possible to conduct a more powerful analysis of thermodynamic changes if infinitesimally small changes of state are considered (represented by prefix d). Hence:

$$dU = dq + dw \tag{3}$$

In practice, work can take forms other than lifting a weight (such as the electrical work of driving a current through a circuit, for example) so can be defined as work of expansion (w_{exp}) and additional work (w_a). As such, Equation (3) can be reformulated as:

$$dU = dq + dw_{exp} + dw_a \tag{4}$$

If a system is kept at a constant volume, then no work can be done on the surroundings ($w_{exp} = 0$). If it is assumed that no additional work can be performed ($w_a = 0$), then the heat added or removed from a system is equal to the change in its internal energy:

$$dU = dq_V \tag{5}$$

where q_V represents the change in heat for a system at constant volume.

If a system is free to change volume against a constant external pressure, then the change in internal energy is not equal to the heat supplied or removed, because the system must convert some of the heat to work against its surroundings. In this case, $dU < dq$. However, it can be shown that (6), at constant pressure, the heat supplied or removed is equal to the change in enthalpy (ΔH) of the system (enthalpy, like internal energy, is a state property and an extensive property):

$$dH = dq_p \tag{6}$$

where q_p represents the change in heat for a system at constant pressure.
    Enthalpy and internal energy are related through:

$$H = U + pV \tag{7}$$

where p is the pressure and V is the volume of the system. Note that as in nearly all cases, unless the experiment is specifically designed otherwise, measurements are performed at constant (atmospheric) pressure, calorimeters measure heat as a change in enthalpy, not a change in internal energy.

Heat Capacities

Both the internal energy and enthalpy of a system change as a function of temperature, a dependence that can be formulated in terms of the heat capacity (C) of the system.
    For a system at constant volume, the heat required to bring about a change in temperature (dT) is given by:

$$dq_V = C_V\, dT \tag{8}$$

where C_V is defined as the heat capacity at a constant volume. Since, as noted above, $dU = dq_v$ then:

$$dU = C_V\, dT \tag{9}$$

If one of the variables is held constant during a change of another, the derivative is known as a partial derivative. The prefix d is replaced by ∂, and the variables held constant are represented by a subscript. Hence:

$$C_V = \left(\frac{\partial U}{\partial T}\right)_V \tag{10}$$

A similar consideration of the change in enthalpy for a system at constant pressure results in the following definition:

$$C_P = \left(\frac{\partial H}{\partial T}\right)_P \tag{11}$$

Second Law of Thermodynamics

The second law deals with the direction of spontaneous change. It is evident in everyday life that some processes occur spontaneously and some do not.

A glass knocked from a table will smash upon contact with the floor, but the glass fragments do not spontaneously reassemble back into the glass. Similarly, hot objects and their surroundings spontaneously equalize to the same temperature, but an object does not spontaneously rise above or below the temperature of its surroundings. Clearly, there is some property of a system that characterizes whether change will occur spontaneously or not; this is termed entropy (*S*). Entropy relates to the degree of disorder of a system. The more disorder is present, the higher the entropy. In simplistic terms, enthalpy and internal energy show which changes are permissible (i.e., those changes where energy is conserved), whereas entropy shows which changes are spontaneous. The second law of thermodynamics can be described as, "the entropy of an isolated system must increase during the process of spontaneous change." Mathematically,

$$\Delta S_{tot} > 0$$

where ΔS_{tot} is the total entropy of all parts of an isolated system.

For a system in mechanical and thermal contact with its surroundings, any change in state is accompanied by a change in entropy of the system (ΔS) and the surroundings ($\Delta S'$). For an irreversible change, the change in entropy must be greater than zero:

$$dS = dS' \geq 0 \quad \text{or} \quad dS \geq -dS'$$

As $dS' = -dq/T$, it follows that for any change:

$$dS \geq \frac{dq}{T}$$

This is known as the Clausius inequality and is important in the derivation of two more fundamental terms, the Helmholtz and the Gibbs functions (see section "Helmholtz and Gibbs Functions").

Third Law of Thermodynamics

The third law states that "if the entropy of every element in its stable state at $T = 0$ is zero, then every substance has a positive entropy which may become zero at $T = 0$ and does become zero for crystalline substances." In other words, the entropy of a perfect crystal becomes zero at absolute zero.

Helmholtz and Gibbs Functions

For a system at a given volume, $dq_v = dU$; the Clausius inequality thus becomes:

$$dS - \frac{dU}{T} \geq 0$$

or

$$T\,dS \geq dU$$

If either the internal energy or entropy is constant then:

$$dU_{S,V} \geq 0 \quad \text{or} \quad dS_{U,V} \geq 0$$

These two terms effectively define the conditions that must be met for change to occur spontaneously; either the entropy must increase or the internal energy must decrease. (This means energy must transfer to the surroundings, thus increasing the entropy of the surroundings.)

Similarly for a system at constant pressure, $dq_p = dH$ and:

$$dS - \frac{dH}{T} \geq 0$$

or

$$T\,dS \geq dH$$

If either the enthalpy or entropy is constant then:

$$dH_{S,P} \geq 0 \quad \text{or} \quad dS_{H,P} \geq 0$$

As before, these two terms define the conditions that must be met for change to occur spontaneously; either the entropy must increase or the enthalpy must decrease (again causing an increase in entropy of the surroundings).

These conclusions are usually expressed more simply by the introduction of two more thermodynamic terms: the Helmholtz function (A) and the Gibbs function (G). Hence:

$$A = U - TS \tag{12}$$
$$G = H - TS \tag{13}$$

and as it is more common to consider changes in the Helmholtz and Gibbs functions at constant temperature:

$$dA = dU - T\,dS \tag{14}$$
$$dG = dH - T\,dS \tag{15}$$

These equations can be written with Δ prefixes rather than d prefixes when the change involved is measurable rather than infinitesimal.

$$\Delta A = \Delta U - T\Delta S \tag{16}$$
$$\Delta G = \Delta H - T\Delta S \tag{17}$$

The Gibbs function and the Helmholtz function simply summarize the changes in entropy and energy that are required if a system is to undergo a spontaneous change. Evidently, from the preceding discussion, if spontaneous change is to occur, then the value of ΔA or ΔG will be negative. As ΔS must be positive, attainment of a negative ΔA or ΔG is most readily achieved if a process involves a decrease in ΔH or ΔU. In this case, the process is said to be enthalpically driven. A positive value of ΔH or ΔU can also lead to spontaneous change, but only with a significant contribution from ΔS. In this case, change is said to be entropically driven. As most experimental measurements are conducted at constant pressure, the Gibbs function is usually used as the determinant of change.

CLASSIFICATION OF CALORIMETERS

Nomenclature

There is a wide range of instruments commercially available; they are based on a number of different operating principles and use a variety of nomenclatures. Unfortunately, there is no common agreement on the naming of instruments and this often leads to confusion (7). Two notable examples are micro- (or nano-) calorimetry and isothermal DSC. In the former case, it is unclear as to whether the instrument is measuring heat on a micro- (or nano-) Watt scale, or if the sample size required for measurement is on the order of micro- (or nano-) grams. The latter case refers to a DSC temperature program that includes isothermal steps, but the name literally means isothermal temperature scanning calorimetry! Of course, such terms are often introduced by instrument manufacturers eager to promote the benefits of a new design (surely a nanoWatt instrument would be better than a microWatt instrument). An instrument may also be named after a designer or an operating characteristic (Black's ice calorimeter, Parr's oxygen bomb calorimeter, gas perfusion calorimetry, and so on), which again reveals no details of its operating principles.

An excellent example of the spread of names used by authors to describe their calorimeters is provided by Hansen (7), who noted that the following techniques were described in a Special Issue of Thermochimica Acta on developments in calorimetry; levitation melting calorimetry, IC, flow calorimetry, water-absorbed dose calorimetry, microcalorimetry, DSC analysis, temperature-modulated DSC (TMDSC), high-temperature calorimetry, and nanocalorimetry. Clearly, unless one is familiar with all the above instruments, these names will cause confusion, not least because it is unclear in most of the cases as to how the actual measurement is made. It has been suggested that, as in the way common names for chemicals have been replaced with IUPAC-sanctioned nomenclature, names for calorimeters and calorimetric procedures must be replaced or supplemented with a systematic nomenclature that gives a clear indication of the method and mode of operation of the calorimeter used (7), although this is some way away, and common or trade names will probably never disappear.

The problems caused as a result of the lack of a common standard nomenclature have been discussed elsewhere (7–10) and shall not be further discussed here. It is, however, important to know the basic designs and operating principles underpinning all modern calorimeters in order to understand the origin of calorimetric signals and to draw comparison between them.

Measuring Principles

There are only three methods by which heat can be experimentally measured:

1. Measurement of the power required to maintain isothermal conditions in a calorimeter, the power being supplied by an electronic temperature controller in direct contact with the calorimeter (power compensation calorimetry).
2. Measurement of a temperature change in a system which is then multiplied by an experimentally determined cell constant (adiabatic calorimetry).
3. Measurement of a temperature difference across a path of fixed thermal conductivity which is then multiplied by an experimentally determined cell constant (heat conduction calorimetry).

Note that all calorimetric measurements therefore require a minimum of two experiments (one for measurement and one for calibration), although further measurements may be needed for blank corrections (such as the determination of a baseline or the correction for dilution enthalpies in a titration experiment). A discussion of the need for, and methods of, calibration (including chemical test reactions) is the basis of Chapter 2.

Power Compensation Calorimeters

In power compensation calorimetry, an electrical element is used either to add heat or remove heat from the calorimetric vessel as the sample reacts, maintaining the sample and vessel at a given temperature. The power output from the sample is thus the inverse of the power supplied by the element. In order to be able to heat and cool, the element is usually based on the Peltier principle. A typical application of this type of calorimetry is power compensation DSC.

Adiabatic Calorimetry

In an ideal adiabatic calorimeter, there is no heat exchange between the calorimetric vessel and its surroundings. This is usually attained by placing an adiabatic shield around the vessel. Thus, any change in the heat content of a sample as it reacts causes either a temperature rise (exothermic processes) or fall (endothermic processes) in the vessel. The change in heat is then equal to the product of the temperature change and an experimentally determined proportionality constant (or calibration constant, ε, which is the effective heat capacity of the system). The proportionality constant is usually determined by electrical calibration.

In practice, true adiabatic conditions are difficult to achieve, and there is usually some heat-leak to the surroundings. If this heat-leak is designed into the calorimeter, the system operates under semi-adiabatic (or isoperibol) conditions and corrections must be made in order to return accurate data. These corrections are usually based on Newton's law of cooling (the most common being the method of Regnault–Pfaundler). These principles are discussed more fully, in the context of solution calorimeters, in the section "Semi-Adiabatic Solution Calorimeters".

Heat Conduction Calorimeters

A heat conduction calorimeter is surrounded by a heat-sink, which acts to maintain the system at a constant temperature. Between the vessel and the heat-sink is a thermopile wall. Any heat released or absorbed upon reaction is quantitatively exchanged with the heat-sink. The thermopiles generate a voltage signal that is proportional to the heat flowing across them; this signal is amplified, multiplied by the cell constant (determined through electrical calibration), and recorded as power versus time. An isothermal system is not limited to reaction processes that reach completion within a short time, as semi-adiabatic instruments are, because it is always (essentially) in equilibrium with its surrounding heat-sink. Furthermore, the greater measuring sensitivity of the thermopiles (as opposed to the thermisters used in semi-adiabatic instruments) means that smaller sample masses can be used.

Single and Twin Calorimeters

In addition to the three types of calorimetry mentioned earlier, which deal with the method by which heat is measured, it is also important to consider the merits of single and twin calorimeters. A single calorimeter has, as its name suggests, only one calorimetric vessel, meaning that a blank experiment is a necessary precursor for any sample experiment in order to subtract out the heat change inherent in simply conducting an experiment. Adiabatic calorimeters are usually single vessels (such as the solution calorimeter discussed in the section "Semi-Adiabatic Solution Calorimeters"). Twin calorimeters have two calorimetric vessels, usually machined to be as closely matched as possible, connected in the opposition (i.e., an exothermic process will produce a positive signal on one side and a negative signal on the other). Such a design automatically corrects for any environmental factors that may affect the data and, if a suitable reference is available, also allows subtraction of the blank signal during an experimental measurement. Usually, twin calorimeters are used for extremely sensitive measurements as they produce considerably less baseline noise than single instruments.

Calorimeter Selection for Pharmaceutical Studies

There is clearly a wide range of calorimetric instrumentation available and it is important to understand the types of sample which each is most suited to study in order that calorimetric measurements are made efficiently and, more importantly, accurately. The ubiquitous nature of heat means that calorimetric measurements are more prone to systematic and operator-induced error than most other analytical techniques; these errors become proportionately more significant as the heat output from the study sample decreases. Moreover, the design of many instruments is a compromise between factors desirable for good calorimetric measurement and factors necessary to ensure a certain measurement function (11), meaning that careful experimental design and suitable calibration (preferentially using chemical test reactions) are essential precursors to accurate sample measurement.

Adiabatic (temperature change) calorimeters are really suited only to the study of events that go to completion within one to two hours because of the inherent difficulties in making the necessary heat-loss corrections over longer time periods (12). This effectively means that this type of calorimeter is not suited to pharmaceutical stability assays, where a rate of degradation of just 1% to 2% per year or less may reasonably be expected. However, adiabatic calorimeters have found widespread application in pharmaceutical preformulation, because they allow the physical characterization of APIs and excipients, and the number of applications of the technique is growing. Examples from the recent literature include its use to detect polymorphs (13), to rank the stability of polymorphs (14), to investigate interactions between species (15–17), to quantify small amorphous contents (18–19), and to study the formation of liposomes (20). In these types of experiment, reaction is initiated by breaking an ampoule, containing (usually) a solid sample, into a reservoir of liquid; they are therefore often referred to as solution calorimeters, and this term will be adopted throughout the rest of the text.

Both power compensation and heat conduction calorimeters are suitable for studying long-term processes. Power compensation instruments have a poorer detection limit but a better capacity to cope with high powers and are therefore usually used to study energetic reactions with large heat rates (21). The power compensation principle also underpins many DSC designs. Heat conduction calorimeters, because they can measure micro- and even nanoWatts are most suited, and most commonly used, for the study of long-term, low-energy reactions, typified by the degradation mechanisms often followed by pharmaceuticals.

OPERATING PRINCIPLES

From the above discussion, it can be seen that there is a great range of calorimetric techniques and that some instruments are more suited to the study of pharmaceutical materials than others. In this section, the operating principles of the most commonly encountered forms of calorimetry in pharmaceutics are explained and it is shown how data are generated and manipulated. Knowledge of how calorimetric data are derived is essential to ensure proper reporting and interpretation of results.

Heat Conduction Calorimetry

In heat conduction IC, a sample and a reference (of similar heat capacity and in equal quantity to the sample) are maintained at a constant temperature in the calorimeter vessel; any heat released (or absorbed) by reaction is quantitatively exchanged with a surrounding heat-sink via an array of thermocouples (a thermo-pile). The thermopile generates an electrical potential (U; note that this term is distinct from internal energy) that is proportional to the heat flowing across it; this is multiplied by an experimentally determined proportionality constant (ε) to give the raw power signal (P_R), which is plotted as a function of time:

$$P_R = \frac{dq}{dt} = \varepsilon U \qquad (18)$$

The value of ε is determined by an electrical calibration (see section "Electrical Calibration"). The total heat quantity released (Q) is given by the product of the electrical potential time integral and the proportionality constant:

$$Q = \varepsilon \int U \, dt \qquad (19)$$

A typical calorimetric arrangement of this type is shown in Figure 1. Note that both the sample and the reference cells will generate an electrical potential; the reference cell is therefore connected in opposition to the sample (i.e., events, which generate a positive potential on the sample side, will generate a negative potential on the reference side and vice versa), because this automatically cor-rects for environmental temperature fluctuations and instrumental noise. Note also that exothermic reactions cause heat to flow toward the heat-sink, whereas endothermic reactions cause heat to flow from the heat-sink. Data can be plotted with exothermic events as either a positive or negative value (depending upon whether the calorimeter records data from the point of view of the instru-ment or the sample), and it is important to define which convention is being used when reporting data. If, as is often the case, exothermic events are recorded as positive signals, then, in order to comply with scientific convention, it is necessary to invert the sign of any thermodynamic data produced.

Although this simple description of a twin cell isothermal calorimeter allows a basic understanding of power–time data, a mathematical description of the instrument is preferable. Such a description starts with the principles of the thermocouples that make up the thermopiles. A thermocouple consists of two strips of dissimilar metals that are coupled at one end. When there is a temp-erature difference between the coupled and uncoupled ends, the thermocouple generates a small electrical potential (known as the Seebeck effect). The output from a thermocouple array (or thermopile) is thus the sum of the electrical potentials produced by the individual thermocouples (U), which is directly

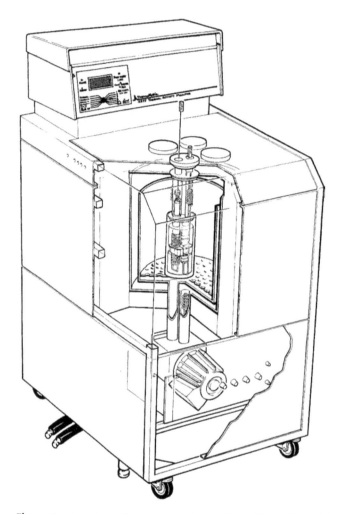

Figure 1 A schematic representation of a Thermal Activity Monitor (TAM). *Source*: Courtesy of Thermometric AB.

proportional to the temperature difference (ΔT) caused by reaction:

$$U = g\Delta T = g(T - T_o) \tag{20}$$

where g is the Seebeck coefficient (V K^{-1}); T_o, the equilibrium temperature of the vessel; and T, the temperature rise (or fall) caused by reaction. The value of g is characteristic for a given type of thermopile; the larger its value the more sensitive the thermopile. From this:

$$T = \frac{U}{g} + T_o \tag{21}$$

And hence the time derivative of dT is given by:

$$\frac{\mathrm{d}T}{\mathrm{d}t} = \frac{1}{g}\frac{\mathrm{d}U}{\mathrm{d}t} \tag{22}$$

If ideal conditions are assumed, then there is no time delay between heat being released (or absorbed) and heat being measured. In that case, the measured electrical potential, multiplied by the proportionality constant, gives power directly, as stated earlier. However, the real case is not usually that simple, and there is a delay between heat change in the sample and heat measurement (known as the dynamic response of the instrument). In order to quantify the dynamic response, it is necessary to have some appreciation of the steps that occur between heat being released (or absorbed) by a sample and that heat causing an electrical potential to be generated by the thermopiles.

For an exothermic process, as reaction occurs, heat is released, which initially accumulates in the sample vessel. This must cause the temperature in the vessel to rise (although this rise is usually regarded to be negligible, because otherwise the experiment would not be conducted under isothermal conditions), the magnitude of the temperature rise being dependent upon the heat capacity of the vessel (C; note here that this heat capacity term is sample-dependent, because the vessel is assumed to mean the calorimetric ampoule and its contents). For simplicity, it is necessary to assume that the heat generated by, and hence the temperature rise of, the sample is uniform. As the temperature of the sample rises, a temperature gradient is formed between the vessel and the heat-sink. Heat must necessarily flow towards the heat-sink in order to restore thermal equilibrium; the rate of heat transfer is dependent on the heat transfer coefficient (k in W K^{-1}) of the thermopiles. As the heat transfers, the thermopiles generate the electrical potential from which the raw data derive. The principle is the same for endothermic processes but will involve a temperature fall and subsequent heat-flow into the vessel. The measured thermal power therefore derives from two events (heat accumulation and heat transfer), and can be represented by a heat balance equation:

$$\frac{\mathrm{d}q}{\mathrm{d}t} = \Phi + C\frac{\mathrm{d}T}{\mathrm{d}t} \tag{23}$$

where the term $C\,\mathrm{d}T/\mathrm{d}t$ represents the temperature change (heat accumulation) in the vessel caused by reaction, and Φ represents the heat flow in or out of the vessel across the thermopiles. (Note that it is assumed that all the heat released or absorbed traverses the thermopiles, which is unlikely. However, this discrepancy is accounted for by the electrical calibration, discussed subsequently.) According to Newton's law of cooling, the rate of heat transfer is proportional to the temperature difference between the vessel (T) and the heat-sink (T_o):

$$\Phi = k\Delta T = k(T - T_o) \tag{24}$$

where k is the heat transfer coefficient.

It follows that where k is the heat transfer coefficient:

$$\frac{dq}{dt} = k(T - T_o) + C\frac{dT}{dt} \tag{25}$$

This is the general heat balance equation for a single calorimeter and shows why single calorimeters are greatly affected by small temperature fluctuations in the heat-sink (because of the term $T - T_o$).

For a twin calorimeter, a heat balance equation can be written for both the sample and the reference sides (denoted by the subscripts S and R respectively):

$$\frac{dq_S}{dt} = k_S(T_S - T_o) + C_S\frac{dT_S}{dt} \tag{26}$$

$$\frac{dq_R}{dt} = k_R(T_R - T_o) + C_R\frac{dT_R}{dt} \tag{27}$$

As stated above, the output from a twin calorimeter is the difference between the sample and reference sides then:

$$\frac{dq}{dt} = \frac{dq_S}{dt} - \frac{dq_R}{dt} = k_S(T_S - T_o) + C_S\frac{dT_S}{dt} - k_R(T_R - T_o) - C_R\frac{dT_R}{dt} \tag{28}$$

Assuming a precision instrument and careful selection of an inert reference with a heat capacity matching the reference then:

$$k = k_S = k_R \tag{29}$$

$$C = C_S = C_R \tag{30}$$

$$T_R = T_o \tag{31}$$

$$\frac{dq_R}{dt} = 0 \tag{32}$$

And hence it can be written that:

$$\frac{dq}{dt} = \frac{dq_s}{dt} = k(T_S - T_R) + C\frac{dT_S}{dt} \tag{33}$$

The reason that twin calorimeters are unaffected by external temperature fluctuations is now clear; the heat flow from the sample can be determined by the measurement of the temperature difference between the sample and reference sides and does not require knowledge of T_o.

Tian Equation

The Tian equation is perhaps the most well-known mathematical description of heat conduction calorimeters and relates the heat balance equations derived earlier to the electrical potential produced by the thermopiles. From the earlier discussion it can be seen that:

$$(T - T_o) = \frac{U}{g} \tag{34}$$

which allows the following equation to be constructed:

$$\frac{dq}{dt} = \frac{k}{g} U + \frac{C}{g} \frac{dU}{dt} \tag{35}$$

Rearrangement gives:

$$\frac{dq}{dt} = \frac{k}{g} \left(U + \frac{C}{k} \frac{dU}{dt} \right) \tag{36}$$

If the following definitions are made:

$$\varepsilon = \frac{k}{g} \tag{37}$$

$$\tau = \frac{C}{k} \tag{38}$$

where ε is the proportionality constant referred to earlier and τ is the time constant of the instrument, then the Tian equation is obtained:

$$\frac{dq}{dt} = \varepsilon \left(U + \tau \frac{dU}{dt} \right) \tag{39}$$

The Tian equation as written above applies to a single vessel, but exactly the same derivation applies to a twin vessel, although in this case k and τ will represent the differential proportionality constant and the differential time constant respectively. The proportionality constant has units of $W\ V^{-1}$ and typically ranges between 2 and 6 $W\ V^{-1}$ for a modern instrument. The time constant has units of s^{-1} and reflects the dynamic response of the instrument. The time constant increases if the heat capacity of the vessel increases or the heat transfer coefficient decreases. The time constant of a commercial instrument with a 5 mL volume is around 100 seconds and with a 20 mL volume is around 500 seconds. If the study reaction occurs over a long time period (hours or days), then the output signal recorded by the instrument approximates to a real-time measurement and the raw data can be used directly to determine the kinetic parameters. For short-term reactions (on the order of minutes), the delay in measurement response can become significant and dynamic correction is needed in order to reconstruct the true power–time profile. Dynamic correction is discussed in detail in the section "Dynamic Data Correction."

Sensitivity

The sensitivity (S; note that this term is distinct from entropy) of a calorimeter is given by the magnitude of the output signal generated for a given amount of heat produced, in this case:

$$S = \frac{\Delta U}{\Delta q} \tag{40}$$

The sensitivity is also affected by the heat capacity of the calorimetric vessel, because the magnitude of U is dependent upon the temperature difference across the thermopile. It follows that the smaller the heat capacity, the larger the value of ΔT (for a given quantity of heat) and hence ΔU and S. However, calorimeters with small heat capacities are more prone to environmental temperature fluctuations than calorimeters with large heat capacities; therefore, a compromise in design is required between calorimetric sensitivity and long-term stability. Many current designs utilize an aluminium heat-sink around the thermopiles with the whole apparatus contained within an air box or water bath.

Dynamic Data Correction

From the preceding discussion, it can (hopefully) be appreciated that, in the ideal case, the raw power (P_R) output from a calorimeter would reflect the true rate of heat generation in the sample, but in the real case, there is a dynamic delay in measurement, which is approximated by the Tian equation and, as such P_R is actually directly proportional to the measured electrical potential. It is thus possible to use the Tian equation to remove the dynamic delay from the raw power signal, resulting in the corrected power (P_C), if the time constant of the instrument is known. This is called dynamic correction and is usually achieved automatically by dedicated instrument software (methods to determine the time constant of instruments are detailed in the section "Determination of Time Constants").

Dynamic correction of calorimetric data becomes important for short-term processes where the kinetic response of the sample is significantly affected by the delay in the measurement response of the instrument because of its time constant. It is also often used in titration experiments (see section "Titration Calorimetry"), where the true rate of heat generation caused by interactions in the sample drops to zero well before the measured response, allowing experiments to be conducted much more quickly.

It is important to note, however, that a number of assumptions are made in order to derive the Tian equation (the major assumption being that there are no temperature gradients within the sample) and that it only approximates the true dynamic delay inherent to the instrument. The corrected data so produced, while much more closely resembling the true response of the sample, often contain artifacts, such as "overshoots," where both endothermic and exothermic events are indicated, even though it may be known that only one event is occurring in the sample. In principle, these artifacts could be removed by manually altering the values of the time constant(s), but this would be time consuming in practice. It is therefore easier to use corrected data to determine reaction enthalpies and reduce the time taken to perform certain experiments, and to note that the use of such data to elucidate kinetic information must be undertaken with caution.

The following section describes briefly the processes involved in dynamic correction so that a user fully appreciates the nature of the data so produced. A fuller account of the steps discussed subsequently can be found in the superb review by Randzio and Suurkuusk (22). It should also be noted that the discussion that follows is only one way of reconstructing calorimetric data and that other methods have been discussed in the literature (23,24).

Starting with the premise (introduced earlier) that:

$$P_R = \varepsilon U \tag{18}$$

then

$$\frac{dP_R}{dt} = \varepsilon \frac{dU}{dt} \tag{41}$$

and the Tian equation can be written as:

$$P_C = P_R + \tau \frac{dP_R}{dt} \tag{42}$$

or

$$P_C - P_R = \tau \frac{dP_R}{dt} \tag{43}$$

The left-hand side represents the difference between the true rate of heat production from the sample and the rate actually measured by the instrument (the degree of correction); as this term approaches the detection limit of the instrument, it becomes impossible to differentiate between the true and the measured response, and dynamic correction is not possible. The right-hand side essentially represents the gradient of the power–time curve. Therefore, if the detection limit and time constant of an instrument are known, it is possible to calculate the magnitude of the slope above which a Tian correction should be applied if the kinetic response of the sample is important.

The Tian equation as written above is predicated on several assumptions that, while making its derivation relatively straightforward and its application easy to follow, in practice means that it is not the most accurate method for data correction. The major assumption in the derivation of the Tian equation is that the calorimetric cell and its contents are at the same temperature. Consideration of how experiments are actually performed (the sample is loaded into an ampoule which is then sealed and placed in the calorimeter) suggests that, initially at least, the temperature of the sample (T_s) and the cell (T_c) will not be the same. Given that the temperature gradient giving rise to the measured signal is external to the calorimetric cell, it is clear that some temperature difference between the sample and the cell will affect the measuring response of the calorimeter. The degree to which the measurement is affected will be governed by the rate of heat transfer between the sample and the cell (k_{sc}) and between the cell and the heat-sink (k_{co}). As before, both of the heat exchange coefficients have units of W K^{-1}. In the same manner as described earlier, the heat balance equations for both the sample and the cell can then be written:

$$\frac{dq}{dt} = k_{sc}(T_s - T_c) + C_s \frac{dT_s}{dt} \tag{44}$$

$$k_{co}(T_c - T_o) + C_c \frac{dT_c}{dt} = k_{sc}(T_s - T_c) \tag{45}$$

where C_c and C_s are the heat capacities of the cell and the sample, respectively. (Note that earlier the heat capacity was defined as that of both the cell and its contents.) The right-hand side of Equation (45) represents the power exchanged between the sample and the cell; this term also appears in Equation (44). In order to remove T_s (which is not directly measured in an experiment), it is necessary to differentiate Equation (45) and then substitute both Equation (45) and its derivative into Equation (44), resulting in:

$$\tau_s \tau_c \frac{d^2 T_c}{dt^2} + (\tau_s + \tau_c) \frac{dT_c}{dt} + \frac{k_{co}}{k_{sc} + k_{co}} T_c = \frac{1}{k_{co} + k_{sc}} \frac{dq}{dt}$$
$$+ \frac{k_{co}}{k_{sc} + k_{co}} \left(T_o + \tau_s \frac{dT_o}{dt} \right) \tag{46}$$

where τ_s and τ_c are defined as:

$$\tau_s = \frac{C_s}{k_{sc}} \tag{47}$$

$$\tau_c = \frac{C_o}{k_{sc} + k_{co}} \tag{48}$$

A similar equation can be written for a reference material, prepared under the same conditions as the sample (the subscript r denoting a property of the reference material):

$$\tau_r \tau_c \frac{d^2 T_c}{dt^2} + (\tau_r + \tau_c) \frac{dT_c}{dt} + \frac{k_{co}}{k_{rc} + k_{co}} T_c = \frac{1}{k_{co} + k_{rc}} \frac{dq}{dt}$$
$$+ \frac{k_{co}}{k_{rc} + k_{co}} \left(T_o + \tau_r \frac{dT_o}{dt} \right) \tag{49}$$

Subtracting the sample and reference equations (which is valid only assuming the reference material has the same heat capacity and heat transfer coefficient to the cell as the sample) yields:

$$\frac{dq}{dt}=k_{co}(T_c-T_r)+(k_{co}+k_{sc})\left[(\tau_s+\tau_c)\frac{d(T_c-T_r)}{dt}+\tau_s\tau_c\frac{d^2(T_c-T_r)}{dt^2}\right] \quad (50)$$

Note here that T_c appears in Equation (50), whereas in the earlier derivations T_s was used; this simply reflects the fact that in the earlier cases T_s and T_c were assumed to be equal and here the actual measured temperature (T_c) is used.

If the rate of heat exchange between the sample and the cell is much smaller than the rate of heat exchange between the cell and the heat-sink (in other words, the calorimeter behaves ideally and the rate-limiting step is heat transfer from the sample to the calorimeter), then k_{sc} is much smaller than k_{co} and:

$$\frac{dq}{dt}=k_{co}\left[(T_c-T_r)+(\tau_s+\tau_c)\frac{d(T_c-T_r)}{dt}+\tau_s\tau_c\frac{d^2(T_c-T_r)}{dt^2}\right] \quad (51)$$

It can be seen that Equation (51) has a form similar to that of the Tian equation but that there appears a second-order time derivative of the temperature difference and the coefficient of the first-order derivative has altered. In practice, it is difficult to know the value of k_{sc}. However, if the following definitions are made:

$$\tau_1+\tau_2=\left|\frac{k_{co}+k_{sc}}{k_{co}}\right|(\tau_s+\tau_c) \quad (52)$$

$$\tau_1\tau_2=\frac{k_{co}+k_{sc}}{k_{co}}(\tau_s\tau_c) \quad (53)$$

Then Equation (51) can be rewritten as:

$$\frac{dq}{dt}=k_{co}\left[(T_c-T_r)+(\tau_1+\tau_2)\frac{d(T_c-T_r)}{dt}+\tau_1\tau_2\frac{d^2(T_c-T_r)}{dt^2}\right] \quad (54)$$

The values of τ_1 and τ_2 differ from the values of τ_s and τ_c and are known as the first and second time constants of the calorimeter, respectively. Using the same definitions of P_R and P_C as described earlier, Equation (54) can be rewritten in the form:

$$P_C=P_R+(\tau_1+\tau_2)\frac{dP_R}{dt}+\tau_1\tau_2\frac{d^2P_R}{dt^2} \quad (55)$$

Equation (55) can be used to reconstruct corrected power data from the raw power signal and is indeed used in some commercial software (for instance, in Digitam, Thermometric AB). In order to apply this correction, the proportionality constant (ε) must be known (as this is required to determine P_R) as must the two time constants. The value of the proportionality constant is determined by electrical calibration while the two time constants are usually determined by a least-squares analysis following an electrical calibration, although graphical methods may also be employed.

Determination of Time Constants

Because modern calorimeters are operated under computer control, time constants are usually determined by fitting power–time data to Equation (55), using a least-squares analysis. However, it is possible to calculate both time constants manually using a graphical method, which, for completeness, is discussed briefly subsequently.

Dynamic parameters can only be determined once the calorimeter has been perturbed from equilibrium; this is usually achieved by applying an electrical heat pulse, which produces a power signal such as that shown in Figure 2. Because the method of analysis requires that no heat is generated within the ampoule (i.e., by a sample) during perturbation, the heat pulse applied is usually large and occurs over a short time period. Before and after the perturbation, the power signal from the calorimeter should be zero (i.e., the instrument should be in a steady state). The response shown in Figure 2 is known as the impulse response of the calorimeter and represents the dynamic delay inherent to the instrument. Four time points are indicated in Figure 2; these are used as reference points for the following discussion. It has been shown that the

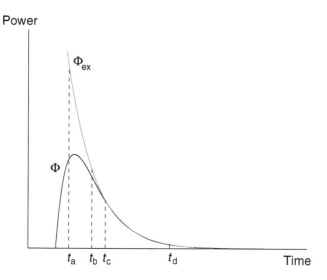

Figure 2 A typical impulse response from an isothermal calorimeter (Φ) and the extrapolated power (Φ_{ex}) calculated using Equation (58). The differences between Φ and Φ_{ex} become apparent between times b and c.

impulse response can be described by the following power function (25):

$$\frac{dq}{dt} = \Phi = Ae^{-t/\tau_1} - Be^{-t/\tau_2} + \cdots \tag{56}$$

If it is assumed that the second and subsequent terms are not contributing to the dynamic delay of the instrument, then Equation (56) can be written in logarithmic form:

$$\ln \Phi = -\frac{1}{\tau_1} t + \ln A \tag{57}$$

Thus, a plot of $\ln \Phi$ versus time should yield a straight line, the gradient of which equals the reciprocal of τ_1, Figure 3 (which is calculated using power values from time t_c to time t_d). Once the value of τ_1 is known, then it is possible to calculate a series of values of Φ, starting from some arbitrary time t_n (Φ_n) near the end of the impulse response and working backwards through times t_{n-1}, $t_{n-2}, \ldots, t_{n-x}$ ($\Phi_{n-1}, \Phi_{n-2}, \ldots, \Phi_{n-x}$), using Equation (58):

$$\Phi_{n-1} = \Phi_n e^{(t_n - t_{n-1})/\tau_1} \tag{58}$$

This produces the series of data termed Φ_{ex} in Figure 2. At some point, the differences between the extrapolated and the measured power data become significant (between times t_b and t_c in Fig. 2); these reflect the influence of the second and higher (which are ignored in practice) time constants. If the differences between the extrapolated (Φ_{ex}) and measured (Φ) data are plotted versus

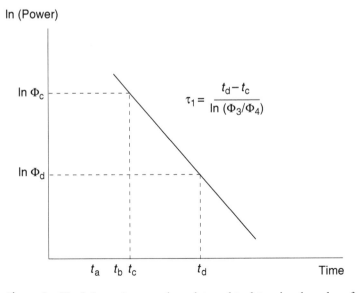

Figure 3   The ln(power) versus time plot used to determine the value of τ_1.

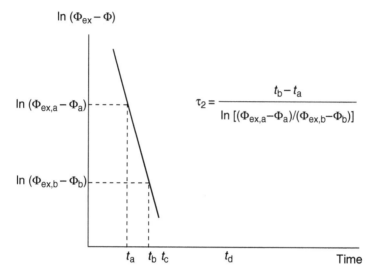

Figure 4 The $\ln(\Phi_{ex} - \Phi)$ versus time plot used to determine the value of τ_2.

time, then τ_2 is given by the reciprocal of the gradient of the resulting straight line (Fig. 4).

Power Compensation Calorimetry

As mentioned earlier, in power compensation calorimetry, the temperature of the sample is maintained at a given temperature by an electrical element that either adds or removes heat as necessary. Because the element needs to be able to remove as well as supply heat, it usually operates on the Peltier principle. The heat supplied or removed by the element is directly measured and it is this signal that is plotted against time in the output data.

Electrical Calibration

As noted earlier, the raw power output (P_R) from a heat conduction calorimeter is given by εU, where ε is the proportionality constant and U is the electrical potential produced by the thermopiles. The value of ε is determined by an electrical calibration. Most calorimeters have a resistance heater located under the sample (and sometimes the reference) ampoule; the resistor produces a power proportional to the current passed through it:

$$\frac{dq}{dt} = I^2 R \tag{59}$$

where I is the current (in Amps) and R is the resistance (in Ohms). If the time for which current is passed (t) is known, then the total quantity of heat produced (Q)

can be calculated:

$$Q = I^2 Rt \tag{60}$$

During calibration, an instrument is loaded with two reference ampoules (i.e., both the reference and the sample sides contain the same reference material, in the same quantity that will be used in any subsequent experiments) and, once thermal equilibrium has been attained, the baseline value is adjusted to zero. A current is then passed through the resistance heater (usually the sample side) for a given length of time. This will cause a deflection in the baseline and, assuming the time period is long enough, a steady state will be reached where the baseline has deflected to a constant value. This value is then adjusted to match the expected power input from the heater [given by Equation (59)]. After the current is switched off, the baseline returns to zero. As a further check, the total area under the curve caused by the calibration can be determined by integration and should match that calculated by Equation (60).

A typical electrical calibration response is shown in Figure 5. In an ideal system, this calibration would be square-form (i.e., there would be no delay in reaching the steady state indicated by the plateau). The shape of this curve is therefore related to the dynamic response of the instrument [and, hence, correction of these data using Equation (55) should result in a square-form trace]. The corrected data are also shown in Figure 5; these data much more closely resemble the ideal case.

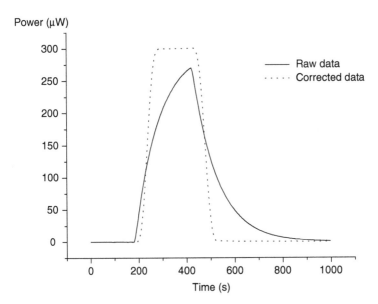

Figure 5   A typical calibration pulse produced by an electrical heater (*raw data*) and the dynamically corrected data produced using Equation (55) (*corrected data*).

Chemical Test Reactions

Electrical calibration is by far the most commonly used method for ensuring the accuracy of calorimetric data, principally because of its ease of use and reproducibility, but there are some notable problems with the procedure. The first is that the accuracy of the calibration is dependent entirely upon knowing the resistance of the heater (which will change with temperature) and the amount of current supplied. This can be mitigated to a certain degree by ensuring the heating resistor is part of a Wheatstone bridge, as this allows determination of its resistance as a function of temperature. The second, and perhaps more significant, issue is that the processes of heat generation and dissipation from a resistance heater do not accurately mimic those that occur during a chemical reaction [for instance, with a resistance heater, all the heat is supplied externally from the base (usually) of the ampoule, whereas in a homogeneous solution phase reaction, heat is generated uniformly inside the ampoule]. This has led to much discussion and attempts to define standard chemical test reactions for isothermal calorimeters, a debate that has recently been summarized in an IUPAC technical report (26).

There are a number of requirements imposed upon a chemical test reaction; it must be chemically robust (i.e., it must give a consistent output over a range of conditions), it must have certified reaction parameters (such as a rate constant or an enthalpy of reaction/solution), and it must be repeatable from laboratory to laboratory and from instrument to instrument. If a chemical test reaction is run in a calorimeter, the user should, through appropriate analysis of the data, obtain values that agree with the literature parameters. This assures that (*i*) the instrument is functioning correctly and (*ii*) the operator is competent in the use of the technique.

It is notable here that chemical reactions are not intended to be used to calibrate instruments (that is still the role of the electrical heater); they are designed to give a reproducible and quantifiable heat output such that analysis in an electrically calibrated instrument returns a universally accepted set of values (be these rate constants, enthalpies etc.). An additional benefit of the use of chemical test reactions is that, because they are prepared by the user externally from the instrument and require loading into ampoules, they can be used for user training as well as allowing a quantitative comparison of the performance of different instruments, whether they are in the same laboratory or on different sites.

A number of test reactions are recommended for IC; proper selection of a test reaction depends upon both the type of instrument being used and the nature of the data being recorded. For instance, if the calorimeter is being used to measure the stability of a formulation over a period of days, then it is desirable to have a chemical test reaction that proceeds over several days (a medium-term test reaction) and that gives a reasonable heat output; this ensures excellent reproducibility of measurement and allows both short-term (hours) and long-term (days and weeks) assessment of calorimetric performance. Conversely, if the calorimeter is being used in a titration mode, then the test reaction should also involve a titration (a short-term test reaction), to allow validation of the heat per injection and the binding constant.

The recommended test reactions for a number of isothermal calorimetric techniques are discussed briefly in the following sections. Further details can be found in the relevant sections of Chapter 2.

Ampoule Calorimetry

Usually, short-term calibration for ampoule experiments is performed using the electrical substitution method outlined earlier. However, if a longer term or a chemical assessment of the performance of the calorimeter is required, then one test reaction that has received a lot of attention in the recent literature is the imidazole catalyzed hydrolysis of triacetin (27–32). In particular, this reaction, which was first proposed as a test reaction for isothermal calorimeters by Chen and Wadsö (33), has been the subject of an extensive inter- and intra-laboratory study (34) as part of an IUPAC trial; the reaction parameters that should be returned if the reaction is conducted under the conditions stated in the trial are $\Delta_R H = -91.7 \pm 3\,\text{kJ mol}^{-1}$ and $k = 2.8 \pm 0.1 \times 10^{-6}\,\text{dm}^3\,\text{mol}^{-1}\,\text{s}^{-1}$.

An alternative reaction that has been discussed in the recent literature is the base-catalyzed hydrolysis of methyl paraben (31,35). When set up at the concentrations specified in the aforementioned papers, the reaction reaches completion within 24 hours at 25°C, following first-order kinetics. The use of this test reaction to troubleshoot instrumental performance has also recently been discussed (36)—work that demonstrated how the use of the test reaction during the installation of an instrument identified a systematic misreporting of instrumental temperature.

Isothermal Titration Calorimetry

The choice of test reaction depends upon the type of interaction being measured. If the aim is simply to quantify an enthalpy of binding, and one component is being held in excess in the ampoule, then the neutralization of NaOH (0.1 M) by HCl (0.1 M) should be used. This returns values of $\Delta_R H = -57.1\,\text{kJ mol}^{-1}$ at 25°C (37). If, additionally, a binding constant is sought, then two-test reactions are recommended by IUPAC; the interaction between Ba^{2+} and 18-crown-6 or the interaction between 2'CMP and Rnase. The former reaction returns values of $\Delta_R H = -31.42\,\text{kJ mol}^{-1}$, $K_B = 5.9 \times 10^3\,\text{mol dm}^{-3}$ (38) whereas the latter reaction returns values of $\Delta_R H = -50\,\text{kJ mol}^{-1}$, $K_B = 1.2 \times 10^5\,\text{mol dm}^{-3}$ (39).

Flow Calorimetry

The imidazole catalyzed hydrolysis of triacetin, which has received recent attention as a potential chemical test reaction for ampoule calorimetry (see section "Ampoule Calorimetry"), has been shown to be unsuitable for flow calorimeters because of its relatively slow rate constant and the small effective thermal volume of flow calorimeter vessels (for a further discussion of this point, see section "Chemical Test Reactions" of Chap. 2) (31). However, the base-catalyzed hydrolysis of methyl paraben has been shown to be a suitable chemical test reaction for flow calorimeters (31), because its faster rate constant produces a larger instantaneous heat signal.

Solution Calorimetry

A number of chemical test reactions for solutions calorimeters have been proposed and discussed, including the dissolution of Tris in 0.1 M HCl (40,41), the dissolution of KCl or NaCl in water (42–44) and the dissolution of propan-1-ol in water (45). The dissolution of sucrose in water can also be used (46,47), although this is not currently recognized as a test reaction.

Of these systems, the dissolution of KCl into water is usually recommended, because it is robust, easy to perform, and a standard reference material is available from NIST (the National Institute for Standards and Technology, U.S.A.); the enthalpy of solution of the NIST certified KCl into water is $17.584 \pm 0.017 \text{ kJ mol}^{-1}$ (43). However, the use of KCl is not without drawbacks. Principally, the value of $\Delta_{sol}H$ varies as a function of the concentration achieved after dissolution, because of the effect of the enthalpy of dilution (Φ_L). Thus, the certified value for the NIST reference material of $17.584 \text{ kJ mol}^{-1}$ applies only if a final concentration of $0.111 \text{ mol kg}^{-1}$ is attained in the calorimetric vessel. This corresponds to a molar ratio of water to KCl of 500 to 1 and is often written as $\Delta_{sol}H$ (500 H_2O, 298.15 K). If measurements are performed under different conditions, then the value obtained ($n H_2O$, 298.15 K) must be corrected to that which would have been recorded at 500 H_2O, in order to draw comparison. These corrections are explained in the certification certificate supplied with the NIST sample (43), although the data supplied there apply only to experiments performed where n varies from 100 to 1000.

The effects of KCl concentration on $\Delta_{sol}H$ have been studied extensively by Kilday (42), who corrected the observed enthalpy values over a range of water ratios ($n = 500$ to $10\,000$) to account for the enthalpy of dilution; this resulted in a value for $\Delta_{sol}H_\infty$, the enthalpy of solution at infinite dilution ($\Delta_{sol}H_\infty = 17.241 \pm 0.018 \text{ kJ mol}^{-1}$).

Modern solution calorimeters often use microgram samples and are capable of detecting very small powers; one consequence of this is that it is not possible to perform the KCl experiment under the NIST certification conditions, because the heat generated would be of a magnitude sufficient to saturate the amplifiers. Because of this, several recent studies have been conducted looking at the applicability of test reactions to modern solution calorimeters (48,49). The outcome from these studies appears to be that, if KCl is to be used, then it is better to correct the data to the enthalpy of solution at infinite dilution, but that sucrose may offer a better and cheaper alternative, especially for heat conduction instruments that use very small (mg) samples.

INSTRUMENTATION

Modern calorimeters are available in a number of different designs; systems are more amenable to study in particular instruments than others. Some of the common designs, the applications of which are discussed in subsequent chapters, are described subsequently.

Ampoule Calorimetry

Ampoule calorimetry simply means placing sample and reference materials in sealed ampoules in the calorimeter. Ampoules can be reusable or disposable, can vary in volume, and can be constructed from a variety of materials (typically, they are either glass or metal). There are many advantages to using ampoules to load samples (as opposed to loading samples directly into a calorimetric chamber; see section "High-Sensitivity DSC" for further discussion of such instruments), the biggest being that it affords the opportunity to study any sample so long as it, or at least a representative fraction of it, fits within the ampoule. Pharmaceutically, this means that it is possible directly to study heterogeneous systems, such as creams and emulsions, as well as solids and liquids; calorimetry is one of the few analytical tools that affords such utility. It also means that the local environment (i.e., RH) in the ampoule can be controlled. Furthermore, it is possible to construct specific apparatus that will fit within the calorimeter for specific analyses (such as a titration unit for instance).

The main drawback of using a removable ampoule is that the thermal contact between the ampoule and the calorimeter will vary each time it is loaded, increasing noise in the raw data signal. There is also the possibility that the sample and reference ampoules will be mismatched, further reducing the quality of the experimental data.

Batch Calorimetry

In a batch calorimeter, the reaction vessel is divided into two compartments that are connected by an air space. Each compartment can be separately charged with sample (either a solid or a liquid). Once thermal equilibrium has been attained, the contents of the two compartments are mixed, usually by rotating the vessel, and the heat of interaction is measured. Batch calorimeters are also known as mixing calorimeters.

Flow Calorimetry

In flow calorimetry, as is implied by its title, a liquid flows through the calorimetric vessel. Solutions are held in a reservoir external to the instrument and a peristaltic pump is used to circulate the liquid. Flow calorimeters operate in two modes, flow-through and flow-mix. In flow-through operation, a solution is pumped from an external reservoir, passes through the calorimetric vessel, and is returned either to the initial reservoir or to waste. If two or more samples are to be mixed, then mixing takes place in the external reservoir. In flow-mix operation, two (usually) liquids are pumped into the calorimetric vessel where they are mixed and then pumped to waste. Thus, in flow-through operation the calorimeter measures the power output from a system after mixing and in flow-mix operation the calorimeter measures the power associated with the mixing process itself.

Titration Calorimetry

In isothermal titration calorimetry (ITC), small aliquots of a titrant solution (held in a reservoir external to the instrument) are added in sequential aliquots to a solution held within the calorimetric vessel and the heat change per injection is recorded. In a typical experiment, up to 30 injections ($\sim$10–15 μL each) are made into the liquid reservoir. Usually, titration calorimetry is used to study the binding interaction between a ligand (a drug or potential drug candidate) and a substrate (typically, a protein, enzyme, or some other biological target), although of course the technique is not limited to this area.

In a typical binding experiment, the area under each peak gives the heat change per injection of titrant and can be plotted against the number of moles of titrant injected (or concentration of titrant in the vessel) to yield a binding isotherm. It is important to recognize that if the interaction under investigation involves binding then, in order to analyze the data properly, it must be ensured that the number of moles of ligand injected is sufficient to ensure all the binding sites on the target are occupied; this is shown by later injections reaching a minimum and consistent value.

It is also important to note that the measured heats from such an experiment comprise contributions from many effects, such as dilution (both of ligand and receptor) and mixing, and care must be taken to ensure that the proper reference experiments have been conducted (for accurate work three are required: solvent into solvent, ligand into solvent, and solvent into receptor). Assuming that both the ligand and receptor are prepared in the same solution (usually a buffer; note that the mixing of samples prepared in different buffers can often cause a large heat effect that arises from the ionization of the buffer species), then the greatest additional contribution to the measured heat is the enthalpy of dilution of the ligand solution. This results from the need to prepare relatively concentrated solutions for injection, because the volumes injected are so small.

If properly constructed, the principal parameter returned from a binding experiment is then the binding constant (K_b). This is usually obtained by non-linear analysis of the binding isotherm. Knowledge of the value of K_b allows the derivation of the Gibbs energy of binding ($\Delta_b G$) via:

$$K_b = e^{(-\Delta_b G)/RT} \tag{61}$$

The binding enthalpy ($\Delta_b H$) can be determined from those injections where total binding is assumed (i.e., the initial injections). Once $\Delta_b G$ is known, and because $\Delta_b H$ is measured directly, it is possible to calculate $\Delta_b S$ through:

$$\Delta_b G = \Delta_b H - T\Delta_b S \tag{62}$$

where the subscript b refers to the enthalpy or entropy change of binding. Thus, an ITC experiment can be used to derive a complete thermodynamic picture of a process.

The drive behind the development of many of the modern ITC instruments was the need to be able to study directly dilute solutions of biological macro-molecules and it is within this area that many applications of ITC are found. However, because the range of reaction parameters that can be derived from ITC data is comprehensive, its use is becoming more widespread in pharmaceutics, especially for characterizing interactions between molecules, and it can used to construct quantitative structure–activity relationships for potential drugs binding to a biological target. Specific application areas and examples are discussed in Chapter 5.

There is a wide range of commercially available ITC instruments and they can be classified according to the type of calorimetric vessel employed; vessels can be fixed in place in the instrument and loaded directly (this limits all samples to being liquids), or vessels can be removed and charged with sample externally from the instrument (which additionally allows the use of suspensions). Instruments can also work on power compensation or heat conduction principles. The specifications of a number of currently available commercial instruments are given in Table 1.

Gas Perfusion Calorimetry

In gas perfusion calorimetry, the flow of a vapor over a (usually solid) sample is regulated; the vapor can be an organic solvent or, more often, humidified air and is selected to interact with the sample in some way. Typical examples of the use of perfusion calorimetry include using the vapor to probe specific binding sites on the surface of the sample (for instance, acidic or basic sites), wetting the sample to induce the formation of hydrates or solvates, and increasing the partial pressure of a vapor to induce recrystallization of an amorphous material. The instruments used can control the relative partial pressure of the vapor in the air space in the ampoule, allowing samples to be held "dry" before the commencement of an interaction. The relative partial pressure of the probe gas can then be increased, either in discrete steps or in a linear ramp. This is usually achieved by using mass flow controllers to mix quantitatively two vapor streams flowing from different sources; one dry and one saturated with vapor. Care must be taken when running a gas perfusion experiment to ensure that the vapor used does not degrade any rubber components in the equipment. Furthermore, the effects of wetting the internal surfaces of the ampoule and associated parts of the instrument must also be considered. It is usually the case that wetting of the ampoule causes a large heat that must be corrected for during subsequent data analysis.

Table 1 Properties of a Number of Commercially-Available Isothermal Titration Calorimetry Instruments

Instrument	VP-ITC	Nano-ITC III	2251 titration ampoule	μ-Reaction calorimeter
Manufacturer	MicroCal, LLC 22 Industrial Drive, East Northampton, MA 01060, U.S.A.	Calorimetry Sciences Corporation 799 E. Utah Valley Drive, American Fork, UT 84003, U.S.A.	Thermometric AB, Spjutvägen 5A, S-175 61 Järfälla, Sweden	THT, 1 North House, Bond Avenue, Bletchley, MK1 1SW, U.K.
Website	www.microcalorimetry.com	www.calscorp.com	www.thermometric.com	www.science.org.uk
Operation	Power compensation	Power compensation	Heat conduction	
Sample volume	1.4 mL	0.75 mL	4 mL	1.5 mL
Temperature range	2–80°C	0–80°C	5–90°C	
Detection limit	Not specified	0.1 μJ	20 μJ[a]	
Ampoule	Hastelloy, fixed in place	24 K gold, fixed in place	Glass or stainless steel, removable	Glass, disposable

[a]When housed in a 2277-201 thermal activity monitor.

Abbreviation: ITC, isothermal titration calorimetry.

Table 2 Relative Humidities Maintained by Various Saturated Salt Solutions at 15°, 20°, and 25°C

Salt	RH maintained		
	15°C	20°C	25°C
Lithium chloride	11.3	11.3	11.3
Magnesium chloride	33.3	33.1	32.8
Potassium carbonate	43.1	43.2	43.2
Sodium bromide	60.7	59.1	57.6
Sodium chloride	75.6	75.7	75.3
Potassium chloride	85.9	85.1	84.3
Potassium sulfate	97.9	97.6	97.3

Abbreviation: RH, relative humidity.
Source: From Ref. 50.

This can be achieved either by flowing a similar vapor through a matched reference ampoule and plotting the differential heat flow or, more usually, by subtracting the data from a blank experiment from the data obtained in the sample experiment.

Controlling the gas flow, using proportional mixing of two vapor lines, allows precise control of the partial pressures of gas in the ampoule and allows the partial pressures to be altered with time, either in discrete steps or in a linear ramp. However, if it is only necessary to maintain the sample under one specific partial pressure, then a much simpler methodology may be employed. This involves the placement of a small glass tube (variously known as a Durham tube, a hydrostat or a hygrostat) holding a small quantity of solvent within an airtight ampoule containing the sample. Saturated salt solutions will maintain a constant RH within a confined space at equilibrium. The specific RH attained is dependent on the ambient temperature and salt used. Table 2 lists the RHs achieved with some common salts (50).

Solution Calorimetry

Solution calorimetry (Sol-Cal) is taken here to mean the measurement of the heat change when a solute (usually a solid but liquids may also be used) is dispersed in a large volume of solvent (to ensure complete dissolution). Usually, this is achieved using "ampoule breaking" instrumentation (wherein the solute is held in an ampoule which is mechanically broken into the solvent reservoir).

There are two types of solution calorimeter design commercially available: instruments that operate on a semi-adiabatic principle (i.e., that record a temperature change upon reaction) and instruments that operate on a heat conduction principle (i.e., that record a power change directly upon reaction). The properties of a selection of commercially available instruments are shown in Table 3. The

Table 3 Properties of a Number of Commercially-Available Solution Calorimeters

Instrument	6755 solution calorimeter	C80	MicroDSC III with mixing-batch vessel	2225 precision solution calorimeter	2265 microsolution ampoule
Manufacturer	211 Fifty Third St, Moline, IL 61265-9984, U.S.A.	Setaram SA, 7 Rue de l'Oratoire F-69300, Caluire, France	Setaram SA, 7 Rue de l'Oratoire F-69300, Caluire, France	Thermometric AB, Spjutvägen 5A, S-175 61 Järfälla, Sweden	Thermometric AB, Spjutvägen 5A, S-175 61 Järfälla, Sweden
Website	www.parrinst.com	www.setaram.com	www.setaram.com	www.thermometric.com	www.thermometric.com
Operation	Semi-adiabatic	Heat conduction	Heat conduction	Semi-adiabatic	Heat conduction
Sample volume	90–120 mL	12.5 mL	1 mL	100 mL	19 mL
Solute volume	20 mL	2.5 mL		1.1 mL	20 μL or 40 μL
Temperature range	10–50°C	Ambient to 300°C	−20–120°C	5–90°C	5–90°C
Detection limit	0.4 J	2–5 μW	0.2–2 μW	1–4 mJ	100 μJ[a]
Ampoule	Silvered glass	Hastelloy C or stainless steel	Hastelloy C	Glass	Glass or stainless steel

[a]When housed in a 2277-201 thermal activity monitor.
Abbreviation: DSC, differential scanning calorimetry.

sensitivities of these instruments, and hence the quantities of solute and solvent required for experiment vary considerably (typically, semi-adiabatic instruments are less sensitive and require much larger sample volumes). Furthermore, it is possible to use dynamic correction on data from heat conduction instruments (see section "Dynamic Data Correction") and hence collect both "raw" and "corrected" data.

Semi-adiabatic and heat conduction solution calorimeters operate on different principles and it is worth considering these differences briefly here, because they impact upon the discussion of the data presented in later chapters.

Semi-Adiabatic Solution Calorimeters

As discussed earlier, in the ideal adiabatic calorimeter, there is no heat exchange between the calorimetric vessel and its surroundings. This is usually attained by placing an adiabatic shield around the vessel. Thus, any change in the heat content of a sample as it reacts causes either a temperature rise (exothermic processes) or fall (endothermic processes) in the vessel. The change in heat is then equal to the product of the temperature change and an experimentally determined proportionality constant (or calibration constant, ε). The proportionality constant is usually determined by electrical calibration. Thus:

$$\Delta T = \frac{q}{\varepsilon} \tag{63}$$

$$\frac{dT}{dt} = \frac{\Phi}{\varepsilon} \tag{64}$$

where Φ represents power. Ideally, the value of ε returned after calibration should equal the heat capacity of the calorimeter vessel (C_v, the vessel including the calorimetric ampoule, block, heaters, thermopiles, and the sample) but in practice, losses in heat mean the value may differ slightly. However, assuming the losses are the same for both sample and reference, the power value returned will be accurate. It is also the case that C_v varies depending on the heat capacity of the sample being studied. The value of C_v affects the measuring sensitivity of the instrument. A small heat capacity results in a large rise in temperature for a given quantity of heat and, consequently, better sensitivity. However, calorimeters with low heat capacities are more sensitive to environmental temperature fluctuations and therefore have lower baseline stabilities. Any calorimeter design therefore results in a compromise between baseline stability and measurement sensitivity.

In practice, true adiabatic conditions are difficult to achieve and there is usually some heat-leak to the surroundings. If this heat-leak is designed into the calorimeter (as the case with the SolCal, Thermometric AB), the system operates under semi-adiabatic (or isoperibol) conditions and corrections must be made in order to return accurate data. These corrections are usually based on Newton's law of cooling (the most common being the method of Regnault–Pfaundler, discussed subsequently).

In the case of the Thermometric SolCal (the principles apply to all similar solution calorimeters), at the start of an experiment, the instrument is held above or below the temperature of its thermostatting bath (typically by up to 200 mK). With time, the instrument will approach the temperature of the thermostatting bath. Data capture is initiated when this approach becomes exponential (this assumption is a necessary precursor to employing the heat-balance equations used to calculate the heat evolved or absorbed by the system contained within the vessel). Thus, the response due to dissolution and any electrical calibrations (usually two are performed: one before and one after the break to ensure the heat capacity of the system has remained constant) must be performed before the instrument reaches thermal equilibrium with the bath. In practice, this limits the technique to studying events that, ideally, reach completion in less than 30 minutes.

Upon completion of an event in a solution calorimeter, a quantity of heat will be recorded. As noted earlier, the heat will be given by the product of the temperature change and the calibration constant (which in the ideal case is the heat capacity of the vessel). This interpretation assumes that the measured temperature change arises solely from the event occurring in the vessel. In practice, other events, such as ampoule breaking, stirring and heat-leakage, all contribute to the temperature change of the vessel. For accurate data analysis, these effects must be removed from the observed temperature change (ΔT_{obs}) to give ΔT_{corr}, the temperature change that would have occurred under ideal conditions. Thus:

$$\Delta T_{obs} = \Delta T_{corr} + \Delta T_{adj} \tag{65}$$

where ΔT_{adj} is defined as the temperature change arising from all the other contributing events in the vessel. Usually, the method of Regnault–Pfaundler, which is based on the dynamics of the break, is used to determine the value of ΔT_{adj} (51). In this case:

$$\Delta T_{adj} = \int_{t_{start}}^{t_{end}} \frac{1}{\tau}(T_\infty - T)\mathrm{d}t \tag{66}$$

where T is the temperature of the vessel and its contents at time t. T_∞ is the temperature that the vessel would attain after an infinitely long time period. t_{start} and t_{end} are the start and end times of the experiment, respectively and τ is the time constant of the instrument. Note that T_∞ is effectively the value of T at t_∞ and is commonly described as the steady-state temperature of the vessel. The time constant has units of seconds and can also be expressed as:

$$\tau = \frac{\varepsilon}{k} \tag{67}$$

where k is the heat exchange coefficient of the vessel.

The values of T_∞ are calculated by analysis of the baseline regions immediately preceding and following the break. These baseline sections will be approaching the temperature of the surrounding heat-sink exponentially and are described by:

$$T = T_\infty + (T_0 - T_\infty)e^{-t/\tau} \tag{68}$$

The data are then fitted to Equation (68), using a least-squares minimizing routine to return values for T_∞ and τ. Once these are known, ΔT_{adj} can be calculated. This value is then used to calculate ΔT_{corr} for the sample break and also for the two electrical calibrations ($\Delta T_{corr, calibration}$). There should be no significant difference in the $\Delta T_{corr, calibration}$ values determined for the two calibrations and an averaged value is used. The calibration constant is then determined from:

$$\varepsilon = \frac{Q_{calibration}}{\Delta T_{corr, calibration}} \tag{69}$$

The heat change for the break is then easily determined:

$$Q_{reaction} = \varepsilon \Delta T_{corr} \tag{70}$$

Isothermal Heat Conduction Solution Calorimeters

The principles for these calorimeters are exactly the same as those described above for isothermal heat conduction calorimeters and hence do not need to be described here. The only difference is that in a solution calorimeter some mechanical arrangement exists, which enables the solute to be added to the solvent. A heat conduction solution calorimeter is not limited to reaction processes that reach completion within 30 minutes as semi-adiabatic instruments are, because it is always (essentially) in equilibrium with its surrounding heat-sink. Furthermore, the greater measuring sensitivity of the thermopiles (as opposed to the thermisters used in semi-adiabatic instruments) means that smaller sample masses can be used.

Differential Scanning Calorimetry

DSC is not, as is implied by its title, an isothermal technique, and one may question its discussion in this text. However, the use of DSC is so widespread in the pharmaceutical arena that the principles of its operation are discussed here in order that the reader is able to understand the important differences between DSC data and isothermal data. Furthermore, in some instances, DSC data can be a valuable aid in the interpretation of phase transitions and other events that may be seen during an isothermal measurement, particularly during preformulation, and a number of its applications are discussed in Chapter 4.

In DSC, the power output from a sample is measured, relative to an inert reference, as it is heated or cooled in accordance with an underlying temperature program. The temperature program can be linear, modulated by some mathematical function or modified by the instantaneous sample reaction rate. DSC data can be presented in a number of ways. Most commonly, power is plotted versus temperature although it is also common to plot heat capacity versus temperature (obtained by dividing the power data by the heating rate). Note that if the area under a peak in a DSC trace is integrated to obtain the heat output (Q, in J), then the data must be plotted as power (J s^{-1}) versus time (s); most DSC software packages automatically take this into account and the user "integrates" the power versus temperature data. Sample masses, ampoules and heating rates are all dependent upon the type of instrument used and are discussed in the specific subsections below.

There are two principal types of DSC design: heat-flux DSC and power compensation DSC.

Heat-Flux Differential Scanning Calorimetry

In heat-flux DSC, the sample and reference materials are heated or cooled from a common furnace and the temperature difference (ΔT) between the sample and reference is recorded. The power change occurring in the sample is directly proportional to the temperature difference.

Power-Compensation Differential Scanning Calorimetry

In power-compensation DSC, the sample and reference materials are heated or cooled from separate furnaces. The instrument varies the power supplied by the two furnaces to maintain the temperature difference between the sample and reference at zero, and the power difference (ΔP) between the sample and reference is measured directly.

Information from Differential Scanning Calorimetry Data

Because DSC data are differential (i.e., they are normalized to the response of a thermally equivalent, inert reference), if the sample does not undergo a process during the experimental run, then the DSC plot will be a horizontal line at $y = 0$ (Fig. 6). If the sample and reference have different heat capacities, but the sample still does not undergo any change during the experiment, then the power plotted will still be a horizontal line, but it will be displaced from zero by an amount proportional to the difference in heat capacity (Fig. 6). Comparison of such data with similar data recorded for an inert reference material of known heat capacity (such as sapphire) allows the heat capacity of a sample to be quantified.

With all DSC instruments, there is a trade-off between resolution (the separation between thermal events) and sensitivity (the magnitude of each thermal event), which is dependent upon the temperature scan rate. To understand this, it is convenient to consider the processes that occur during a thermal event, such as melting. Before the melt, the sample and reference are being heated at the same rate and the power being supplied to both is the same (assuming the sample and reference materials are present in equal quantity and have the same heat capacities). (This discussion is based on using a power compensation instrument but the case for a heat-flux instrument is analogous.) Therefore ΔP is zero and a baseline is plotted. As the sample starts to melt, the energy being

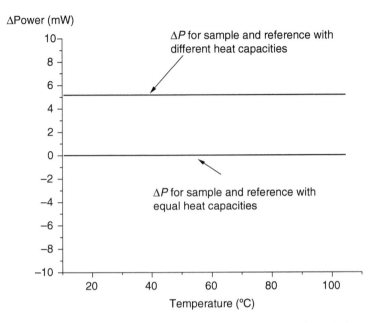

Figure 6 A schematic representation of the differential scanning calorimetry traces that would be obtained for sample and reference materials with equal or different heat capacities.

supplied by the furnace is utilized to break bonds rather than raise temperature and the sample momentarily remains at a constant temperature. The reference, conversely, continues to rise in temperature. The instrument responds to drive the temperature difference between the sample and reference back to zero by increasing the amount of power supplied to the sample side. As a consequence, ΔP is not zero and a peak is seen on the DSC trace. Once the melt is completed, the sample temperature equalizes to that of the reference, and a baseline is once again seen in the DSC trace. These processes are represented graphically in Figure 7. If the heat capacity of the sample has changed in going through the event, then the baselines before and after the peak will be different.

The effect of scan rate is now clear. The faster the scan rate, the greater the difference in temperature between sample and reference during the event and the greater the amount of power needed to maintain ΔT at zero. This gives high sensitivity but the resolution is poor, because the peaks become much broader (because the time period over which the system acts to restore ΔT to zero increases). Conversely, at slow scan rates, the temperature difference between sample and reference is reduced, leading to poor sensitivity but good resolution. These effects are represented graphically in Figure 8. Selection of the proper scan rate for an experiment is therefore critical. It must also be ensured that an

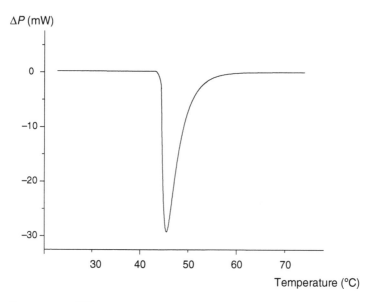

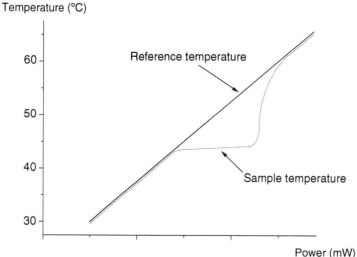

Figure 7   A schematic representation of a melting event as recorded by a differential scanning calorimetry (*top*) and the corresponding temperature profiles followed by the sample and reference materials (*bottom*).

instrument is calibrated (using the appropriate IPUAC standard materials) at the same scan rate at which any subsequent experiments are run.

By analyzing the data from a typical DSC experiment, it is possible to determine a number of useful thermodynamic parameters, including the calorimetric enthalpy (H_{cal}), the van't Hoff enthalpy (H_{vH}), the onset temperatures of

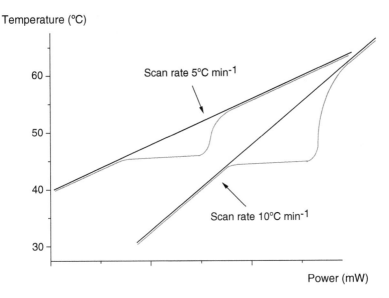

Figure 8 The effect of scan rate on the differential scanning calorimetry signal. At 5°C min^{-1}, the maximum difference between the sample and reference powers is smaller than at 10°C min^{-1}. This results in greater data resolution (separation of events) at slower scan rates but greater sensitivity (ultimate peak size) at higher scan rates.

phase transitions (T_o), glass transition temperatures (T_g), and changes in heat capacities.

Although DSC is a highly sensitive technique, typical instruments do not possess the sensitivity that can be attained with isothermal microcalorimeters, a constraint imposed by the relatively small heat capacity of the calorimetric block that is required to enable rapid changes in temperature. A typical isothermal calorimeter is capable of running separate experiments over a range of temperatures and can attain a baseline sensitivity of ± 0.1 μW, but it may require 24 hours or more to change the temperature of the instrument. A DSC, however, can operate at temperature scan rates of 200°C min^{-1} or higher (52). Because of the instrument design constraints imposed to attain such versatility, this had led to instruments with a sensitivity, in practice, of ± 10 μW.

There have been a number of recent developments in DSC experimental design, leading to three new derivatives of the technique: fast-scan DSC, high-sensitivity DSC (HSDSC), and TMDSC. The principles of these three techniques are briefly discussed in the following sections.

Fast-Scan Differential Scanning Calorimetry

In fast-scan DSC, the sample-heating rate is typically of the order of 150°C min^{-1} or higher. This will produce a data set, which, from the discussion above, will exhibit very high sensitivity but poor resolution. However, the use of fast heating rates has two principal benefits. Firstly, it becomes much easier to see glass transitions, if they are present, because of the greater difference in the baseline before and after the event. It is therefore preferable to test solid pharmaceuticals for amorphous nature by fast-scan DSC than conventional DSC (although isothermal techniques can often be more sensitive for detection of small amorphous contents; see chap. 6). Secondly, the sample is heated so rapidly that the data effectively represent a "snapshot" of its form at the start of the experiment. (In other words, if a sample contains different polymorphs, for instance, then during a conventional DSC experiment there is time for different forms to interconvert and one cannot be certain of the number and ratios of forms present in the sample initially. However, in a fast-scan DSC experiment, the melting temperature of a higher melting form is reached before it has time to convert to a different form. A further discussion of this application is deferred until Chapter 4. An additional benefit of fast-scan DSC is that most power compensation instruments can operate at these scan rates with no hardware modification.

High-Sensitivity Differential Scanning Calorimetry

HSDSC instruments are characterized by operating at very slow scan rates (typically up to $1°C \, min^{-1}$), producing data that are very well resolved. However, if high-sensitivity instruments used the same architecture as conventional DSC instruments, then the sensitivity of the signal would be very poor. This problem is overcome by using much larger sample sizes (typically 1 mL or larger). Moreover, the larger calorimetric vessels have a greater heat capacity which acts to increase instrumental sensitivity, and most HSDSC instruments are capable of detecting signals of $\pm 0.5 \, \mu W$. The increase in sensitivity over standard DSC instruments and ability to study reactions occurring in solution directly mean that HSDSC may be applied to the study of a range of systems not amenable to study by standard DSC. Typical examples include the denaturation of proteins, phase changes in lipid bilayers, phase transitions in dilute polymer solutions, and changes in the structure of creams and emulsions.

There are two basic designs of HSDSC: fixed cell instruments and batch cell instruments.

With fixed cell instruments, such as the VP-DSC/ITC (Microcal Inc.), samples must be loaded directly into the calorimetric chamber and the need for a metal pan or batch cell is eliminated. Such a system has many advantages in terms of sensitivity because, as the cells are never removed from the calorimeter, each experiment is conducted under conditions where the thermal contact between the cell and the instrument is identical, increasing the baseline repeatability of the instrument. Furthermore, because the cells are never removed, there is very little chance of damaging either the cells or the instrument, and the mechanical stability imparted to the cells reduces the baseline noise. Usually, to facilitate loading and cleaning of the cells, any samples studied must be liquids. A typical fixed cell design allows a constant volume of sample to be loaded, minimizing errors that might occur by using a balance to determine the mass of a sample loaded into a removable pan. The drawbacks of the fixed cell design include the need to use liquid samples and the difficulty that may arise in cleaning the cells after an experiment. The sample and reference cells are fixed within a cylindrical adiabatic shield, and are loaded through access tubes. As the temperature of the cell changes during the course of an experiment, the sample is free to expand and contract and, within the measuring space, a constant sample volume is achieved.

With batch cell instruments, such as the MicroDSC III (Setaram), samples are loaded into external, reusable cells and these cells are then placed in the measuring chamber of the instrument. Because the cells are removable, they are much easier to load and clean than the fixed cells discussed earlier and, hence, can be used to study liquids, solids, or heterogeneous samples. This means that it is possible to investigate, directly, the behavior of pharmaceutical products, such as creams or powders. Perhaps, the biggest advantage of the batch cell design, however, is that the sample cell can be replaced with different vessels designed to study specific systems. It is possible to obtain vessels designed to study a range of systems, which include circulating liquids, mixing circulating liquids, and mixing of liquids or solids and liquids. There is also the possibility of designing a specific vessel for a specific type of reaction. However, there are some drawbacks to the batch cell design. The cells must be loaded into the instrument, and the thermal contact between the cells and the instrument may vary between experiments. The cells are loaded to a constant weight rather than a constant volume, and it difficult to load liquid samples without a vapor space. It is also possible to damage cells though misuse or accident.

While most fixed cell instruments are power compensation instruments, because it is possible to place heaters on the base of cells that are not removable, batch cell instruments are available as either power compensation or heat-flux designs. Irrespective of whether an instrument employs fixed cells or batch cells, some design considerations are common to all HSDSC instruments. To measure small powers accurately, it is necessary to ensure good baseline stability throughout the course of an experiment. This is usually achieved by maintaining very accurate control of the calorimetric block temperature and ensuring that the properties of the sample and reference cells such as geometry, cell volume, local environment, thermal conductivity and conduction pathways, and heating rate are identical.

Temperature-Modulated Differential Scanning Calorimetry

The principles of TMDSC were first described by Reading et al. (53). Briefly, the underlying linear temperature program is modulated by some periodical function. The modulation can have any form but is typically sinusoidal or sawtooth. These are the modulations used commercially by Thermal Analysis (TA) Instruments and Perkins–Elmer Thermal Analysis (PETA), respectively. In the case of a sinusoidal modulation (similar principles apply to other modulations), the dependence of temperature of time is described by:

$$T = T_0 + \beta t + \beta \sin(\omega t) \tag{71}$$

where T_0 is the starting temperature, β is amplitude of the modulation, and ω is the frequency of the modulation. Temperature modulation means that the heating rate experienced by the sample at any particular time (the instantaneous heating rate) is constantly varying. Mathematically, the instantaneous heating rate is described by:

$$\frac{dT}{dt} = \beta + A_T \omega \cos(\omega t) \tag{72}$$

where A_T is the amplitude of the modulation. The utility of the technique is predicated on the fact that the measured power signal comprises two types of component:

1. a signal dependent on the rate of temperature change (related to the heat capacity of the sample),
2. a signal dependent on the absolute temperature (related to the kinetic changes in the sample).

The measured power can thus be described by the sum of these components:

$$\frac{dq}{dt} = C_P \frac{dT}{dt} + f(T, t) \tag{73}$$

Note that this is a simplified equation where $f(T, t)$ represents some function that describes the power from the kinetic processes in the sample; for further details, refer to Reading et al. (53).

Analysis of TMDSC data allows the separation of these components, because data are effectively recorded at two heating rates in a single experiment (the underlying heating rate and the sinusoidal heating rate).

SUMMARY

Calorimetry offers a set of qualities that makes it unique among analytical instrumentation and it is ideally suited to the study of pharmaceuticals, because it is invariant to physical form, does not destroy samples, and is sensitive enough to be able to detect degradation directly under storage conditions. In this chapter, the basic principles of the technique have been explored and the types of instrumentation commercially available have been discussed. The remainder of this book is devoted to two main areas: the principles of good experimental design and data analysis and the discussion of the applications of the technique to specific subject areas.

RESOURCES

There is a wide range of information sources available to the user of IC. The following (by no means exhaustive) sections summarizes some of these resources.

Books

There are a number of excellent books that provide information on pharmaceutical thermal analysis but many do not include sections on IC. Probably the most comprehensive textbook is *The Handbook of Thermal Analysis. Vol. 1: Principles and Practice*, edited by ME Brown, Elsevier Science (Amsterdam), 1998, which comprehensively covers all aspects of thermal analysis. A more specialized, but now out of print, text is *Biological Microcalorimetry*, edited by AE Beezer, Academic Press (London), 1982. An excellent chapter can be found in *Principles of Thermal Analysis and Calorimetry*, edited by PJ Haines, RSC (Cambridge), 2002. The subject of pharmaceutical uses of DSC is more widely covered with excellent texts including *Biocalorimetry*, edited by JE Ladbury and BZ Chowdhry, Wiley (Chichester), 1998 and *Pharmaceutical Thermal Analysis*, edited by JL Ford and P Timmins, Ellis Harwood, 1989.

Journals

Most research papers concerned with pharmaceutical calorimetry can usually be found in the pharmaceutical journals *International Journal of Pharmaceutics* (Elsevier Science Publishers, BV, Amsterdam) and *Pharmaceutical Research* (Plenum Publishing Corporation, New York) while more fundamental thermodynamics work will be found in *Thermochimica Acta* (Elsevier Science Publishers BV, Amsterdam) or *The Journal of Thermal Analysis and Calorimetry* (Klewer, New York).

Periodically, reviews of the pharmaceutical applications of IC appear in the literature. Recent examples include

Gaisford S. Stability assessment of pharmaceuticals and biopharmaceuticals by isothermal calorimetry. Curr Pharm Biotech 2005; 6:181–191.

Gaisford S, Buckton G. Potential applications of microcalorimetry for the study of physical processes in pharmaceuticals. Thermochimica Acta 2001; 380: 85–198.

Phipps MA, Mackin LA. Application of isothermal microcalorimetry in solid state drug development. Pharm Sci Tech Today 2000; 3:9–17.

Buckton G. Applications of isothermal microcalorimetry in the pharmaceutical sciences. Thermochimica Acta 1995; 248:117–129.

Koenigbauer MJ. Pharmaceutical applications of microcalorimetry. Pharm Res 1994; 11:777–783.

Buckton G, Russell SJ, Beezer AE. Pharmaceutical calorimetry: a selective review. Thermochimica Acta 1991; 193:195–214.

Manufacturers

Information about specific instruments and application notes can be obtained by writing directly to thermal equipment manufacturers or by visiting their websites. The details of selected manufacturers include

1. Calorimetry Sciences Corporation: 515 East 1860 South, PO Box 799, Provo, Utah, 84603-0799, U.S.A. www.calscorp.com.
2. Microcal Inc: 22 Industrial Drive East, Northampton, Massachusetts 01060, U.S.A. www.microcalorimetry.com.
3. Parr Instrument Company: 211 Fifty Third Street, Moline, Illinois 61265-9984, U.S.A. www.parrinst.com.
4. Setaram: 7 rue de l'Oratoire, F-69300 Caluire, France. www.setaram.com.
5. Thermal Hazard Technology: 1 North House, Bond Avenue, Bletchley, MK1 1SW, U.K. www.thermalhazardtechnology.com.
6. Thermometric AB: Spjutvägen 5A, S-175 61 Järfälla, Sweden. www.thermometric.com.

Groups and Conferences

The IUPAC sets the standards and conventions used in calorimetry and publishes the journal Pure and Applied Chemistry. IUPAC also organizes a biannual international conference. Further information can be found from www.iupac.org.

There is an annual conference held in North America organized by the Calorimetry Conference, which usually has a pharmaceutical session.

In the United Kingdom, the Thermal Methods Group (part of the Royal Society of Chemistry) holds the Thermal Analysis and Calorimetry Conference (a two-day meeting) as well as a subject-specific one-day meeting annually. More details are available from www.thermalmethodsgroup.org.

REFERENCES

1. Lavoisier A, Laplace PS. de Histoire de l'Academie Royale des Sciences 1780, 355.
2. Wadsö I. Neither calorimeters nor calorimetrists are what they used to be. Thermochimica Acta 1997; 300:1–5.
3. Skinner HA, Sunner S, Sturtevant JM. The design and operation of reaction calorimeters. In: Skinner HE, ed. Experimental Thermochemistry. Vol. 2, London: Interscience, 1962:157.
4. Wadsö I. Isothermal microcalorimetry near ambient temperature: an overview and discussion. Thermochimica Acta 1997; 294:1–11.
5. Moore WJ. Physical Chemistry. 5th ed. Harlow: Longman, 1987:48–49.
6. Atkins PW. Physical Chemistry. 4th ed. Oxford: OUP, 1990:38–39.
7. Hansen LD. Toward a standard nomenclature for calorimetry. Thermochimica Acta 2001; 371:19–22.
8. Hemminger W, Höhne G. Calorimetry: Fundamentals and Practice. 1st ed. Weinheim: Verlag Chemie,1984:131–133.

9. Kemp RB. Nonscanning calorimetry. In: Brown ME, ed. Handbook of Thermal Analysis and Calorimetry. Vol. 1. Principles and Practice. 1st ed. Amsterdam: Elsevier 1998:577–586.
10. Hansen LD, Hart RM. The art of calorimetry. Thermochimica Acta 2004; 417: 257–273.
11. Wadsö I. Recent developments in reaction and solution calorimetry. Thermochimica Acta 1990; 169:151–160.
12. Hansen LD. Instrument selection for calorimetric drug stability studies. Pharmaceutical Technology 1996; 20:64–74.
13. Souillac PO, Dave P, Rytting JH. The use of solution calorimetry with micellar solvent systems for the detection of polymorphism. Int J Pharm 2002; 231:185–196.
14. Willson RJ, Sokoloski TD. Ranking of polymorph stability for a pharmaceutical drug using the Noyes-Whitney titration template method. Thermochimica Acta 2004; 417:239–243.
15. Arnot LF, Minet A, Patel N, Royall PG, Forbes B. Solution calorimetry as a tool for investigating drug interaction with intestinal fluid. Thermochimica Acta 2004; 419:259–266.
16. Chadha R, Kashid N, Jain DVS. Microcalorimetric evaluation of the in vitro compatibility of amoxicillin/clavulanic acid and ampicillin/sulbactam with ciprofloxacin. J Pharm Biomed Anal 2004; 36:295–307.
17. Chadha R, Kashid N, Kumar A, Jain DVS. Calorimetric studies of diclofenac sodium in aqueous solution of cyclodextrin and water-ethanol mixtures. J Pharm Pharmacol 2002; 54:481–486.
18. Harjunen P, Lehto V-P, Koivisto M, Levonen E, Paronen P, Järvinen K. Determination of amorphous content of lactose samples by solution calorimetry. Drug Dev Ind Pharm 2004; 30:809–815.
19. Hogan SE, Buckton G. The quantification of small degrees of disorder in lactose using solution calorimetry. Int J Pharm 2000; 207:57–64.
20. Barriocanal L, Taylor KMG, Buckton G. A study of liposome formation using a solution (isoperibol) calorimeter. Int J Pharm 2004; 287:113–121.
21. Hansen LD, Eatough DJ. Comparison of the detection limits of microcalorimeters. Thermochimica Acta 1983; 70:257–268.
22. Randzio SL, Suurkuusk J. Interpretation of calorimetric thermograms and their dynamic corrections. In: Beezer AE, ed. Biological Microcalorimetry. 1st ed. London: Academic Press, 1980:311–341.
23. Weber H. Isothermal Calorimetry for Thermodynamic and Kinetic Measurement. European University Papers, Frankfurt: Peter Lang, 1973.
24. Zielenkiewicz W, Margas E. Theory of Calorimetry. 1st ed. Dordercht: Klewer Academic, 2002.
25. Laville G. Théorie générale du microcalorimètre Calvet. C.R. Hebd Séanc Acad Sci Paris 1955; 240:1060–1063.
26. Wadsö I, Goldberg RN. Standards in isothermal microcalorimetry (IUPAC technical report). Pure App Chem 2001; 73:1625–1639.
27. Willson RJ, Beezer AE, Hills AK, Mitchell JC. The imidazole catalysed hydrolysis of triacetin: a medium term chemical calibrant for isothermal microcalorimeters. Thermochimica Acta 1999; 325:125–132.
28. Hills AK, Beezer AE, Mitchell JC, Connor JA. Sources of error, and their correction, in the analysis of isothermal heat conduction microcalorimetric data: applications of a newly developed test reaction. Thermochimica Acta 2001; 380:19–26.

29. Hills AK, Beezer AE, Connor JA, Mitchell JC, Wolf G, Baitalow F. Microcalorimetric study of a proposed test reaction—the imidazole catalysed hydrolysis of triacetin: temperature and imidazole concentration dependence. Thermochimica Acta 2001; 386:139–142.

30. O'Neill MAA, Beezer AE, Deal RM, et al. Survey of the effect of fill volume on the values for the enthalpy and rate constant derived from isothermal microcalorimetry: applications of a newly developed test reaction. Thermochimica Acta 2003; 397:163–169.

31. O'Neill MAA, Beezer AE, Labetoulle C, et al. The base catalysed hydrolysis of methyl paraben: a test reaction for flow microcalorimeters used for the determination of both kinetic and thermodynamic parameters. Thermochimica Acta 2003; 399:63–71.

32. Parmar MK, Beezer AE, Willson RJ. An isothermal microcalorimetric study of the imidazole catalysed hydrolysis of triacetin and the observed rate constant dependence of triacetin concentration. Thermochimica Acta 2004; 424:219–222.

33. Chen A-T, Wadsö I. A test and calibration process for microcalorimeters used as thermal power meters. J Biochem Biophys Methods 1982; 6:297–306.

34. Beezer AE, Hills AK, O'Neill MAA, et al. The imidazole catalysed hydrolysis of triacetin: an inter- and intra-laboratory development of a test reaction for isothermal heat conduction microcalorimeters used for determination of both thermodynamic and kinetic parameters. Thermochimica Acta 2001; 380:13–17.

35. Skaria CV, Gaisford S, O'Neill MAA, Buckton G, Beezer AE. Stability assessment of pharmaceuticals by isothermal calorimetry: two component systems. Int J Pharm 2005; 292:127–135.

36. Finnin BA, O'Neill MAA, Gaisford S, Beezer AE, Hadgraft J, Sears P. Performance validation of step-isothermal calorimeters: application of a test and reference reaction. J Therm Anal Cal 2006; 83:331–334.

37. Stark JG, Wallace HG. Chemistry Data Book. 2nd ed. London: John Murray, 1982:56.

38. Briggner L-E, Wadsö I. Test and calibration procedures for microcalorimeters, with special reference to heat-conduction instruments used with aqueous systems. J Biochem Biophys Methods 1991; 22:101–118.

39. Straume M, Freire E. 2-Dimensional differential scanning calorimetry—simultaneous resolution of intrinsic protein structural energetics and ligand-binding interactions by global linkage analysis. Anal Biochem 1992; 203:259–268.

40. Irving RJ, Wadsö I. Use of tris(hydroxymethyl) aminomethane as a test substance in reaction calorimetry. Acta Chem Scand 1964; 18:195–201.

41. Hill JO, Öjelund G, Wadsö I. Thermochemical results for "tris" as a test substance in solution calorimetry. J Chem Thermodyn 1969; 1:111–116.

42. Kilday MV. The enthalpy of solution of SRM 1655 (KCl) in H_2O. J Res Nat Bur Stand 1980; 85:467–481.

43. Uriano GA. National Bureau of Standards Certificate. Standard Reference Material 1655, Potassium Chloride, KCl (cr) for Solution Calorimetry. 1981.

44. Archer DG, Kirklin D. NIST and standards for calorimetry. Thermochimica Acta 2000; 347:21–30.

45. Olofsson G, Berling D, Markova N, Molund M. The dissolution of propan-1-ol and dilution of 10 wt% propan-1-ol solution in water as calibration and test reactions in solution calorimetry. Thermochimica Acta 2000; 347:31–36.

46. Gao D, Rytting JH. Use of solution calorimetry to determine the extent of crystallinity of drugs and excipients. Int J Pharm 1997; 151:183–192.

47. Salvetti G, Tognoni E, Tombari E, Johari GP. Excess energy of polymorphic states or glass over the crystal state by heat of solution measurement. Thermochimica Acta 1996; 285:243–252.

48. Yff BTS, Royall PG, Brown MB, Martin GP. An investigation of calibration methods for solution calorimetry. Int J Pharm 2004; 269:361–372.

49. Ramos R, Gaisford S, Buckton G, Royall PG, Yff B, O'Neill MAA. A comparison of chemical reference materials for solution calorimeters. Int J Pharm 2005; 299:73–83.

50. Greenspan L. Humidity fixed of binary saturated aqueous solutions. J Res Nat Bur Stand 1977; 81A:89–96.

51. Wadsö I. Calculation methods in reaction calorimetry. The LKB Inst J 1966; 13:33–39.

52. Noble D. DSC balances out. Anal Chem 1995; 67:323A–327A.

53. Reading M, Luget A, Wilson R. Modulated differential scanning calorimetry. Thermochimica Acta 1994; 238:295–307.

2

Good Experimental Practice

INTRODUCTION

In Chapter 1 the fundamental principles of isothermal calorimetry and the basic design of modern calorimeters were outlined. It was highlighted there that modern calorimeters can achieve a high level of sensitivity, are nonspecific in their operation (i.e., they record all heat flow regardless of source), and can yield high-quality, quantitative data for a wide range of chemical and physical processes (1–4). If these advantages are to be exploited to their fullest potential, it is essential that the calorimeter operator is fully aware of the possible sources of error in their data, either through poor experimental practice or method development and/or poor calibration. It is important that the operator is able to minimize this potential for error and have some means for identifying and correcting systematic errors, should they be introduced. As calorimetric methods become more widespread and if the data obtained from calorimetric experiments are to be fully utilized, it is vital that there is some means of validating data obtained from different instruments and different operators.

This chapter will discuss the fundamentals of good experimental practice for a variety of instruments, covering the main range of application (ampoule, isothermal titration, solution, flow, and perfusion) of calorimetry. Good experimental practice is, of course, of the utmost importance for any scientific study but it is particularly so for calorimetry. Due to the high sensitivity and nonspecificity of calorimetric instruments there is the ever-present risk of introducing erroneous powers to the system which may give rise to misleading information being derived for a system under study. Some of the most important factors which must be considered when performing a calorimetric experiment include choice of ampoule, choice of reference material, and location of the instrument. These and other considerations are discussed.

For any scientific instrument, calibration and validation are essential to ensure the quality of the data obtained. The calibration and validation of calorimetric instruments is a much-debated topic (5–9), and hence attention will be paid to this subject with specific reference to the International Union of Pure and Applied Chemistry (IUPAC) guidelines for calibration and validation of calorimeters (10).

GOOD EXPERIMENTAL PRACTICE

Before any calorimetric study is undertaken, there are a number of factors to consider before even one datum is recorded; any decisions made at this point can impact severely on the returned calorimetric data. Such factors include, for example: location of the calorimeter, choice of ampoule material (stainless steel or glass), choice of reference material, quantity of reacting material, and the time allowed for the system to reach thermal equilibrium with the instrument. All of these factors require careful consideration before any experiment is performed. These factors and others are discussed in the following sections.

Sample Preparation

As heat is ubiquitous, it is very easy to corrupt an experiment through poor sample preparation. As a general rule, 90% of the effort in running a calorimetric experiment is spent on sample preparation. In simple terms, this means ensuring that the sample is pure (or, at least, not contaminated with any unexpected material), the ampoule is clean and properly sealed, there are no interactions likely to occur between the sample and the ampoule, and no dust or other debris is present in the calorimeter. There are other specific sample considerations depending upon the type of experiment to be run and these are discussed in the following relevant sections.

Location of the Instrument

The physical location of the calorimeter is very important. The quality of the data returned from the instrument and its high sensitivity derive from having accurate and precise control of temperature. If the environmental conditions fluctuate significantly then the instrument's temperature control may be compromised, resulting in poor or misleading data. Generally, the calorimeter will be housed in a temperature-controlled environment designed to operate within specified limits. For example, the TAM (Thermal Activity Monitor, Thermometric AB) will function within the manufacturer's stated limits, operating temperature ± 1 mK, with a baseline stability of ± 0.1 μW over 24 hours if the environmental temperature is within $\pm 10°C$ of the operating temperature of the calorimetric unit. As noted earlier, the calorimeter is nonspecific in its operation, and therefore it is essential to minimize the risk of introducing any external, erroneous powers to the system. As for all sensitive pieces of equipment, the calorimeter should not be subject to any movement while data are being collected. Even small vibrations can induce a relatively large signal, giving rise to poor data. Also important to note is that the calorimeter should be protected from electrical surges/spikes from the electricity supply (the populace switching their kettles on at 8 a.m., for example) by the use of appropriate electrical circuit–protection equipment.

Choice of Ampoules

Calorimeters, unless they are operating in flow-mode, require the use of ampoules to contain the sample. Some manufacturers offer a variety of disposable and non-disposable ampoules, usually constructed of glass or stainless/hastelloy steel. There is, therefore, a choice to be made when designing the experiment. Each type of ampoule has a number of advantages and disadvantages. For example, glass ampoules are usually disposable (hence the risk of cross contamination between experiments is minimized) and they are impervious to most types of chemical attack. Steel ampoules, on the other hand, are susceptible to corrosion and are relatively expensive while alloy ampoules, such as those made of hastelloy, can be susceptible to corrosion from strong acids and bases. They are, however, reusable, easy to clean, and very robust. In all cases, it is highly desirable to have as closely matched sample and reference ampoules as possible in order to minimize systematic errors arising from discrepancies in heat capacities and differential thermal contacts with the thermopiles.

Most instrument manufacturers supply disposable and nondisposable ampoules engineered to high degrees of precision specifically to minimize such sources of error. Having removable ampoules allows a wide range of experiments to be performed (as well as allowing ancillary apparatus such as titration or solution units to be used) but it can introduce nonsystematic errors for each experiment. Some instrument manufacturers employ fixed ampoules to overcome this limitation but as always this is a compromise. Fixed ampoules may limit the range of experiments one can perform but offer the significant advantage of having a defined system, which does not change with each new experiment. While there is little choice over the material from which fixed ampoules are constructed, there is some choice over the materials from which removable ampoules are manufactured; the choice of such can impact on the outcome of a calorimetric experiment and hence deserves some consideration.

Glass Ampoules

Glass ampoules offer a major benefit in that they are impervious to most chemicals and hence any erroneous power associated with corrosion is generally not an issue. Uniformity between experiments is an important factor to consider, however, but modern mass production methods generally mean that inter- and intra-batch variability is minimal. Glass ampoules do suffer the limitation that the long-term costs are higher than those associated with reusable ampoules.

One area where the use of glass ampoules can be problematic is where studies on proteins or other biologicals are conducted. Enzymes, for example, are well known for their affinity for surfaces and the Si-O terminal groups found on glass surfaces are well suited to bind and immobilize enzymes, often with a concomitant change in the enzyme activity. This effect can be ameliorated through the use of a suitable surface-coating agent, such as a silicone-based product like Repelcote[a], for example.

It is necessary to have a tightly fitting lid because any evaporation from the ampoule, apart from changing the system under study, will give rise to a very large endothermic signal which will affect or even completely mask the thermal events of interest. Glass ampoules are usually sealed by means of a crimped lid and a rubber seal. This arrangement may not always give a hermetic seal and so great care must be taken when sealing such crimp-fit lids.

Steel Ampoules

Steel ampoules, as they are reusable, offer the advantage of being cost effective over a number of experiments and they are usually a screw-top fit, which allows for a more effective seal to be achieved. They are, however, susceptible to corrosion, especially by halides. This means that they are unsuitable for use in experiments in which high halide concentrations are present (mineral acids, salt solutions, and so on) or for long experimental studies with low concentrations of halide. Salt solutions, in particular, are often used to maintain a specific relative humidity (RH) in an ampoule and great care must be taken when using these in steel ampoules. As noted in Chapter 1, most chemical changes give rise to a change in heat, and corrosion is no exception. If data are being collected against a background power associated with corrosion, then the returned data will be a summation of the power of interest and the power associated with corrosion, thus rendering those data effectively useless.

Being steel, the ampoules have a higher thermal mass relative to glass ampoules and hence there are some considerations to be taken into account when designing the experiment. For example, steel ampoules will take longer to reach thermal equilibrium with the calorimeter and hence important data may be lost (this is discussed in more detail later in this chapter).

Choice of Reference

Modern calorimeters, on the whole, operate as differential instruments; that is, the power from the sample side is recorded as the difference between it and an inert reference side. This, in principle, maximizes the signal-to-noise ratio and minimizes any erroneous signals arising from environmental temperature fluctuations, as described in the theoretical treatment in Chapter 1. The reference material should be inert and as closely matched to the sample material as possible, both in terms of the heat capacity (and hence, mass) and the nature of the sample material. For example, if a solution phase reaction is to be studied then the reference material would be an identical volume of solvent (usually deionized water),

---

[a]Repelcote is a 2% v/v solution of dimethyldichlorosilane in octamethylcyclotetrasiloxane.

while for solid-state studies dried talc is usually sufficient. With closely balanced ampoules and with an inert reference comes confidence that any displacement of the signal from zero is the result of chemical or physical processes occurring in the sample ampoule and not from other factors.

Attaining Thermal Equilibrium with the Calorimeter

A limitation of calorimetric instruments in which removable ampoules are used is the requirement that the ampoules and system to be studied must be allowed to reach thermal equilibrium with the calorimeter before any meaningful data can be collected. The procedure varies between instruments, but often the ampoules are placed in an equilibration position for a set time ($\sim$30 minutes) before being placed in the measuring position. The time taken for thermal equilibrium to be achieved is directly related to the heat capacity of the ampoules and their contents so, for example, steel ampoules generally require a longer equilibration time than glass ampoules. It is usual for experiments performed using removable ampoules that the reaction is initiated outside of the calorimeter and hence the reaction precedes the commencement of data capture.

If the ampoules are lowered into the measuring position, a frictional heat will be generated, which must dissipate before any useful data can be recorded. These limitations pose a severe problem if reactions with short half-lives are to be studied because a significant proportion of the data is lost (typically one hour from reaction initiation to useful data collection). For long, slow reactions this is less of an issue providing that the mechanism is constant over that initial period of time. These effects can be minimized to some extent by prethermostatting the ampoules and reagents to the operational temperature of the instrument and by controlled lowering of the ampoules (11). A further difficulty associated with this is the fact that it is impossible to know truly what the calorimetric signal is at time zero (a value required for some of the calculations). This is discussed in more detail in Chapter 3.

CALIBRATION AND VALIDATION

The accuracy of any calorimetric measurement performed is essentially determined by the reliability of the calibration routine of the instrument and can be no better than the accuracy of the calibration. It is, therefore, of primary importance that the calorimeter be calibrated on a regular basis and certainly after any modifications to the system have been performed (such as changing the temperature or ampoule type). Most modern calorimeters are calibrated by means of an integral internal electrical calibration heater located in close proximity to the calorimetric channel. Internal electrical calibration can be performed using either the integral calibration heater or by means of an insertion heater. Internal electrical calibration relies on generating a known quantity of heat from a resistance heater located in close proximity to the sample measuring position.

Electrical Calibration

The heat released or absorbed by a system under study is detected by exchange of that heat through a series of thermocouples (collectively known as the thermopile) located around the perimeter of the ampoule. The consequence of a power flowing across the thermopile is that an electrical potential (U) is generated; the potential is quantitatively proportional to the power flowing across it. The raw calorimetric signal is found from multiplying U by a proportionality (calibration) constant (ε), which is determined experimentally. The raw power P_R are then plotted as a function of time to give the characteristic form of calorimetric data dq/dt, Equation (1):

$$P_R = \frac{dq}{dt} = \varepsilon U \tag{1}$$

In electrical calibration, a precisely known current, I (Amps), is passed through a wire of precisely known resistance, R_w (Ohms):

$$\frac{dq}{dt} = I^2 R_w \tag{2}$$

Consequently, the total heat released, Q (J), by the calibration heater can be calculated for a given period of time, t (seconds):

$$Q = I^2 R_w t \tag{3}$$

This is a convenient method in that it is relatively easy to perform.

In broad terms, there are two types of calorimetric experiment. Those in which the reaction under investigation has some kinetic dependence (i.e., chemical degradation) and those in which there is little or no kinetic dependence (i.e., the reaction can be considered to be essentially instantaneous, ligand–substrate binding, for example). This distinction is required because of the limitation of all calorimeters record data instantaneously [the delay in measurement being related to the time constant(s) of the instrument]. The main principles of these design limitations are discussed in more detail in Chapter 1.

To recap, the calorimeter records the heat released, or absorbed, by a reacting system via a thermopile arrangement located close to the sample site. If the calorimeter were to operate ideally, then the heat change from the sample would be detected instantaneously and the calorimetric signal would be derived directly. However, such a perfect calorimeter does not and cannot exist and consequently there is always a delay between heat generation and heat transfer, the so-called dynamic response of the calorimeter. For modern heat conduction calorimeters this dynamic response is of the order of 100 to 500 seconds. As a general rule, for systems where the half-life of reaction is significantly >500 seconds, the raw calorimetric signal is essentially the true response and the dynamic response of the calorimeter can be ignored. However, for reactions in which the half-life is less than or approaching 500 seconds, the dynamic response becomes an important factor in the returned data and hence steps must be taken to compensate for its effects, such as the dynamic correction method discussed in Chapter 1.

Limitations of Electrical Calibration

Although useful, electrical calibration does have some limitations. At least two conditions must be met for electrical calibrations to be valid; the first is that all of the heat must be generated by the heater (there must be no other heat source) and the second is that all of that heat must be transferred to the calorimeter. In practice, these conditions can be never met; the electrical heater must be connected to the calorimeter surroundings by wires that have both finite thermal conductivity and electrical resistance. The requirements of reducing both thermal conductivity and electrical resistance are diametrically opposed, and as a consequence, any heater design must be a compromise of the two. Another important issue to consider is that many different research groups use calorimetry as a technique for deriving kinetic and thermodynamic information for various systems. It is possible that nominally identical calorimeters may perform differently and, hence, in order for information to be directly comparable with those reported by other groups, it is necessary to have some accepted standard by which all values are measured.

The use of electrical calibration heaters can be unsatisfactory for several other reasons. It is, in principle, possible for the behavior of the internal calibration heaters to vary over a period of time, being affected by changes in the resistance of the heater itself, its electrical connectors, or a malfunction in the heater supply resulting in the introduction of systematic errors to the recorded data. Misreporting of the calorimeter-operating temperature would also not necessarily be detected by electrical calibration (12).

A further criticism of electrical calibration is that it is difficult to generate power patterns identical to those produced by a real process, and this may again contribute toward systematic error (6,13). For an ideal heat conduction calorimeter, all heat exchange between the ampoule and the heat-sink takes place through the thermopiles. However, for practical reasons, this is not possible and alternative power pathways are introduced; for example, air gaps, support structures, and electrical leads. As a consequence, it is not uncommon for the proportion of heat, which is unrecorded to approach 50% (13) and even greater for some designs of flow calorimeter (14). This is compounded by other factors such as nonideal design of the thermopiles, vessel and/or the electrical heater, the position of the heater, the heat conductance of the ampoule material, the contents of the ampoule (liquid or solid), and so on. If these errors are to be minimized, it is essential that the calibration routine imitates accurately the power expected for the system under investigation. This does not mean that electrical calibration should not be performed; it simply means the results should not necessarily be taken at face value.

It is not necessary to calibrate every time an experiment is performed if a series of similar experiments are to be conducted. For instance, if a number of solution phase reactions in 3 mL glass ampoules are to be carried out, then a single calibration at the start is usually sufficient for at least one or two weeks. If, however, the experimental set-up is altered, if steel ampoules or inserts are used, for example, then it is necessary to perform a new calibration. If an insert such as a perfusion unit is used, the calorimeter must be calibrated with the insert and a reference ampoule in place. While every effort should be made to have the sample and reference sides as closely matched as possible, it is inevitable that, unless an identical insert is used in the reference side, the sample and the reference will not be balanced. If a calibration has previously been run on a balanced set of ampoules, then the zero point will not be the same and a new calibration will have to be run. If the type of reaction being run is of a similar nature to the substances used in the calibration, then the recorded thermal power is the true value of that produced by the reaction under investigation.

Chemical Test Reactions

Most isothermal calorimeters are calibrated by means of a precisely controlled electrical heater, but as already noted in some instances it is difficult to conduct electrical calibration experiments which mimic closely the power produced by the reaction or process under investigation, leading potentially to systematic errors being introduced to the recorded data. In such instances, the use of chemical test reactions has been proposed (8,13,14) as a complementary technique to alleviate some of these issues.

The use of chemical test reactions for isothermal calorimeters, specifically for ampoule heat conduction instruments, was proposed some time ago by Chen and Wadsö (8). The principles put forward have been debated extensively in the literature (15–20) and extended to incorporate other types of calorimeter, culminating in the publication of recommendations for chemical test reactions for a variety of isothermal calorimeter types by IUPAC (10).

Chemical test reactions serve purposes other than as a validation for isothermal calorimeters; because the chemical standards used are well characterized they can be employed as training routines for new users allowing a check on their experimental technique. Chemical test reactions also allow troubleshooting of instruments (12,18) suspected of performing out of specification, because any malfunction will manifest itself through the values derived for the known parameter.

For a chemical test reaction to be useful, it must fulfil several criteria (10). It must be readily available in a quality that does not require extensive purification or analysis and, wherever possible, expensive, hazardous, and unstable materials should be avoided. If the calibration is to be used as a training exercise, the reaction should also be relatively easy to perform. Based upon these requirements, the IUPAC committee (6) proposed several different test reactions covering the main range of applications of calorimetry. In order to appreciate fully the need for chemical test reactions, it is necessary to consider aspects of operation of some general calorimeter designs. What follows is an overview of standard operating procedures for titration, solution, batch/ampoule, and flow calorimeters with appropriate discussion as to the recommended chemical test reactions.

GOOD PRACTICE FOR AMPOULE/BATCH CALORIMETRY

Ampoule or batch calorimetric experiments (those in which the sample remains static) are generally the simplest experiments to design and perform, as they require no external intervention, unlike perfusion or solution calorimetry, for example. This simplicity of operation permits the study of a wide variety (21) of systems including those which are homogeneous (22) in nature and those which are heterogeneous (23,24) (suspensions, semisolids, etc.). It is also possible to study living biological systems such as microbial (25,26) species as well as insects and plant material (27). The only requirement is that the sample, or at least a representative fraction of it, can be contained within the calorimetric ampoule. This versatility in the systems amenable to study does carry some limitations. For example, in some instruments the capacity to stir the reaction medium is not available. While this is not necessarily a difficulty for homogeneous systems it clearly can present serious difficulties for heterogeneous samples. Ampoule/batch experiments are truly noninvasive and nondestructive as they merely monitor the sample changing as a function of time; essentially, when the experiment is concluded, the only change that has occurred to the sample is that which it would have undergone anyway, had it been stored under the same conditions.

The majority of ampoules are removable and, therefore, particular attention must be paid to ensure the ampoule is hermetically sealed after loading. As soon as the sample components are mixed or, in the case of single-compound studies, sealed within the ampoule, the time should be noted. The loaded ampoules should then be allowed to reach thermal equilibrium with the calorimeter, following which data collection is initiated. The time between mixing or sealing and the commencement of data capture should be recorded and must be appended to the start of the time column before any analysis is performed.

Calibration of Ampoule/Batch Calorimeters

Electrical Calibration

Electrical calibration should be performed before the first use of the apparatus, following a change in amplifier range or loss of power supply and/or on a weekly basis.

Chemical Test Reactions

As with solution calorimetry, ampoule/batch calorimeters are versatile in their application; that is, solids, semisolids, liquids, or suspensions can all be studied using these instruments but, broadly speaking (in terms of requirements for chemical test substances), they can be classified as either solid-state or solution phase.

Modern calorimeters are now sensitive enough, and have sufficient long-term stability, to monitor extremely slow reactions with lifetimes of many years and because of the form of the returned data it is possible to derive not only thermodynamic but also kinetic information. The ability to derive kinetic information is vastly important for studies on systems for which stability is an issue. Due to the invariance to physical form, stability studies can be conducted on both homogenous and heterogeneous solution phase and solid-state systems (single component or formulations, for example). As a consequence, there is also a requirement for test reactions that mimic these long-term processes.

Solid-state test reactions: As interest in calorimetry grows within the pharmaceutical sector, the application of calorimetric methods to the study of long-term physical and chemical stability of solid compounds (actives) or formulations (tablets, for example) will also increase. There exist many examples of the use of calorimetry for such studies (2) although, to date, no test reaction has been proposed for solid-state systems. There is a need for different test reactions to assess calorimeter performance depending upon the type of system and the nature of the change expected; that is, it is necessary to have a test reaction for assessment of physical change and a second for assessment of chemical change. Even with this apparent need for suitable test reactions, there are no recommended systems. The issue is somewhat clouded by the inherent variability and heterogeneity usually encountered in solid systems from factors such as particle size/surface area, defects, and impurities all of which may give rise to unacceptable variability in even the most carefully collected data.

Solution phase test reactions: There are many instances where it is desirable to monitor a system for days, weeks, or even months in a calorimeter. In such instances, a test reaction which will release a predictable quantity of heat over extended periods of time is required. The imidazole-catalyzed hydrolysis of triacetin (ICHT) was proposed (8–13) some time ago as a test reaction to fulfil this purpose. The ICHT has been the subject of several recent research articles (16–19), including an inter- and intra-laboratory study, which resulted in values for ΔH and k of -91.7 ± -3 kJ mol^{-1} and $2.8 \times 10^{-6} \pm 0.097$ mol^{-1} s^{-1} dm^3, respectively.

Photocalorimetric test reactions: The use of ampoule calorimeters for the study of photo-induced reactions has been reported by several groups (28–31). In order to compare results from different experiments and between operators and/or different laboratories, it is essential that the light energy (intensity, W m^{-2}) being delivered to the sample can be quantified (actinometry). There are two methods by which this can be achieved. The first is through spectroradiometry, which can give a precise value for the number of photons produced by the light source but, like the electrical calibration routine, it does not provide any information on how the calorimeter might respond in the presence of a sample undergoing reaction.

The second is to use a chemical actinometer. The IUPAC-recommended system is the photoreduction of potassium ferrioxalate $(0.15 \text{ mol dm}^{-3})$ in aqueous sulfuric acid solution $(0.05 \text{ mol dm}^{-3})$. The recommended value for ΔH is $-52.6 \pm 0.8 \text{ kJ mol}^{-1}$. The use of potassium ferrioxalate as a test reaction has been criticized by Morris (31) because of its sensitivity to oxygen and hence the requirement that it be prepared under an inert atmosphere and in a deoxygenated solvent; it has also been postulated that the reaction may in fact be complex, comprising of several reaction steps, further reducing its usefulness as a test reaction.

The photoreduction of 2-nitrobenzaldehyde has been suggested as an alternative (31). This reaction does not suffer the same limitations as those for the photoreduction of potassium ferrioxalate, as it is less sensitive to oxygen and, hence, can be prepared under atmospheric conditions; it also conforms to simple zero-order kinetics, making data analysis straightforward, although currently there are no recommended standard values for the enthalpy and rate of reaction.

Data Interpretation

Ampoule calorimeters are most commonly used for the monitoring of stability and/or compatibility, often of complex systems. The parameters sought are therefore usually related to the kinetic nature (rate constant, Arrhenius values, and so on) of the system. Perhaps uniquely of the instrumentation described here, there are so many approaches to data analysis, because of the complex nature of the data returned, that this subject forms the basis of a separate chapter (Chap. 3).

GOOD PRACTICE FOR TITRATION CALORIMETRY

Titration calorimetry is used to study the interaction between two entities in solution or suspension, and directly measures the interaction enthalpy (32). In a typical titration experiment, a known and fixed amount of liquid (typically ~ 1 mL) is held in the ampoule (the titrand) and is continuously stirred. A second preequilibrated liquid (the titrant) is injected via an accurately controlled syringe pump and cannula into the ampoule. The titrant can be introduced by continuous or periodic injections of fixed volumes of liquid. On injection of the titrant to the titrand the calorimeter measures any accompanying heat changes associated with physical (such as dissolution and/or dilution) and chemical (such as ligand–substrate binding) change. The versatility of such instruments is demonstrated by the range of experimental studies reported in the literature (33–36).

It is evident that one species must be selected as the titrand and the other as the titrant. In principle, the arrangement is not important and indeed repeating the experiment with the reverse arrangement can be a useful check of the experimental results. There are, of course, factors which may influence the decision as to which is the titrand and which is to be the titrant.

1. The volume of sample in the cell is typically larger than the volume of solution injected.
2. The molar ratios of titrand and titrant must be selected carefully, to ensure that binding saturation occurs during the experiment.
3. The titrand solution is usually quite concentrated (because it is desirable to inject small quantities of material to minimize thermal shocks and heat-capacity effects), so the solubility of the species is important.

Sample preparation is then the same as for any other calorimetric experiment; care must be taken to ensure that materials are pure and, especially if biological materials are involved, that buffers are used. Note here that it is imperative that if buffers are used as the solvent, then the same buffer is used for both the titrand and the titrant solutions because the enthalpy of mixing/dilution of buffers can be significant. It should also be noted that if a reaction produces a product which causes one or more of the buffer components to ionize the heat of ionization can be considerable and should be corrected for. It is also important to ensure that, if the same interaction is studied in more than one buffer, the ionic strength is maintained at a constant value.

The general objective of a calorimetric titration experiment is to find the binding enthalpy for an interaction and, ideally, the equilibrium constant. In order to achieve this, sufficient titrant needs to be added to the titrand to span the reaction stoichiometry (equivalence preferably being attained at the midpoint of the experiment). This is usually ensured by proper calculation of the concentrations of the two solutions, resulting in an experiment consisting of 10 to 25 injections. The time between the injections must be sufficient for the system to return to equilibrium (i.e., all reactions must be complete before the next injection is made). Usually the gap is on the order of minutes.

For an instrument with a batch cell (such as a TAM), the cell is typically filled to approximately 75% of its full capacity to maximize the signal without the risk of overflowing when the stirrer and titrant are added. For a dedicated titration calorimeter (such as a Microcal VP-ITC), the cell is actually overfilled and a constant volume is maintained throughout the experiment (and, consequently, a correction must be applied to account for the dilution, and loss of material, from the cell as the titration proceeds). If the calorimeter has a reference cell, this is filled with either the same substance as the sample cell or buffer alone (if this has similar heat capacity to the sample). To ensure the highest quality of data, it is essential that the titrand and titrant are mixed efficiently. To ensure this, a variety of stirrer designs such as propellers and turbines can be employed (37). The stirrer blades should be positioned just above the base of the cell (to avoid friction) and, for good mixing, should be completely covered by liquid. For results to be comparable between experiments, the stirrer speed should be maintained at some constant value.

One final consideration is with the positioning of the cannula through which the titrant solution is injected. In the case of Microcal instruments, the cannula is below the surface of the sample liquid. For TAM experiments, the cannula can be placed above or below the surface of the sample liquid. In general, it is preferable that the cannula be placed below the liquid surface, because this obviates the significant problem of a drop forming, but not detaching, from the end of the cannula during an injection. However, there is always the possibility of diffusion of liquid from the cannula into the solution during equilibration/calibration; this usually results in the first peak in a titration experiment being smaller than expected, and this must be corrected for in any subsequent calculation.

As with all calorimeters, it is necessary to protect against unwanted interactions between the chemical entities and the exposed components of the calorimeter. If nondisposable ampoules are to be used (as is usually the case with titration calorimeters), then cleaning of the cell becomes vitally important.

Calibration of Titration Calorimeters

Electrical Calibration

Perhaps uniquely for the systems discussed in this chapter, titration calorimeters are often affected by the dynamic response caused by the time constant(s). This is because most experiments involve measurement of the heat change when small aliquots of solution are injected into the ampoules, and this usually occurs over a period of minutes.

Therefore, two types of electrical calibration can be indicated for titration experiments. Following the attainment of thermal equilibrium, a normal electrical calibration (as already described) is performed. This is used to set the position of the baseline and the full-scale deflection prior to experimental measurement. Subsequently (and optionally, depending upon the type of instrument being used), a dynamic calibration (i.e., one that corrects for the dynamic response of the instrument) may be performed. The principles of dynamic correction are discussed more fully in Chapter 1.

Chemical Test Reactions

Chemical test reactions for titration calorimeters can be designed to validate the measurement of enthalpy only or to validate the measurement of enthalpy and the binding constant. In the former case, acid/base neutralizations are often used, while in the latter there are two IUPAC-indicated binding interactions.

Acid/base reactions: The standard test and reference reaction for acid/base reactions is the neutralization of 0.1 M HCl with a large excess of 0.1 M NaOH. The recommended value for the protonation of hydroxyl ions at 298.15 K and infinite dilution is $\Delta_{sol}H_m^\infty = -55.81$ kJ mol^{-1} (10). The experimental values can be corrected for the enthalpy of dilution of the concentrated injected solution.

Neutralization reactions often have problems, which may render them effectively useless as test reactions. For example, in reactions where one component is present in large excess, it is likely that the reaction will proceed to completion, even if the mixing efficiency is poor. In such an instance, a correct value can be returned for the calibration reaction but, because of the poor mixing efficiency, false results may be returned for other reactions. There is also the issue of CO_2 contamination, a particular problem for dilute NaOH solutions, and hence it is necessary to prepare the solutions in CO_2-free water and prevent uptake of CO_2 during the experiment. If the reaction vessel, or the injection tube, is made from stainless steel, erroneous powers may also be introduced by corrosion of the vessel by the acid.

Ligand-binding test reactions: There are two ligand-binding experiments used routinely in the calibration of titration calorimeters. These permit the simultaneous determination of the enthalpy change ($\Delta_b H$) and the equilibrium-binding constant (K_b).

The first reaction is the binding of Ba^{2+} to 18-crown-6 (1,4,7,10,13,26-hexaoxacyclooctadecane) (10,13,38). The values for the enthalpy and equilibrium constant (at 298.15 K), when run under the conditions specified in the IUPAC report, are 31.42 ± 0.20 kJ mol^{-1} and $5.90 \pm 0.20 \times 10^3$ mol^{-1} dm^3, respectively (10). Again, it is normal to apply a correction for the enthalpy of dilution of the injected barium salt solution.

The second reaction is the binding of 2′ CMP to RnaseA (10). The derived thermodynamic properties for this reaction are highly dependent on pH, ionic strength, temperature, and RnaseA concentration, and hence, its use as a test reaction is limited. The IUPAC committee has proposed standard conditions (potassium acetate, 0.2 mol dm^{-3}; potassium chloride, 0.2 mol dm^{-3}; RnaseA, 0.175 mol dm^{-3} at pH 5.5) for which values of the enthalpy and equilibrium constant of -50 ± 3 kJ mol^{-1} and $1.20 \pm 0.05 \times 10^5$ mol^{-1} dm^3, respectively, should be attained (10).

Data Interpretation

The enthalpy change for each injection is found by integrating the power–time curve relative to a suitable baseline. However, titration experiments always involve a compromise in design and a minimum of three blank experiments must be conducted: the dilution of the titrant, the dilution of the titrand, and the dilution of the solvent. Assuming the controls are proper, the area under each peak is therefore the enthalpy of interaction. Initially, it is assumed that complete binding occurs; this allows estimation of $\Delta_b H$ in J mol^{-1} because the number of moles of titrand injected is known. As the experiment progresses and the titration end-point is reached, the measured heat falls. A plot can then be constructed of the enthalpy per injection (J) versus number of moles of titrand injected (equivalent to a binding isotherm). The binding constant is calculated by fitting the binding isotherm to (usually) a two-state binding model using least-squares minimization.

As is always the case with calorimetric experiments, care must be taken when designing ITC experiments such that the measured heat derives solely from the interaction of interest and not from an artefact. For instance, it is difficult to use ITC to characterize binding reactions with very high-binding affinities, because once the binding affinity increases beyond approximately $10^8 - 10^9$ M^{-1}, isotherms will lose their characteristic curves and become indistinguishable (39). Recently, an elegant solution to this problem, which had previously been a significant limitation of the technique, has been presented (40). In this approach, a high-affinity ligand is titrated into protein that has been prebound to a weaker inhibitor. Over time, competitive displacement results in the stronger ligand replacing the weaker. The approach requires three titrations: a titration with the weak inhibitor in order to characterize its binding thermodynamics, a titration with the high-affinity ligand, and the displacement titration. Once the binding characteristics of the weak inhibitor are known, under a given set of conditions, this titration does not need to be repeated. The titration with the high-affinity ligand confirms the binding enthalpy and allows a more rigorous analysis of the displacement titration.

This methodology has allowed the analysis of systems previously outside the scope of ITC, such as the inhibition of HIV-1 protease (41). Furthermore, selection of a weak inhibitor that has a binding enthalpy of opposite sign to the high-affinity ligand amplifies the calorimetric signal. If binding is accompanied by a change in protonation, then the interaction will be pH dependent and the measured binding enthalpy will be dependent upon the ionization enthalpy of the buffer system in which the reaction is occurring (42). The ITC experiment must then be carefully constructed in order that this effect is correctly compensated for. Initially, this is done by performing titrations in buffers with different (and known) ionization enthalpies ($\Delta_i H$). The measured enthalpy ($\Delta_e H$) is then given by:

$$\Delta_e H = \Delta H + n_H \Delta_i H \tag{4}$$

where n_H is the number of protons absorbed or released.

GOOD PRACTICE FOR SOLUTION CALORIMETRY

A solution calorimeter records the heat change when a sample (solid or liquid) is dissolved in a large volume of solvent. Two types of solution calorimeter are commonly used:

1. adiabatic instruments (i.e., calorimeters that record a temperature change upon reaction) and
2. heat conduction instruments (i.e., that record a power change directly upon reaction).

The operating principles of these instruments were discussed fully in Chapter 1. The solute and solvent can be combined by various means. For example, the C80 (Setaram) utilizes a batch cell. The cell is rotated end over end, allowing the solute and the solvent to mix. An alternative is to seal the solute in a glass ampoule, which is broken into the solvent to initiate reaction (used in the 2225 precision solution calorimeter, Thermometric AB). Another option is to hold the solute in reusable metal canisters, which are broken into the solvent (used in the 2265 20 mL microsolution ampoule, Thermometric AB).

In order to generate precise values for the desired parameters, it is essential that the mass of the sample to be studied is measured accurately. Note that for optimal accuracy, the quantity of sample used should be freely soluble in the volume of solvent used and for semi-adiabatic instruments the sample should dissolve completely within the 30-minute window discussed earlier. Once loaded, the ampoule should then be stored under a dry atmosphere (e.g., in a desiccator over P_2O_5 or in a vacuum oven). This is particularly important for amorphous or partially amorphous samples. Once the sample is dry, the ampoule and contents should be weighed again. At this point, it may be necessary to correct for air buoyancy (this can affect the weight by as much as 0.05%, depending on sample density) especially if high-precision measurements are to be made. If the sample is prone to oxidization, it may be necessary to flush the ampoule headspace with an inert gas before sealing. Solution phase and liquid samples require a similar routine but obviously the drying step is not necessary.

Calibration

Electrical Calibration

For semi-adiabatic instruments it is necessary to perform two (usually) electrical calibrations (one before and after the mixing) in order to convert the raw data (temperature) to power. For heat conduction instruments, this calibration is not necessary but the instrument should be calibrated on a regular basis, as described already. As discussed earlier, for other designs of calorimeter it is necessary to calibrate after temperature changes, loss in power, and so on.

Chemical Test Reactions

The many different types of system which can be studied by solution calorimetry mean there are a myriad of test reactions which exist to cover these applications. The main types are discussed individually in the following sections.

Mixing of two liquids: The dissolution of propan-1-ol in water has been proposed (13) as a test reaction for experiments in which small volumes of one liquid are injected into a larger volume of solvent. If the final molality of propan-1-ol is <0.017 mol kg^{-1}, then the solution can be essentially considered as being infinitely dilute and the reported enthalpy of solution at 298.15 K is -10.16 ± 0.02 kJ mol^{-1} (10). If the experiment is to be conducted at a temperature other than 298.15 K, then the expected enthalpy of dilution can be calculated from:

$$\Delta_{sol}H_m^{\infty} = -15.7484 + 0.237901(T - 273.15) - 5.7674$$
$$\times 10^{-4}(T - 273.15)^2$$

(5)

where T is the experimental measurement temperature (K).

However, the sensitivity of modern calorimeters is such that, for experiments conducted at ambient temperatures, the heat evolved in the dissolution of propan-1-ol often saturates the calorimeter amplifiers. In such instances, the dilution of aqueous propan-1-ol solutions may be more appropriate. The molar enthalpy of dilution can be calculated from McMillan–Mayer theory:

$$\Delta_{dil}H_m(m_i \rightarrow m_f) = h_a(m_f \rightarrow m_i) + h_b(m_f \rightarrow m_i)^2$$

(6)

where m_i and m_f are the initial and final molalities, and h_a and h_b are the virial coefficients for the enthalpy. At 298.15 K, these values are $h_a = 558 \pm 9$ J kg mol^{-2} and $h_b = 158 \pm 8$ J kg^2 mol^{-3}.

Sparingly soluble organic liquids: In some instances, one of the liquids may be only sparingly soluble in the aqueous phase. Benzene was initially proposed (13) as a reference material, but its use was criticized because of its highly toxic nature, and toluene (17) has been proposed as an alternative. A calorimetric value for the dissolution of toluene at 298.15 K has been reported (10); $\Delta_{sol}H_m^{\infty} = 1.73 \pm 0.04$ kJ mol^{-1}. However, the temperature coefficient is high and, hence, several other test reactions have been proposed including the dissolution of octan-1-ol and several esters. The IUPAC report (10) states that no suitable test and reference reaction have been proposed at present for dissolution of sparingly soluble organic liquids in water.

Dissolution of solid compounds: A number of chemical test reactions for the dissolution of solids have been proposed and discussed, including the dissolution of KCl or NaCl in water (43) as a suitable test reaction for endothermic enthalpies of solution and the dissolution of Tris in HCl as a test reaction for exothermic heats of solution (44). The dissolution of sucrose in water has also been discussed (45,46), although this is not currently recognized as a test reaction. The application of these reactions to solution calorimeters has recently been discussed, for both adiabatic (47) and heat conduction instruments (48).

Of these, the dissolution of KCl into water has been recommended by the IUPAC committee as the preferred method because it is robust, easy to perform, and a defined, standard reference material is available from the National Institute for Standards and Technology (NIST, U.S.A.); the enthalpy of solution of the NIST certified KCl into water is 17.584 ± 0.017 kJ mol^{-1} (43,44,49). However, this test reaction does suffer from some limitations, which were discussed in Chapter 1.

The dissolution of Tris in dilute HCl solution was first suggested as a test reaction for solution calorimeters by Irving and Wadsö (44) [and subsequently characterized by Hill et al. (50)]. The reaction gives a value of $\Delta_{sol}H$ of -29.744 ± 0.003 kJ mol^{-1} at 25°C (50). Although it has been widely used as a test reaction, there are potential problems with its use in steel vessels, which are susceptible to acidic attack, and it is recommended that the reaction be carried out only in glass ampoules.

Data Interpretation

The most commonly desired parameter is the enthalpy of solution $\Delta_{sol}H$. This parameter is readily available from the calorimetric data by simple integration, provided the instrument is designed and performed carefully. Effects from breaking the ampoule and sample stirring (most easily compensated for by subtracting a blank experiment from the sample experiment), changes in the solvent activity because of solute dissolution, changes in the rate of evaporation of solvent into the headspace, and changes in volume upon mixing of the two phases all must be taken into account. The last three effects can be assumed to be negligible if an ideal (dilute) solution is formed; if this is not the case, a fact that would be confirmed by obtaining different values of $\Delta_{sol}H$ with different quantities of sample, then the data must be extrapolated to infinite dilution to give $\Delta_{sol}H^{\infty}$.

Analysis of data can be performed either using the instrument dedicated software package or by exporting the data into a data-fitting package such as Origin® and manually integrating the power–time data. These methods will be dealt with in more detail in Chapter 3.

GOOD PRACTICE FOR PERFUSION CALORIMETRY

Sorption reactions (adsorption, absorption, and desorption of vapors and solutes onto solids) are extremely important and have found particular use in the pharmaceutical industry (51).

In gas perfusion, as implied by its name, a gas (of defined vapor pressure, often RH) flows over a solid sample. This is normally achieved using a "wet" gas line (i.e., saturated with vapor) and a "dry" gas line; proportional mixing of the two gas streams using mass-flow controllers results in any desired atmosphere in the ampoule. In order to operate most effectively, the following must be ensured:

1. The gas supply (nitrogen usually) is dry (most easily achieved through the use of an in-line desiccant before the gas stream is split).
2. The reservoirs for the wet-line are full.
3. There are no leaks in the system (connecting a gas-flow meter to the outlet is the simplest check).
4. If a solvent other than water is used, that the solvent does not degrade the o-rings used to maintain the system air-tight.

5. That the external parts of the apparatus are held approximately 25°C above the temperature of the experiment (to prevent condensation of any vapor).
6. That the flow rate of the gas is sufficient that the rate of vapor supply is not rate limiting (usually between 120 and 150 mL h^{-1} for a 5 mL ampoule volume).

It is advisable to use mass-flow controllers to regulate the flow of the wet and dry lines, as this ensures proportional mixing (the alternative being to flow wet gas and then dry gas for different time periods; on average, the RH attained in the ampoule would be that desired but in practice the sample is exposed to either dry air or saturated air).

Assuming the sample is a solid, there are a number of factors that must be taken into consideration when preparing a sample for perfusion calorimetry. First, the moisture (or solvent) content of the sample must be controlled. This is most easily achieved either by storing the sample in a desiccator prior to use or by flowing a dry gas over the sample during the initial phase of the experiment. This step is particularly important for samples with small amorphous contents, because small amounts of moisture can cause recrystallization of the amorphous regions, which will then not be registered during any subsequent measurement.

Second, the wetting response will be affected by both the particle size distribution and the total mass of the sample. The smaller the average particle size, the greater the area to be wetted and the greater the wetting signal. The kinetic response of the sample will also be affected in this case. The best option here is to sieve the sample prior to any experiment and ensure that a consistent particle size fraction is used for each measurement. A large mass of sample results in a column of powder in the ampoule, which is then difficult to wet; again, this will affect the kinetic response of the sample and it is best to ensure that the smallest amount of sample (but that which gives a measurable signal) is used.

Calibration

The calibration of gas perfusion instruments is, in the main, the same as for the ampoule/batch instruments already described. However, in order to obtain quantitative information, the RH produced in the ampoule must be known precisely. One method for validating the generated RH is to use selected inorganic salts (commonly used as saturated solutions to control RH in a sealed environment), which have a critical RH at which deliquescence occurs. An example of this is a saturated solution of NaCl which will, at 25°C, produce an RH of 75.3% (Table 1, Chap. 1). If a saturated solution of NaCl is placed in a perfusion ampoule and a "wet" gas is passed through the ampoule, then at RHs <75.3% water will evaporate from the NaCl solution and at RHs >75.3% water will condense from the gas. These changes will be readily observed as endothermic and exothermic responses, respectively.

Electrical Calibration

Electrical calibration should be performed before the first use of the apparatus, following a change in amplifier range or loss of power supply and/or on a weekly basis.

Chemical Test Reactions

There are no recommended chemical test reactions for gas perfusion calorimetry, but it is possible to perform an RH validation by placing a small quantity of a saturated salt solution in the ampoule, as described. The difficulties in finding a suitable chemical test reaction for perfusion calorimeters lie in finding a material with suitable properties. Buckton (3) defined four criteria that a test substance for perfusion calorimeters should have:

1. physical and chemical stability,
2. a water uptake which falls within the range encountered in the majority of systems (for crystalline organic powders this is ~0.2% at 90% RH),
3. a uniform surface area per unit mass, and
4. no batch-to-batch variability.

In practice, a material which fulfils all of these requirements is extremely difficult to find. A proposed solution is to consider uniform-size fractions of well-defined polymeric materials, which do not absorb water but do have a suitable level of water absorption. At this time, there is no well-defined material proposed for use as a standard for perfusion calorimeters.

Data Interpretation

The usual quantity to be obtained is the enthalpy change upon wetting. This is given by the area under the peak following each RH change (for a stepped-RH experiment) or the total area under the curve (for a ramp-RH experiment). The value can be obtained either through the dedicated instrument software or by exporting the data to a suitable software package and integrating manually. The area (J) should be divided by the sample mass (g) to give the enthalpy change ($J\ g^{-1}$) or number of moles to give the enthalpy change ($J\ mol^{-1}$). It should be noted that with a change in RH the entire apparatus will wet (RH increase) or dry (RH decrease), an event that will manifest itself in the measured signal. It is therefore necessary to perform a blank experiment (using the same RH program but without the sample in the ampoule) and subtract these data from the experimental data.

GOOD PRACTICE FOR FLOW CALORIMETRY

One of the disadvantages encountered in some types of calorimeters is the incapacity to stir the contents of the ampoule. One way to overcome this problem is to utilize flow calorimetry. These instruments allow the reaction medium (held externally to the calorimeter) to be pumped through the calorimetric cell. This circumvents the requirement to stir inside the cell, as it is possible to stir the reaction medium while in its holding chamber. Flow calorimetric techniques have proved to be useful in variety of situations; for example, if the sample is heterogeneous or is susceptible to sedimentation. Alternatively, if some property of mixing of two solutions is to be followed, then flow techniques offer a convenient way to follow these interactions.

As with ampoule calorimeters, if kinetic and thermodynamic information are to be obtained from the data then it is important to know accurately the concentration of the reactant(s). The sample should be covered to prevent evaporation/condensation as well as contamination and a volume sufficient to allow complete filling of the lead-in and lead-out tubing and the calorimeter cell should be used. The reaction medium should be stirred to ensure thorough mixing and to prevent sedimentation if the sample is in suspension form.

The reaction system must be preequilibrated to the operating temperature of the calorimeter by means of an external water bath and must be kept at that temperature throughout the experiment lifetime. The reaction medium flows through the calorimeter usually by means of a peristaltic pump, which must also be calibrated for flow rate (mass or volume delivered per unit time). It is desirable to achieve a flow which is as constant as possible, and so care should be taken to ensure that the peristaltic pump used does not introduce a significant pulsed flow.

The ideal reference for a flow calorimetric experiment would be an inert liquid (the solvent, for example) flowing through the reference cell at the same rate as the sample. In practice, this is near impossible to achieve and so careful consideration must be given as to the reference arrangement used. The optimum arrangement can vary from calorimeter to calorimeter and can only be determined experimentally (52).

Calibration

Electrical Calibration

Electrical calibration should be performed before the first use of the apparatus, following a change in amplifier range or loss of power supply and/or on a weekly basis.

Chemical Test Reactions

The design of flow calorimeters means that it is often the case that the actual operating volume of the calorimetric cell is significantly different from the physical volume; in addition, the operating volume changes as a function of flow rate. This imposes a severe restriction on the applicability of electrical calibration to flow calorimeters and, hence, there is an absolute necessity for a robust chemical test reaction. In some instances, it is possible that the calorimeter being used may not have sufficient sensitivity (52) to follow the solution phase test reaction already discussed (ICHT) (which has a relatively slow rate of reaction); it, there-fore, becomes necessary to use an alternative. This was highlighted by O'Neill et al. (52,53) for the LKB (now Thermometric) flow calorimeter, which was found to be too insensitive to follow the ICHT, a consequence of the combination of a relatively slow rate of reaction and a relatively small enthalpy of reaction, resulting in the generated signal falling outside the detection limit of that calorimeter. As a compromise, a secondary test reaction has been proposed; the base-catalyzed hydrolysis of methyl paraben (BCHMP) (15). The values for k and ΔH for the BCHMP are $3.2 \times 10^{-4} \pm 1.1 \times 10^{-5}\,\mathrm{s}^{-1}$ and $-50.5 \pm 4.3\,\mathrm{kJ\,mol}^{-1}$, respectively.

Data Interpretation

As with ampoule/batch calorimetry, flow calorimeters are commonly used for the monitoring of stability and/or compatibility. The parameters sought are, there-fore, usually related to the kinetic nature (rate constant, Arrhenius values, and so on) of the system. There are a variety of different ways to elucidate these values from flow calorimetric data; presently, the most efficient way is to calculate the desired parameters using an appropriate kinetic equation, although novel methods are being developed. There is also the added burden of dealing with the flow rate dependence when analyzing flow calorimetric data, and the kinetic equations used must be adapted to reflect this. The flow rate affects the thermal volume (the volume "seen" by the calorimeter) and so its effects must be accounted for if thermodynamic information is required from the data. An in depth discussion of the analysis of flow calorimetric data can be found in Chapter 3.

APPLICATIONS OF TEST AND REFERENCE REACTIONS

While chemical test reactions play an essential role in the validation of any calori-metric study, they also offer other potential uses. First, they can be used as robust training routines for new calorimeter operators. This versatility arises from the knowledge of accurate values for particular parameters, thus allowing the per-formance of a new user to be monitored through their ability to perform the experiment and return the correct values for the desired parameters. Any error in constructing the experiment or set-up of the instrument will be highlighted by poor recovery of those values.

    A second use is for troubleshooting; for instance, when an instrument appears to be malfunctioning (18) or where the upper and lower calibration limits have been poorly assigned. Chemical test reactions have also been shown to be useful in the identification of sources of systematic error (12), which may not be revealed by electrical calibration alone (such as instruments in which the temperature control is compromised in some way). Chemical test reactions have also been used to probe various physical aspects of calorimeter design, such as finding the optimal fill volume for batch calorimeters and optimal flow rate for flow calorimeters, and thus can be used to aid in experimental design (53).

These particular applications of chemical test reactions are explained with reference to simulated data and, where appropriate, real experimental studies in the following sections.

Incorrect Assignment of Lower and Upper Power Limits

An electrical calibration is conducted to achieve two aims: the first is to set the baseline of the instrument to zero and the second is to set the full-scale deflection through the input of a known amount of energy from an electrical heater. If either of these two steps is not performed correctly, either through operator error or an instrument malfunction, then the recorded calorimetric data will contain systematic errors. The effects of these potential sources of error can be highlighted through an exploration of simulated data for a hypothetical solution phase reaction (18,54).

Incorrect Assignment of a Baseline

The first step in electrical calibration is to set the zero point (baseline) of the calorimeter. If this baseline is incorrectly set, either to the positive or negative, by a fixed amount, x μW, where x is the magnitude of the offset in the baseline, then the recorded data will be shifted by that same amount. Consider the case where the baseline has been incorrectly set by $+2$ μW, and hence the recorded data are all 2 μW greater than their actual value (Fig. 1). The data were generated from Equation (7), using values of -94 kJ mol^{-1} and 2.76×10^{-6} dm^3 mol^{-1} s^{-1} for the enthalpy and rate constant, respectively. If these data are analyzed for the rate constant and enthalpy via iterative procedures, it is found that a satisfactory fit to the model is not generated and hence incorrect values for k and ΔH (2.3×10^{-6} dm^3 mol^{-1} s^{-1} and 338.7 J mol^{-1}, respectively) are returned. Although this is clear for an error of 2 μW, it would not necessarily be so obvious in the case where the error is small, $+0.4$ μW for instance. In this case, the fit of the line appears to be satisfactory but inspection of the returned values reveals that the enthalpy is in error by approximately 3 kJ mol^{-1} (3.2%) and the rate constant by approximately 10×10^{-8} dm^3 mol^{-1} s^{-1} (2.6%). The error in the baseline can be determined by comparing random points along an expected (true) output with equivalent points along the observed output. It is found that for all data points the observed output is greater than the expected output by 0.4 μW. Alternatively, it is possible to modify Equation (7) to incorporate a parameter which takes into account the shift in the data and allows that shift to be quantified [Equation (8)]:

$$\frac{dq}{dt} = -\Delta HVk\left(\frac{[A_0]}{1 + k[A_0]t}\right)^2 \tag{7}$$

$$\frac{dq}{dt} = -\Delta HVk\left(\frac{[A_0]}{1 + k[A_0]t}\right)^2 + R \tag{8}$$

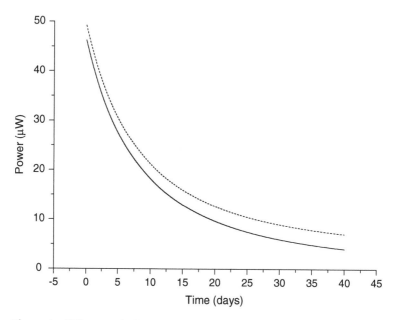

Figure 1   Effect on calorimetric power from incorrectly assigning the zero point (*dotted line*) compared with expected output (*solid line*).

where R is the correction factor (in μW). Once the shift in baseline has been quantified, the data can then be adjusted accordingly and the desired parameters calculated.

Incorrect Assignment of an Upper Power Limit

When calibrating using the electrical calibration heater, it is necessary to assign the upper limit in accordance with the amplifier setting. If the upper limit of the amplifier range is incorrectly defined, such that the zero point is correct but all other values are incorrect by some percentage (e.g., if the calorimeter is zeroed and then electrically calibrated to $100 \mu W$ but through operator or instrument error the calorimeter is actually calibrated to $90 \mu W$), this would lead to falsely low readings for the recorded thermal power. If the system under investigation yields exactly $100 \mu W$, for example, then the effect would be that the calorimeter reads $90 \mu W$. The difference in the output, produced by the calorimeter, and the true output would be greatest at the start of reaction, and as the reaction progresses, this difference would decrease until at completion of the reaction the difference would be zero. This effect can be illustrated by the consideration of the output for the hypothetical solution phase reaction illustrated in Figure 2. As the value of the power output decreases, so does the numerical difference between the observed output and the true output, although the percentage difference remains the same. For each data point, the true power is 90% of the observed heat flow.

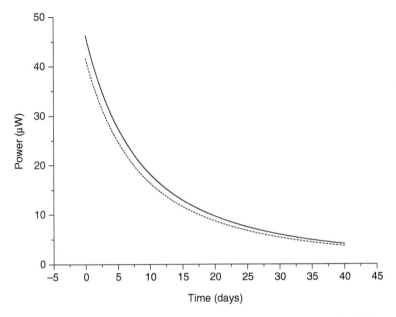

Figure 2 Effect on calorimetric power from incorrectly assigning the full-scale deflection (*dotted line*) compared with expected output (*solid line*).

From Figure 2, the fit of the data reveals the correct value for the rate constant that is used to generate the original data. However, the value for the derived enthalpy is incorrect ($-104.5\,\text{kJ}\,\text{mol}^{-1}$ compared with $-94\,\text{kJ}\,\text{mol}^{-1}$). As the correct value for the reaction enthalpy is known, it is possible to correct the erroneous data set and to calculate the magnitude of the error in the upper limit:

$$\frac{dq_{\text{true}}}{dt} = \Delta H_{\text{true}}kV\left([A]_0 - \frac{q}{\Delta H_{\text{true}}}\right)^n \tag{9}$$

$$\frac{dq_{\text{false}}}{dt} = \Delta H_{\text{false}}kV\left([A]_0 - \frac{q}{\Delta H_{\text{false}}}\right)^n \tag{10}$$

when $q = 0$

$$\frac{dq_{\text{true}}}{dt} = \Delta H_{\text{true}}kV[A]_0^n \tag{11}$$

$$\frac{dq_{\text{false}}}{dt} = \Delta H_{\text{false}}kV[A]_0^n \tag{12}$$

Dividing Equation (11) by (12) and rearranging for dq_{true}/dt:

$$\frac{dq_{\text{true}}}{dt} = \frac{\Delta H_{\text{true}}}{\Delta H_{\text{false}}} \times \frac{dq_{\text{false}}}{dt} \tag{13}$$

Therefore, the erroneous data set can be corrected by multiplying each data point by the ratio $\Delta H_{true}/\Delta H_{false}$ to generate the expected output. Alternatively, this ratio may also be used to calculate the magnitude of the error in the assigned upper limit [Equation (14)]:

$$\frac{\Delta H_{true}}{\Delta H_{false}} \times upper\ limit_{expected} = upper\ limit_{actual} \qquad (14)$$

The observation that a correct value for the rate constant is returned whereas the enthalpy of reaction is not is a strong indicator that the upper limit has been incorrectly assigned.

Nonuniform Error

The potential sources of error discussed in the previous two sections were examined individually in a uniform manner. It is possible that, if the calorimeter is not performing to its specification, both the upper and lower limits may be in error. Figure 3 illustrates a situation where the error in the zero point is $3\ \mu W$ and the error in the upper limit is $6\ \mu W$. Again, a satisfactory fit to Equation (7) is not achieved for these data. Even though the error in the calibration is nonuniform, it can be broken down into its constituent parts and analyzed independently. Therefore, the error in the baseline can be quantified from Equation (8) (the test being a correct value for k and an incorrect value for ΔH), and thus the data can be corrected by shifting each point by the returned value of R

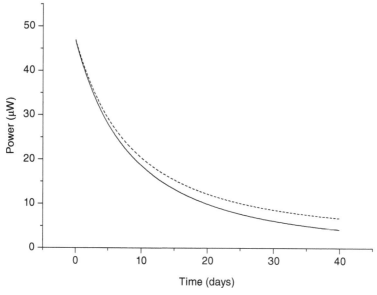

Figure 3   Effect on calorimetric power from incorrectly assigning the zero point and upper limit (*dotted line*) compared with expected output (*solid line*).

(3 μW in this instance). The error in the upper limit can then be identified from the derived (incorrect) value for enthalpy, as described in the previous section.

Time-Dependent Error

So far, the potential sources of error discussed have been time-independent (i.e., the errors have been constant, fixed values across the entirety of the data set). It is more likely, however, that after calibration the baseline and/or upper limit may drift over a period of time. Indeed, this is a recognized issue for all calorimeters and is, of course, one of the reasons why regular calibration is essential. In the case, where a reaction is followed over a significant period of time, it is not possible to calibrate at regular intervals. If the drift in the baseline is large, then the observed output will be significantly affected, especially if the output is small. Figure 4 illustrates the effect that a drift in the baseline would have on the observed output from an initially correctly calibrated instrument.

Fitting the data in Figure 4, using Equation (7), does not yield the correct values. If Equation (7) is modified to take into account a time-dependent change in signal, Equation (15) is obtained:

$$\frac{dq}{dt} = -\Delta H V k \left(\frac{[A_0]}{1 + k[A_0]t} \right)^2 + Pt \tag{15}$$

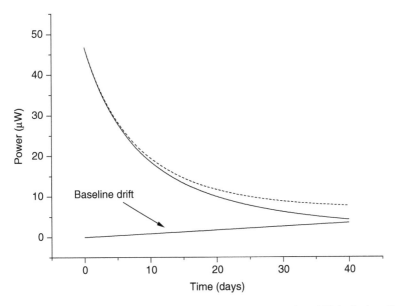

Figure 4 Effect on calorimetric power from a time-dependent drift in the baseline value (*dotted line*) compared with expected output (*solid line*). The baseline drift with time is also shown.

where the constant P (μW s^{-1}) dictates the magnitude of the drift at any time t. Fitting the observed data to Equation (15) yields the correct values for the reaction parameters. It also yields a value for the constant P, in this case 1×10^{-6} μW s^{-1}.

If the calorimeter is used to study short-term reactions, then this effect can be considered negligible. However, if the reaction is long-term and the thermal power is small, then a drift in the baseline could cause significant errors in the results obtained from the calorimeter. Data, where the baseline is drifting, can be corrected using Equation (16):

$$\frac{dq_{\text{true}}}{dt} = \frac{dq_{\text{obs}}}{dt} - Pt \tag{16}$$

Of course, if the baseline can drift over time, then it is just as likely that the full-scale deflection may also drift or, indeed, they may both drift at the same time giving rise to a more complex situation. Dealing with a drift in the full-scale deflection only first, Figure 5 shows the effect of a drift of 1×10^{-3} μW s^{-1}. As the zero point is not affected, the numerical difference between the expected and the observed outputs is greater for lower power values than for higher power values (i.e., the error is propagated as a percentage error) and, hence, cannot be corrected by simple addition or subtraction of a set value. Again, Equation (7) does not satisfactorily return the correct values for

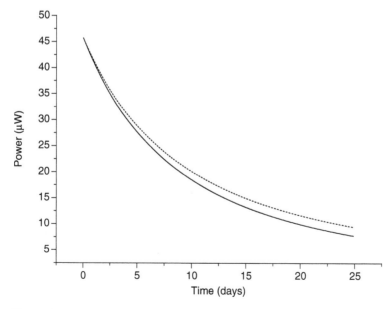

Figure 5 Effect on calorimetric power from a time-dependent drift in the full-scale deflection value (*dotted line*) compared with expected output (*solid line*).

k and ΔH. Equation (7) can be modified to quantify the magnitude of the drift per unit time, S:

$$\frac{dq_{obs}}{dt} = \Delta H V k \left(\frac{[A]_0}{1 + k[A]_0 t}\right)^2 \frac{U + St}{U} \tag{17}$$

Hence, the calorimetric output can be corrected by:

$$\frac{dq_{true}}{dt} = \frac{dq_{obs}}{dt} \frac{U}{U + St} \tag{18}$$

For the example illustrated in Figure 4, the instrument was calibrated between 0 and 100 μW and hence $U = 100$ in this case.

It is not uncommon for both the upper limit and the baseline to drift to different extents at the same time. This can be dealt with simply by introducing the term Pt from Equation (16) to yield:

$$\frac{dq_{obs}}{dt} = \Delta H V k \left(\frac{[A]_0}{1 + k[A]_0 t}\right)^2 \frac{U + St}{U} + Pt \tag{19}$$

Thus far, the sources of error discussed have been dealt with separately, and those which vary as a function of time have been assumed to do so in a linear fashion. Of course, this may not always be the case; in fact, it is likely that the drift in the baseline or the upper limit (or both) may proceed in a nonlinear fashion.

Nonlinear Drift

This can be dealt with by again modifying Equation (18) into a form which contains polynomial functions:

$$\frac{dq_{obs}}{dt} = \frac{dq_{true}}{dt} \frac{U + (a + bt + ct^2 + \cdots + it^{i-1})}{U}$$
$$+ (z + yt + xt^2 + \cdots + jt^{j-1}) \tag{20}$$

The constants a and z take into account the initial error in the calibration and baseline, respectively. The constants b and y reflect the rate of linear drift of the upper limit and baseline, respectively. If the upper limit and baseline do not drift in a linear fashion this will be accounted for by the higher polynomial functions. Hence, Equation (19) can then be used to derive Equation (21), thus allowing the data for the hypothetical system to be corrected:

$$\frac{dq_{obs}}{dt} = \Delta H V k \left(\frac{[A]_0}{1 + k[A]_0 t}\right)^2 \frac{U + (a + bt + ct^2 + \cdots + it^{i-1})}{U}$$
$$+ (z + yt + xt^2 + \cdots + jt^{j-1}) \tag{21}$$

Error Caused by the Thermostat

Systematic errors are not limited to inadequacies of an electrical calibration routine; they may also be introduced via routes beyond the control of the operator. One such example is incorrect reporting of operating temperature of the calorimeter.

The rate of a chemical reaction is directly proportional to the temperature of the system. Hence, if the temperature of the system is misrepresented then the rate of reaction would appear to be in error while the enthalpy of reaction would be returned correctly (if it is assumed that the enthalpy is temperature independent). In this instance, it is possible to quantify the error in the reported temperature through a consideration of the Arrhenius equation.

If the activation energy of a reaction is known, then that value can be used to calculate the rate constants for any temperature using the Arrhenius equation. Assuming that the error in the thermostat is unknown, and hence the temperature of the reaction is unknown, then knowledge of the rate constant for the test reaction at any given temperature and at the unknown temperature will allow calculation of the operating temperature of the calorimeter.

According to the Arrhenius equation:

$$\ln k_1 = \ln A - \frac{E_a}{R}\frac{1}{T_1} \tag{22}$$

and

$$\ln k_2 = \ln A - \frac{E_a}{R}\frac{1}{T_2} \tag{23}$$

Rearranging both equations in terms of $\ln A$ (which is equal in both cases) and setting them equal results in:

$$\ln k_1 + \frac{E_a}{R}\frac{1}{T_1} = \ln k_2 + \frac{E_a}{R}\frac{1}{T_2} \tag{24}$$

letting:

$$\ln k_1 + \frac{E_a}{R}\frac{1}{T_1} = X \tag{25}$$

allows T_2 to be calculated from:

$$\frac{E_a/R}{X - \ln k_2} = T_2 \tag{26}$$

Thus, the actual operating temperature of the calorimeter can be calculated.

The importance of employing chemical calibration routines has been high-lighted by Finnin et al. (12). They report the results of a study employing the BCHMP as a calibration routine for a newly developed calorimetric instrument. They note that, on delivery, the instrument performed within the manufacturer's specified limits but that the performance fell below these standards after updating the firmware for the system. The system was validated using the BCHMP as the test reaction where it was noted that the data from the instrument returned the correct reaction enthalpy but the rate constant fell outside these accepted limits. They postulated that, given the correct enthalpy but incorrect rate constant, the error could be attributed to a misreporting of the temperature of the instrument.

Using the technique (and the knowledge of the "true" rate constant and its temperature dependence) outlined already, they predicted that the operating temperature was approximately 2°C higher than the reported temperature. On inspection by an engineer, this was indeed found to be the case and the temperature discrepancy predicted from the Arrhenius method matched almost exactly that actually found. Had this instrument been calibrated solely by electrical means, it is entirely possible that this misreporting of temperature may have been missed.

A Caution

It should be noted that when fitting data to equations such as Equation (21), via iterative procedures, care must be taken to obtain a suitable fit using as few parameters as possible. There is a danger that the value of chi-squared may be reduced (indicating a better fit), solely on the basis of having many parameters and thus clearly would not represent the true nature of the problem. Consequently, when fitting data obtained from a test reaction one should initially use the simplest form of the equation [Equation (7)]. Only if a poor fit is obtained, or incorrect values for the parameters are returned, then a modified form containing a correction parameter should be used. This is continued until a satisfactory fit and correct values for the required parameters are returned. In some instances, it is possible to gain an insight into the potential source of error (and hence the appropriate modified equation to use) from the values of the returned parameters.

EXPERIMENTAL DESIGN

It has been shown that chemical test reactions, potentially, are useful for operator performance testing, troubleshooting, and identification/correction of systematic errors. The following sections highlight how these test reactions can be used to probe the design of certain aspects of the calorimeter and how they can be used to help design calorimetric experiments more efficiently.

Effect of Fill Volume

Thus far, the discussion has assumed that the experiments have been performed using ampoules which are essentially entirely filled with solution phase reaction medium. In reality, it is likely that the material under investigation might be scarce, for instance material from the discovery laboratory, or for practical reasons the sample may be spread thinly on the base of the ampoule (some solid-state experiments require a thin layer of material to allow exposure of the whole sample to a humid atmosphere, for example).

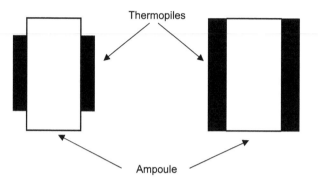

Thermopiles

Ampoule

Figure 6   Schematic of thermopile arrangements.

The design of the thermopile arrangement in some calorimeters is such that the thermopiles only cover a percentage of the ampoule surface (Fig. 6). There are regions at the top and bottom of the ampoule, which are not covered by the thermopiles. Alternatively, the thermopiles may extend over the height of the ampoule and surround it entirely, except for the top and base (Fig. 6). Note that the calibration heater is usually located at the base of the ampoule. This, by its very design, must be an excellent thermal conductor and, therefore, is likely to provide an alternative thermal pathway other than through the thermopiles. There is, therefore, concern as to what effects the ampoule fill volume will have on the measured data.

The effects of fill volume on the recovered parameters, in both the 4 mL (which can accommodate both 3 and 4 mL ampoules) and 20 mL channels of the Thermometric TAM (utilizing the ICHT test and reference reaction) have been investigated (53). The results of the study provide an interesting insight into the performance of the calorimeter over a range of fill volumes.

For the 20 mL ampoule, the rate constant values are recovered consistently across almost the whole fill volume range. The same is not true for the recovered values for the reaction enthalpy; these have been shown (53) to vary at fill volumes <50% fill (Fig. 7). With different fill volumes (20, 5, and 2.5 mL), the recovered mean value for ΔH decreases with fill volume. It is postulated (53–55) that this may be related to the changes in heat conduction pathways involving the sides, base, and lid of the ampoule.

The effect of decreasing fill volume for the 3 and 4 mL ampoules is more marked (Figs. 8 and 9) than that observed for the 20 mL ampoule. In both instances, the reliability of the recovered values for k and ΔH diminishes at <50% fill volume. As with the 20 mL ampoule, the values recovered at and around 100% fill volume all fall within the accepted limits.

Clearly, such observations are of cause for concern and should be taken into account when designing calorimetric experiments, especially those in which material is limited because of its rarity or because of technical issues such as

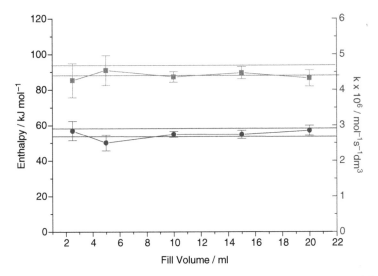

Figure 7 Effect of fill volume on the accuracy and precision of the enthalpy (*square*, left axis) and rate constant (*circle*, right axis) values for data recorded in a 20 mL ampoule. The solid horizontal lines represent the upper and lower accepted limits for the plotted parameters.

those encountered in studies conducted on solid-state systems. These problems may be alleviated by careful experiment design. For example, it is possible to manufacture inserts for the ampoule which allow the sample to be positioned in the center of the area covered by the thermopiles and thus maximize the

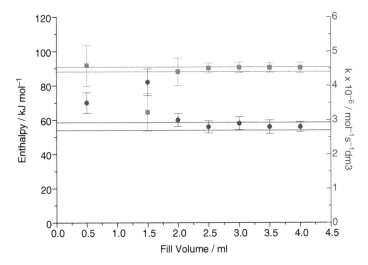

Figure 8 Effect of fill volume on the accuracy and precision of the enthalpy (*square*, left axis) and rate constant (*circle*, right axis) values for data recorded in a 3 mL ampoule. The solid horizontal lines represent the upper and lower accepted limits for the plotted parameters.

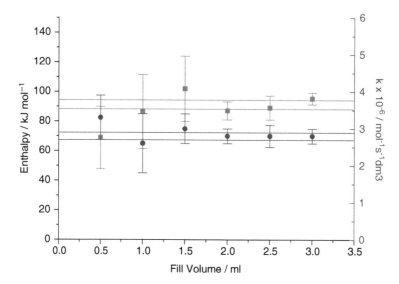

Figure 9   Effect of fill volume on the accuracy and precision of the enthalpy (*square*, left axis) and rate constant (*circle*, right axis) values for data recorded in a 4 mL ampoule. The solid horizontal lines represent the upper and lower accepted limits for the plotted parameters.

capture of any heat produced by that sample. As with any modification to the system, care must be taken to ensure that erroneous heat flow is not introduced to the system by poor choice of material for the inserts (e.g., Teflon is porous to many substances) and that the sample and reference sides are as closely matched as possible. This approach has been tentatively examined by Morris (31), but no definitive conclusion has been reached as to the merits of such an approach.

SPECIAL CONSIDERATIONS FOR FLOW CALORIMETERS

The chemical test reactions discussed in the previous sections all pertain to systems in which the reactants are essentially static. Although this is the most common arrangement, it can prove to be unsatisfactory in some instances. For example, if the system under investigation has a short half-life, the reaction may be complete before the preparatory work for static systems is carried out. Another example is that sedimentation may be an issue in suspensions. To counter these issues flow calorimeters have been developed, some specifically as dedicated flow calorimeters such as those manufactured by Thermometric and Microscal. Other manufacturers have developed their instruments such that they can be adapted for use as flow calorimeters through application of specially modified calorimetric ampoules (the Setaram Micro-DSC III being a good example).

Flow calorimeters operate on the principle that an in-flow of a reacting solution at constant flow rate, F, will give rise to a constant power in the calorimetric cell. If left long enough, this will eventually reach steady-state conditions where the heat generated in the cell will, per unit time, be equal to the heat exchanged from the cell to the heat-sink. It is widely acknowledged that even under ideal conditions some power will be lost. As described earlier, heat loss is possible through air gaps, electrical wires, and so on. In addition to these losses, flow calorimeters can also lose heat with the effluent as it leaves the cell. The effects of these losses (particularly that from the effluent) are significant and can cause serious systematic errors to be introduced to the recorded data.

Flow calorimeters, as discussed earlier, have two modes of operation, flow-mix and flow-through. In both instances, the reagents are preequilibrated outside the calorimeter. In a flow-mix experiment the reagents are mixed inside the cell and the reacting solution is flowed to waste (or retained for further analysis). In a flow-through experiment, the reaction is initiated outside (lead times from the external reaction reservoir to the calorimetric cell are of the order 3 to 15 minutes) the calorimeter and the reacting solution is recycled or again retained for further analysis. For flow-mix systems where instantaneous (i.e., reactions which are rapid relative to the time constant of the instrument) reactions are initiated in the flow-measuring vessel, the thermodynamic parameters for the reaction are calculated through the use of the calibration constant derived from electrical methods determined at the experimental flow rate. In contrast to this simple calibration, a flow-through calorimetric experiment involves a reaction in which the rate of reaction changes continuously with time. A simple electrical calibration is not appropriate for a flow-mix vessel in which a time-dependent reaction takes place. This is because the observed calorimetric signal describes the mean extent of reaction that occurs in the flowing reacting system over the vessel residence time. The extent of reaction is the integral of the reaction rate over this residence time (this is dealt with in more detail in Chap. 3).

Determination of Thermal Volume

The significance of the effect of flow rate on the observed calorimetric signal has been discussed previously. Bakri et al. (56) claimed that the output is not dependent on flow rate and hence the data could be treated as though they were derived from a static instrument. Poore and Beezer (57), and subsequently O'Neill et al. (15,52), have shown this not to be the case. If the reaction has some kinetic dependence (i.e., it is not instantaneous), then the calorimetric signal is a reflection of the average rate of reaction over the residence time in the flow cell and hence the flow rate must have some effect on the observed calorimetric signal. Even though this was known to be a limitation, there was no accurate means of quantifying the effect; thus, Beezer et al. (58) proposed a modification of the traditional flow equations (defined in Chap. 3) to sidestep the requirement of knowledge of residence time. This modification was unsatisfactory as it applied only to the equation which describes a first-order kinetic regime and did not deal with zero- or higher-order reactions. The modification also required knowledge of a thermal volume and they were forced to make the assumption that the effective thermal volume was the same as that of the physical volume of the cell. As discussed later, this assumption was not valid and led, in some instances, to inaccurate values for the reaction enthalpy being reported.

Flow Calorimetric Equations

Equations (27) and (28) (for a derivation, see Chap. 3) describe first- and zero-order reactions in flow calorimeters, respectively:

$$\Phi = -FCH(1 - e^{-k\tau})e^{-kt} \tag{27}$$

$$\Phi = -kV_cH \tag{28}$$

where F is the flow rate (dm^3/sec), C is the concentration of reactant, H is the reaction enthalpy, k is the rate constant, τ is the residence time (the time the reacting solution spends in the calorimetric vessel) and V_c is the thermal volume (the volume that the calorimeter detects as the solution flows through the cell). As both equations (27) and (28) show, in order to elucidate the rate constant, and hence reaction enthalpy, using these equations requires knowledge of both the residence time and the thermal volume:

$$\tau = \frac{V_c}{F} \tag{29}$$

Flowing systems necessarily result in transport of heat from the detection area and thus, while the physical volume of the cell can be calculated from its physical dimensions, this volume may not be the effective thermal volume nor indeed the zero-flow rate volume of the cell (Fig. 10).

If accurate values for the desired thermodynamic and kinetic parameters are to be found, then it is of vital importance that the thermal volume is known, or can be calculated for any experimental flow rate. The determination

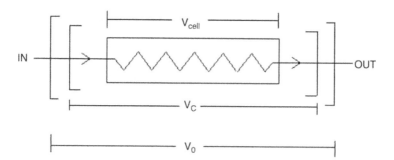

Figure 10 Schematic of a flow calorimeter cell.

of an effective or thermal volume is, therefore, a generic problem associated with all flow calorimeters. The thermal volume is affected by several different factors but the most important is the flow rate of the solution because, as the reacting solution flows through the vessel, a proportion of the heat will be carried out of the calorimetric cell retained in the solution; this is related to the heat capacity of the system. As the flow rate is increased, more heat will be carried out by the flowing solution and hence a smaller thermal volume will be observed.

Conversely, as the flow rate is decreased, less heat will be lost such that as the flow rate approaches zero, the apparent thermal volume will approach a limiting value. It is clear therefore that, if quantitative kinetic and thermodynamic information are to be gleaned from flow calorimetric data, there needs to be some means of validating the data via a chemical test routine.

Secondary Test Reactions for Flow Calorimeters

The IUPAC-recommended chemical test reaction for flow calorimeters operating in flow-through mode is the ICHT; however, for some flow calorimetric instruments this is unsuitable because of sensitivity issues. It then becomes necessary to employ an alternative or secondary test reaction (i.e., one for which the derived parameters can be traced back to, and validated by, a test reaction in a separate instrument).

One such alternative is the BCHMP, discussed previously. This reaction has been proposed as a suitable alternative for the ICHT, as the rate of reaction is greater than that for ICHT and, hence, as the calorimetric output is governed in part by the reaction rate the signal falls within the range of less-sensitive instruments. Values for the rate constant and reaction enthalpy have been calculated from data obtained from isothermal batch calorimeters, which have been validated with the accepted ICHT test reaction (17). These values have then been used to explore the physical characteristics of flow calorimeters.

Flow rate dependence: The flow rate dependences of three different flow calorimeters (LKB 10-700, a Thermometric 2277-201, and a modified 2277-201) have been demonstrated through the use of the BCHMP test reaction (53). The thermal volume varies with flow rate such that at high flow rates a large proportion of the heat generated by the reacting solution is carried out of the calorimetric cell undetected; hence, as the flow rate increases, the thermal volume approaches zero and, conversely, at low flow rates, more heat is detected and the thermal volume approaches the zero-flow rate volume of the calorimetric cell.

The flow rate dependency of thermal volume of the LKB instrument is demonstrated in Figure 11. The zero-flow rate volume of the cell can, in principle, be derived from this plot, by extrapolating the curve to zero-flow rate and taking the intercept value. However, as all extrapolations are subjective, such a determination of the zero-flow rate thermal volume could potentially lead to an inaccurate value. A better alternative is to calculate the thermal volume directly from experimentally attained data. The flow rate dependency of the thermal

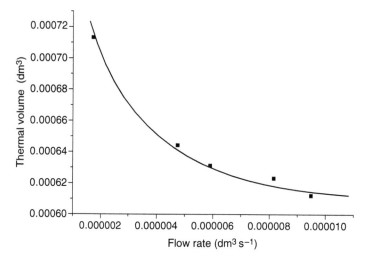

Figure 11 A plot of thermal volume versus flow rate.

volume V_c can be expressed as:

$$V_c = V_0 - FI \tag{30}$$

where I is the incremental change in residence time per unit flow rate.

Note from Figure 11 that the thermal volume approaches its maximum when the flow rate approaches zero, hence the negative dependence of V_c with flow rate. If the terms in Equation (30) are divided by the flow rate, F, then Equation (31) is obtained:

$$\frac{V_c}{F} = \frac{V_0}{F} - I \tag{31}$$

Consideration of Equations (29) and (32) mean the following must be true:

$$\tau = \frac{V_0}{F} - I \tag{32}$$

Therefore, a plot of τ versus $\frac{1}{F}$, will yield a straight line with slope equal to the zero flow rate volume and a *y*-intercept equal to I (Fig. 12). This relationship, therefore, allows a statistically more accurate value for the value of V_0 to be obtained and hence statistically more accurate values for V_c are possible.

For the LKB instrument used in the study from which these data are taken, the calorimetric cell is a gold coil of diameter 1 mm and length 60 cm. These physical dimensions give rise to a nominal volume of approximately 0.45 mL. The zero-flow rate volume for this instrument was calculated to be 0.74 mL. The incremental change in residence time per unit flow rate was calculated as −16.2 s. It is

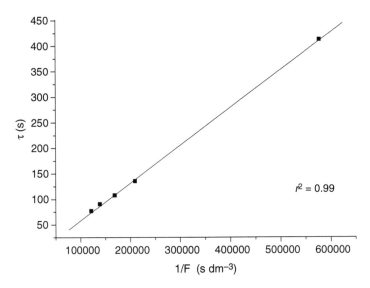

Figure 12   A plot of residence time versus reciprocal flow rate.

important to note here that this value of the zero flow rate thermal volume, V_0 is approximately 60% higher than the nominal (engineered), physical volume.

If the physical volume is used as the defining value (as was the case in early experiments), then it is clear that serious errors in the calculated reaction enthalpy would result. A startling example of this can be found in a comparison of identical studies, on the urea/urease enzyme system, performed by Beezer et al. and O'Neill et al. (58,59) using the same calorimeter some 30 years apart. Beezer et al. (58) analyzed their calorimetric data using the only assumption available to them at that time, that the nominal physical volume (0.47 mL) of the cell was essentially identical to the thermal volume for any given flow rate. In fact, this value for the thermal volume should have been approximately 0.65 mL (for their applied flow rate). This assumption led them to values of $4.8 \times 10^{-4} \pm 1.4 \times 10^{-5} \, \text{s}^{-1}$, 0.05 M, and $-33 \pm 1 \, \text{kJ mol}^{-1}$ for the first-order rate constant, Michaelis constant, and reaction enthalpy, respectively. In an identical study by O'Neill et al. (59), but now with defined values for V_c, the returned values for the first-order rate constant and Michaelis constant are consistent with those reported earlier but the reaction enthalpy is significantly different $-10.01 \pm 1.02 \, \text{kJ mol}^{-1}$. This discrepancy must in part be caused by the incorrect values used for thermal volume and hence incorrect residence time. A more detailed analysis of this particular application of flow calorimetry can be found in Chapter 3.

Experimental Arrangement

The experimental arrangement of the system may also have a bearing on the extent of the flow rate dependence. The thermometric flow module operates as a differential calorimeter (i.e., there are sample and reference chambers). There

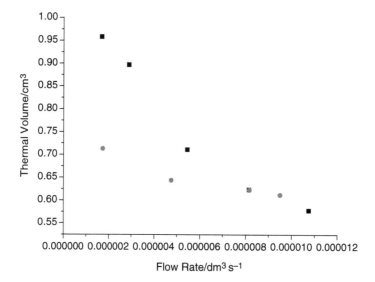

Figure 13   A comparison of the effects of the reference ampoule arrangement on the flow rate dependency of thermal volume. •, filled reference ampoule; ■, empty reference ampoule.

are several arrangements in which this calorimeter can be operated; for example, the reference side can contain no ampoule, an empty ampoule, or an ampoule filled with the solvent. And because of the positioning of the flow cell around the ampoule housing in the sample side, the same arrangements are also possible for the sample side. The arrangement used has been shown to be critical in determining the dependence on flow rate (52). Figure 13 shows a comparison of the effects of different experimental arrangements on the flow rate dependence of thermal volume.

It is evident from Figure 13 that inserting an empty ampoule into the sample side can effect a marked reduction in the flow rate dependence of the thermal volume. For the arrangement with no ampoule present, the thermal volume varied from 0.56 to 0.98 cm^3 over the flow range studied, while V_0 was calculated to be 1.04 cm^3.

Effect of Flow Rate on Residence Time

Knowledge of V_0 and I allows calculation for τ and, hence, Φ at any flow rate. If the maximum possible signal for a first-order reaction (i.e., at initiation of the reaction when $t = 0$) is considered, then Equation (27) can be rewritten as:

$$\Phi_0 = -FCH(1 - e^{-k\tau}) \tag{33}$$

It is now possible to consider the effects of flow rate, enthalpy, and rate constant on a theoretical chemical reaction studied in a flow calorimeter.

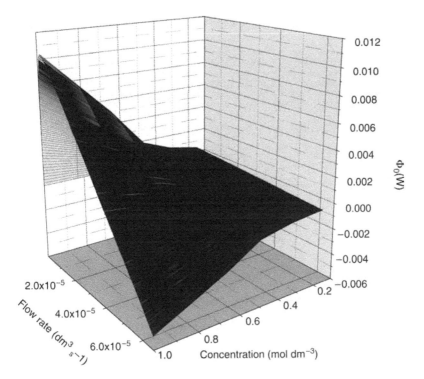

Figure 14 Variation of the initial signal Φ_0 with flow rate and initial concentration.

Note that it is possible to have control over the flow rate and some control over the rate constant (by changing the temperature). However, it is not possible to alter the enthalpy (although it is possible, in some circumstances, to magnify the enthalpy by careful choice of an appropriate buffer system). Figure 14 demonstrates how the initial signal Φ_0 varies with flow rate for a fixed rate constant and varying initial concentration of reagent. It is clear from Figure 14 that if a flow rate is selected and the initial concentration $[A_0]$ is increased then, as expected, Φ_0 increases in direct proportion to $[A_0]$. If $[A_0]$ is now held constant and the flow rate F is varied, it becomes apparent that the dependence of Φ_0 on flow rate is complex.

The calculated values for Φ_0 increase in a linear fashion from zero to a maximum value (governed by the value of $[A_0]$), and then decrease in a different linear fashion back to zero. This is readily explained by a consideration of Equation (27). As the flow rate tends to zero, τ will increase and hence the value of $e^{-k\tau}$ will tend to zero. Thus, the term $(1 - e^{-k\tau})$ will become equal to one. Therefore, at low flow rates, for fixed $[A_0]$ and enthalpy, the thermal output is dependent only on the flow rate. As the flow rate increases, the term RCH has a greater significance for the value of Φ_0, and the initial signal

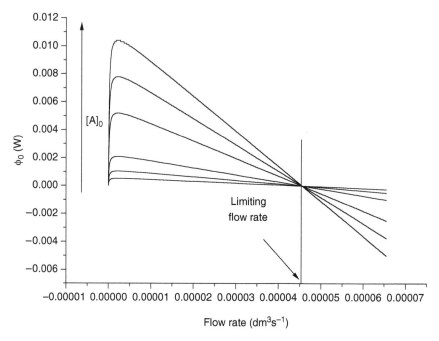

Figure 15 A three-dimensional plot to show the effect of increasing flow rate on the observed initial power for a fixed initial concentration and variable rate constant.

increases. However, as flow rate increases the residence time, τ, will decrease and the term $(1 - e^{-k\tau})$ will tend to zero; hence, the observed output will also tend to zero. The relative effects of these terms can be more clearly seen in Figure 15, where it can be seen that the value of Φ_0 (the maximum possible signal for a single reaction step) decreases as a function of flow rate, eventually reaching zero at a limiting flow rate.

A similar study of the effect of rate constant can also be carried out. Figure 16 describes the effect of varying the rate constant for a given $[A_0]$ with constant enthalpy. Consider the situation where the flow rate remains constant but the rate constant is increased. It becomes apparent that Φ_0 also increases. However, in this instance, the relationship is not directly proportional. Again, this can be readily explained by a consideration of Equation (27). In this case, the rate constant does not have a direct multiplicative effect on Φ_0; the relationship is embedded in an exponential term and will not be linear. Again, this flow rate dependence is seen more clearly in Figure 16.

An important observation is that Φ_0 will always go to zero at a limiting flow rate (Fig. 17). This is an important result because it shows that regardless of the thermokinetic parameters of the system, all the heat generated within the cell will be carried out by the solution when flowing at the limiting flow rate. This limiting flow rate can be determined from a rearrangement of Equation (30). If Φ_0 is equal

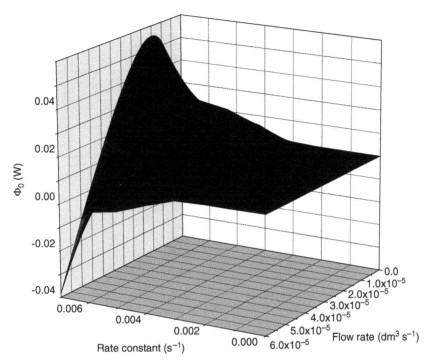

Figure 16 A three-dimensional representation of the effect of varying flow rate and rate constant on the observed initial power for a fixed initial concentration.

to zero then the thermal volume, V_c, must also be zero. Hence:

$$V_0 - FI = 0 \tag{34}$$

$$\frac{V_0}{I} = F \tag{35}$$

Therefore, the limiting flow rate at which Φ_0 goes to zero is approximately $165 \, mL \, h^{-1}$. These observations show the need for caution when investigating unknown reaction systems. If a reaction appears to be zero-order in the flow calorimeter, for instance, it may just be the operating conditions under which the experiment is conducted, or the thermokinetic parameters may give rise to a signal that is outside the detection limits for the calorimeter.

The product of the rate constant and reaction enthalpy for the ICHT reaction falls into this category. It has been reported (55) that the ICHT, a known second-order reaction, yields a zero-order output in the LKB 10-700 calorimeter even when the system is optimized for its detection. With knowledge of the flow rate dependence on the calorimetric output, a theoretical study can be conducted to determine whether the experimental observation is a consequence of the

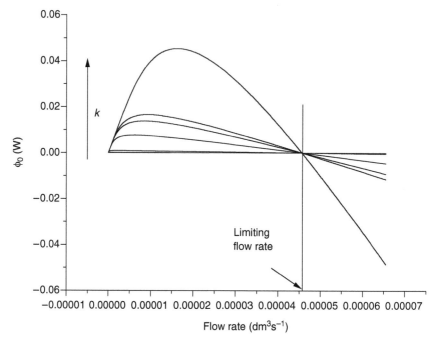

Figure 17   A plot to show the effect of increasing flow rate on the observed initial power for a fixed initial concentration and variable rate constant.

thermokinetic parameters being unsuitable or whether the flow rate used is inappropriate for the instrument. Figure 18 shows the range of theoretical values for Φ_0 as a function of flow rate for the ICHT reaction and, for comparison, the BCHMP reaction. The vertical line represents the experimental flow rate at which the ICHT reaction was studied.

It is obvious that for the ideal condition, where Φ_0 is at a maximum, the methyl paraben reaction yields a signal that is approximately 10-fold greater than that for the ICHT reaction. Note also that the maximum possible signal for the ICHT reaction is almost at the limit of detection for the instrumental settings used, and thus eloquently highlights the effect that the combination of rate constant and reaction enthalpy values has on the calorimetric output.

Attempts have been made to minimize the effects of flow rate on the thermal volume of the calorimetric cell. Guan et al. (60) designed a modified flow insert specifically for the study of dilute suspensions of Chinese hamster ovary cells. The flow characteristics of such cellular suspensions are known to be significantly different from homogenous solution phase systems. To overcome these differences, wide-bore tubing and relatively fast, downward flow-through rates were employed along with a calorimetric vessel of 1 cm^3 nominal (engineered) volume. They also realized the importance of determining the

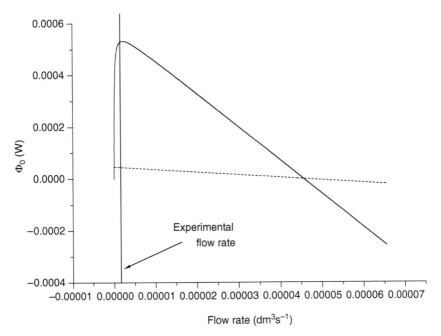

Figure 18 Range of theoretical values for ϕ_0 for the imidazole-catalyzed hydrolysis of triacetin (*top line*) and the base-catalyzed hydrolysis of methyl paraben (*bottom line*).

effective thermal volume of their modified insert in order to derive accurate thermodynamic information.

However, they made no attempt initially to quantify the effect of varying flow rate on the thermal volume. It should be noted that even in this study the thermal volume was found to be significantly different (at their experimental flow rate) from the nominal physical volume. A subsequent study using the BCHMP verified their earlier findings and, moreover, for their modified instrument it was shown that there was no apparent flow rate dependency on the thermal volume (the limiting flow rate for this insert was found to be 13.91 dm^3 h^{-1} over the range studied) confirming the assertion that the design was such that flow effects were minimized (Fig. 19). This observation was attributed by Guan et al. (60) to the fact that more attention was paid by the manufacturers to the geometry and hydrodynamics of the flow vessel, the chamber housing it, and the positioning of the heat detector assembly to minimize the flow dependency of the heat detection system. Nevertheless, there will be a limit to the minimization of such flow consequences that can be achieved by design modifications.

It is still desirable to know the thermal volume of the cell even if its value does not change. The thermal volume at zero flow rate was calculated to be 1.3 mL (nominal engineered value is 1 mL). A possible reason for this

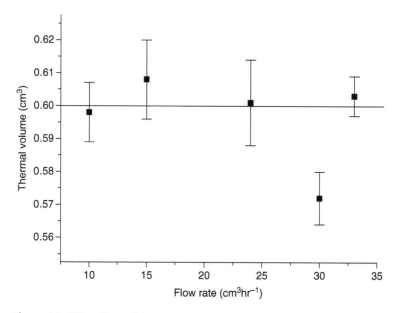

Figure 19 The effect of flow rate on thermal volume for the modified Thermometric 2277 flow calorimeter. *Source*: From Ref. 60.

incongruity is that there may be some contribution to the detected heat from the lead-in and lead-out tubing of the calorimetric cell.

It should also be noted that the designs of the two instruments discussed in the examples given already are different. The LKB instrument consists of a coiled gold tube with Peltier elements adjacent to it. The thermometric instrument essentially is identical to the static ampoule cell, but now with a coiled gold tube running around the outside of the calorimetric well on the sample side, and in the case of the modified insert this is replicated on the reference side. However, it is important to note that the test reaction has identified that the effective thermal volume is significantly different from the physical volume and hence suggests that this chemical calibration routine should be performed for all flow calorimeters if accurate data are to be obtained.

SUMMARY

Isothermal calorimetry is fast becoming an important tool for the study and characterization of a plethora of systems. With this popularity, comes the need for robust calibration and validation routines. Historically, calibration by electrical means has been the method of choice because of its simplicity and the possibility of high-precision measurements. Even though electrical calibration methods are very useful, they can be misleading. There is a real danger that the power patterns generated by such methods do not always mimic those seen in real systems, and hence this must be taken into consideration when inferring conclusions from experimentally obtained data. In light of these limitations, there is a growing trend to use electrical calibration in conjunction with chemical test reaction methods. Several well-characterized systems have been proposed which cover a wide range of calorimetric applications. There are some applications, however, which still need to be addressed; for example, no chemical test reactions have been proposed for solid-state reaction systems or for experiments performed using perfusion calorimeters. A widely accepted fact is that for a test reaction to be truly useful it must be robust and relatively easy to perform. On these grounds the accepted test reaction for photocalorimeters has been challenged by Morris (31) on the basis that the ferrioxalate reaction requires special handling (exclusion of O_2, in particular) in order to obtain reproducible results and hence is not sufficiently robust to be an effective chemical test reaction.

The kinetic and thermodynamic information already given for the various chemical test reactions are accurate within the stated uncertainty limits. It is important, however, that these values be verified and improved by further measurement using a variety of analytical techniques.

REFERENCES

1. Buckton G, Beezer AE. A microcalorimetric study of powder surface energetics. Int J Pharm 1988; 41:139–145.
2. Phipps MA, Mackin LA. Application of isothermal microcalorimetry in solid state drug development. Pharm Sci Tech Today 2000; 3:9–17.
3. Buckton G. Isothermal microcalorimetry water sorption experiments: calibration issues. Thermochim Acta 2000; 347:63–71.
4. Beezer AE. From guinea pigs to cutting fluids—a microcalorimetric journey. Thermochim Acta 2000; 349:1–7.
5. Archer DG, Kirklin DR. NIST and standards for calorimetry. Thermochim Acta 2000; 347:21–30.
6. Wadsö I. Needs for standards in isothermal microcalorimetry. Thermochim Acta 2000; 347:73–77.
7. Bunyan PF. An absolute calibration method for calorimeters. Thermochim Acta 1999; 327:109–116.
8. Chen A, Wadsö I. A test and calibration process for microcalorimeters used as thermal power meters. J Biochem Biophys Methods 1982; 6:297–306.
9. Hansen LD, Hart RM. The art of calorimetry. Thermochim Acta 2004; 417: 257–273.
10. Wadsö I, Goldberg RN. Standards in isothermal microcalorimetry (IUPAC Technical Report). Pure App Chem 2001; 73:1625–1639.
11. Urakami K, Beezer AE. A method to reduce the equilibration time prior to data capture in ampoule-based isothermal microcalorimetric studies. Thermochim Acta 2004; 410:109–117.

12. Finnin BA, O'Neill MAA, Gaisford S, Beezer AE, Hadgraft J, Sears P. The calibration of temperature for step isothermal calorimeters: Application of a test and reference reaction. J Therm Anal Cal 2006; 83:331–334.

13. Briggner L-E, Wadsö I. Test and calibration processes for microcalorimeters, with special reference to heat-conduction instruments used with aqueous systems. J Biochem Biophys Methods 1991; 22:101–118.

14. Marsh KN. Role of reference materials for the realization of physicochemical properties. Past, present and future. Pure App Chem 2000; 72:1809–1818.

15. O'Neill MAA, Beezer AE, Labetoulle C, et al. The base catalysed hydrolysis of methyl paraben: a test reaction for flow microcalorimeters used for the determination of both kinetic and thermodynamic parameters. Thermochim Acta 2003; 399:63–71.

16. Willson RJ, Beezer AE, Hills AK, Mitchell JC. The imidazole catalysed hydrolysis of triacetin: a medium term chemical calibrant for isothermal microcalorimeters. Thermochim Acta 1999; 325:125–132.

17. Beezer AE, Hills AK, O'Neill MAA, et al. The imidazole catalysed hydrolysis of triacetin: an inter- and intra-laboratory development of a test reaction for isothermal heat conduction microcalorimeters used for determination of both thermodynamic and kinetic parameters. Thermochim Acta 2001; 380:13–17.

18. Hills AK, Beezer AE, Mitchell JC, Connor JA. Sources of error, and their correction, in the analysis of isothermal heat conduction microcalorimetric data: Applications of a newly developed test reaction. Thermochim Acta 2001; 380:19–26.

19. Parmar MK, Beezer AE, Willson RJ. An isothermal microcalorimetric study of the imidazole catalysed hydrolysis of triacetin and the observed rate constant dependence of triacetin concentration. Thermochim Acta 2004; 424:219–222.

20. Sabbah R, Xu-wu A, Chickos JS, Planas Leitão ML, Roux MV, Torres LA. Reference materials for calorimetry and differential thermal analysis. Thermochim Acta 1999; 331:93–204.

21. Kemp RB, Lamprecht I. La vie est donc un feu pour la calorimétrie: Half a century of calorimetry—Ingemar Wadsö at 70. Thermochim Acta 2000; 348:1–17.

22. Beezer AE, Mitchell JC, Colegate RM, Scally DJ, Twyman LJ, Willson RJ. Microcalorimetry in the screening of discovery compounds and in the investigation of novel drug delivery systems. Thermochim Acta 1994; 250:277–283.

23. Willson RJ, Beezer AE, Mitchell JC. Solid state reactions studied by isothermal microcalorimetry; The solid state oxidation of ascorbic acid. Int J Pharm 1996; 132:45–51.

24. Zaman F, Beezer AE, Mitchell JC, Clarkson Elliot J, Nisbet M, Davis AF. The stability of benzoyl peroxide formulations determined from isothermal microcalorimetric studies. Int J Pharm 2001; 225:135–143.

25. O'Neill MAA, Vine GJ, Beezer AE, et al. Antimicrobial properties of silver-containing wound dressings: A microcalorimetric study. Int J Pharm 2003; 26:61–68.

26. Montanari MLC, Beezer AE, Montanari CA. QSAR based on biological microcalorimetry. The interaction of Saccharomyces cerivisiae with hydrazides. Thermochim Acta 1999; 328:91–97.

27. Criddle RS, Breidenbach RW, Hansen LD. Plant calorimetry: How to quantitatively compare apples and oranges. Thermochim Acta 1991; 193:67–90.

28. Teixeira C, Wadsö I. A microcalorimetric system for photochemical processes in solution. J Chem Thermodyn 1990; 22:703–713.

29. Wagnera T, Munzarb M, Krbala M, Kasapb SO. Photocalorimetric measurement of the heat flow during optically and thermally induced solid state reaction between Ag and $As_{33}S_{67}$ thin films. Thermochim Acta 2005; 432:241–245.
30. Wadsö I. Microcalorimetric techniques for the investigation of living plant materials. Thermochim Acta 1995; 250:285–304.
31. Morris AC. Photocalorimetry: Design, development and test considerations. Ph.D. dissertation, University of Greenwich, UK, 2004.
32. Christensen T, Toone EJ. Calorimetric evaluation of protein–carbohydrate affinities. Methods Enzymol 2003; 362:486–504.
33. Jelesarov I, Leder L, Bosshard HR. Probing the energetics of antigen–antibody recognition by titration microcalorimetry. Methods 1996; 9:533–541.
34. Holdgate GA. Making cool drugs hot: Isothermal titration calorimetry as a tool to study binding energetics. J Chem Soc Faraday Trans 1998; 94:2261–2267.
35. O'Brien R, Ladbury JE. Isothermal titration calorimetry of biomolecules. In: Harding SE, Chowdhry BZ, eds. Protein–Ligand Interactions: Hydrodynamics and Calorimetry, a Practical Approach. Oxford, UK: Oxford University Press, 2001.
36. Blandamer MJ, Cullis PM, Engberts JBFN. Titration microcalorimetry. J Chem Soc Faraday Trans 1998; 94:2261–2267.
37. Bäckman P, Bastos M, Briggner LE, et al. A system of microcalorimeters. Pure App Chem 1994; 66:375–382.
38. Mizoue LS, Tellinghuisen J. Calorimetric vs. vant Hoff binding enthalpies from isothermal titration calorimetry: Ba^{2+}—crown ether complexation. Biophys Chem 2004; 110:15–24.
39. Wiseman T, Williston S, Brandts JF, Lin LN. Rapid measurement of binding constants and heats of binding using a new titration calorimeter. Anal Biochem 1989; 179: 131–137.
40. Sigurskjold BW. Exact analysis of competition ligand binding by displacement isothermal titration calorimetry. Anal Biochem 2000; 277:260–266.
41. Velazquez-Campoy A, Kiso Y, Freire E. The binding energetics of first- and second-generation HIV-1 protease inhibitors: Implications for drug design. Arch Biochem Biophys 2001; 390:169–175.
42. Gomez J, Freire EJ. Thermodynamic mapping of the inhibitor site of the aspartic protease endothiapepsin. J Mol Biol 1995; 252:337–350.
43. Kilday MV. The enthalpy of solution of SRM 1655 (KCl) in H_2O. J Res Nat Bur Stand 1980; 85:467–481.
44. Irving RJ, Wadsö I. Use of tris(hydroxymethyl) aminomethane as a test substance in reaction calorimetry. Acta Chem Scand 1964; 18:195–201.
45. Gao D, Rytting JH. Use of solution calorimetry to determine the extent of crystallinity of drugs and excipients. Int J Pharm 1997; 151:183–192.
46. Salvetti G, Tognoni E, Tombari E, Johari GP. Excess energy of polymorphic states or glass over the crystal state by heat of solution measurement. Thermochim Acta 1996; 285:243–252.
47. Yff BTS, Royall PG, Brown MB, Martin GP. An investigation of calibration methods for solution calorimetry. Int J Pharm 2004; 269:361–372.
48. Ramos R, Gaisford S, Buckton G, Royall PG, Yff B, O'Neill MAA. A comparison of chemical reference materials for solution calorimeters. Int J Pharm 2005; 299:73–83.

49. Uriano GA. National bureau of standards certificate. Standard Reference Material 1655, potassium chloride, KCl (cr) for solution calorimetry, 1981.
50. Hill JO, Öjelund G, Wadsö I. Thermochemical results for "tris" as a test substance in solution calorimetry. J Chem Thermodyn 1969; 1:111–116.
51. Jakobsen DF, Frokjaer S, Larsen C, Niemann H, Buur A. Application of isothermal microcalorimetry in preformulation. I. Hygroscopicity of drug substances. Int J Pharm 1997; 156:67–77.
52. O'Neill MAA, Beezer AE, Vine GJ, et al. Practical and theoretical consideration of flow-through microcalorimetry: Determination of "thermal volume" and its flow rate dependence. Thermochim Acta 2004; 413:193–199.
53. O'Neill MAA, Beezer AE, Deal RM, et al. Survey of the effect of fill volume on the values for the enthalpy and rate constant derived from isothermal microcalorimetry: Applications of a newly developed test reaction. Thermochim Acta 2003; 397: 163–169.
54. Hills AK. Theoretical and experimental studies in isothermal microcalorimetry. Ph.D. dissertation, University of Kent at Canterbury, UK, 2001.
55. O'Neill MAA. Calorimetric investigation of complex systems: Theoretical developments and experimental studies. Ph.D. dissertation, University of Greenwich, UK, 2002.
56. Bakri A, Janssen LHM, Wilting J. Determination of reaction rate parameters using heat conduction microcalorimetry. J Therm Anal 1988; 33:185–190.
57. Poore VM, Beezer AE. Systematic errors in the measurement of heat by flow microcalorimetry. Thermochim Acta 1983; 63:133–144.
58. Beezer AE, Steenson TI, Tyrrell HJV. The measurement of kinetic and thermodynamic parameters for reacting systems by flow microcalorimetry and application of a method to the urea–urease system in a sodium phosphate buffer. In: Peters H, ed. Protides of the Biological Fluids. Oxford, UK: Pergamon Press, 1972.
59. O'Neill MAA, Beezer AE, Mitchell JC, Orchard JC, Connor JA. Determination of Michaelis–Menten parameters obtained from isothermal flow calorimetric data. Thermochim Acta 2004; 417:187–192.
60. Guan YH, Lloyd PC, Kemp RB. A calorimetric flow vessel optimised for measuring the metabolic activity of animal cells. Thermochim Acta 1999; 332:211–220.

3

Quantitative Data Analysis

INTRODUCTION

From the general discussion given in Chapter 1, it can be seen that although isothermal calorimetry (IC) offers huge potential to the pharmaceutical sciences, its use has so far been relatively limited by the lack of methods for analyzing quantitatively the data obtained. However, developments over the past 15 years have begun to redress this balance, and there are now a number of approaches that can be used to analyze calorimetric data. This chapter provides a theoretical overview of these developments and underpins many of the practical examples given in the latter part of the book. For the benefit of tuition, all of the techniques shown are illustrated using simulated data. Simulated data (which, as suggested by the name, are created using a mathematical worksheet) are ideal; in other words, they are free from the noise, random errors, and other artifacts inherent in real data. They are useful for testing the validity of a model (or the analyst's use of a model) because the correct reaction parameters, and reaction mechanism, are known in advance. A further discussion of the use and construction of simulated data is given in the following sections. The remainder of this chapter is subdivided according to the type of analysis methodology, which are defined as follows:

1. *Empirical fitting.* The use of a generic equation that fits, but does not describe, the data. This allows calculation of the percentage reaction that has occurred over a given time period.
2. *Kinetic modeling.* The construction of equations based on solution- or solid-phase kinetics. These fit and describe the data and can determine reaction parameters via iteration. This requires prior knowledge of the reaction mechanism.
3. *Direct calculation.* Allows the direct determination of reaction parameters without the need for iteration. Requires prior knowledge of the reaction mechanism.

4. *Chemometrics*. The use of principal component analysis (PCA) to determine and deconvolute the individual processes contributing to the overall signal. No prior knowledge of the system is required.

EMPIRICAL FITTING

The simplest approach to modeling calorimetric data is to fit the data to a generic equation; that is, an equation that conveniently fits the data but does not attempt to describe the reaction processes occurring. A simple example would be to use an exponential decay model such as that shown in Equation (1):

$$y = y_0 + A\,e^{-x/t} \tag{1}$$

For example, some calorimetric data are represented in Figure 1. The exact process that gave rise to these data is not important, but it shall be assumed that they represent the heat output of a partially completed reaction. The data can be fitted to Equation (1), by least-squares minimization, to determine the equation parameters that describe them. Once these values have been determined, it is a simple matter to extend the data to the time at which the power signal falls to zero (shown by the dotted line in Fig. 1). The area under the dotted line thus represents the total heat, Q, which would be generated by the reaction if it went to completion. From this information, it is easy to determine the percentage of

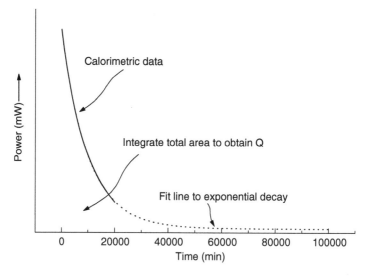

Figure 1   The fit of calorimetric data to an exponential model and the subsequent extrapolation to power = 0.

reaction completed at any time, t, by taking fractional areas. Hence, some useful information can be derived by using nonspecific models.

KINETIC MODELING

For a simple, solution phase, $A \rightarrow P$ reaction, a kinetic expression may be written that describes the rate of disappearance of reactant A, or the build up of product P [Equation (2)]:

$$\frac{dx}{dt} = k (A_0 - x)^n \tag{2}$$

where dx/dt is the rate of reaction, k is the rate constant, A_0 is the initial quantity of reactant A that is available for reaction, x is the quantity of reactant A reacted at time t, and n is the order of reaction. It should be noted that n may have any value, integral or nonintegral. For any given reaction that has gone to completion, the total heat evolved during the course of the reaction, Q, must be equal to the product of the enthalpy of reaction, ΔH, and the number of moles of material reacted, A_0:

$$Q = A_0 \Delta H \tag{3}$$

It follows that:

$$q = x \Delta H \tag{4}$$

where q is the heat evolved at time t.

Substituting $q/\Delta H$ for x and $Q/\Delta H$ for A_0 in Equation (2) and rearranging gives:

$$\frac{dq}{dt} = \Phi = k \Delta H^{1-n} (Q - q)^n \tag{5}$$

where Φ is the calorimetric power (in Watts). Integration of Equation (5) gives:

$$(Q - q) = [k t \Delta H^{1-n} (n - 1) + Q^{1-n}]^{1/1-n} \tag{6}$$

This expression may be substituted into Equation (5) to give:

$$\Phi = k \Delta H^{1-n} [k t \Delta H^{1-n} (n - 1) + Q^{1-n}]^{n/1-n} \tag{7}$$

Equation (7) describes calorimetric data that derive from reactions that follow the general rate expression given in Equation (2) (i.e., any single-step, solution-phase reaction). Calorimetric data from such a reaction may be entered into a suitable mathematical package and, by a process of iteration, may be fitted to Equation (7). This process was first described by Bakri (1) and was later extended by Willson et al. (2,3).

A further consideration of these equations results in methods to calculate directly the parameters of interest. Knowing these values reduces the burden on the fitting program. Methods for calculating these parameters are described later in this chapter.

A Practical Example

The application of these equations is best illustrated by the use of some power–time data. In order to prove the validity of the method when analyzing a reaction, it is necessary to know precisely the values of the parameters describing the reaction, to ensure that the values determined from the fitting procedure are correct. At present, there are insufficient data available in the literature for real systems so, to illustrate the use of the method, simulated calorimetric data will be used. As mentioned briefly earlier, simulated data have many advantages when compared with real data. As the equation and parameters describing the data are known, it is possible to know immediately if the values obtained through the fitting procedure are correct. In this way, it is possible to train an operator in the correct use of the method. Simulated data are also free of the imperfections that might be expected in real calorimetric data such as inaccuracies in solution concentrations, baseline instability, and so on.

However, care should also be exercised when using simulated data to test model equations. When fitting real calorimetric data, it is important that there are no preconceptions about the values that are returned, such that the fitting procedure is not unjustly biased, and the same approach should be adopted when fitting simulated data. It is often beneficial to construct many different sets of simulated data, and to choose one randomly for fitting. Only after the fitting procedure has been completed, should the values obtained be compared with the true answers.

In order to construct simulated data, it is necessary to use a suitable mathematical worksheet and, for the data constructed for this text, the program used was MathCad®. The following example shows how data may be constructed to conform to a simple, one-step model. Similar techniques will be used to construct data for all the methodologies presented in this chapter.

Construction of simulated data starts with the integrated rate expression for the reaction under study. For a one-step scheme the data are described by Equation (7) (integrated rate equations for other schemes are presented in the following sections). The equation and the reaction parameters are entered into the worksheet, which generates the data (Fig. 2). The data can then be imported into the analysis program (in this case, Origin) and treated as if they derived from a real experiment.

Origin uses a least-squares minimization routine to fit data to a model. The statistical measure of fit is chi^2 (χ^2). The smaller the value of χ^2, the better the fit. Although χ^2 values are of some use, their worth is somewhat limited because χ^2 values determined for different data sets are not always comparable and care must be taken when comparing fits to different data sets with each other. If different models are used to fit the same data, then the values of χ^2 become more important, as they allow the fit lines of the different models to be compared directly, the

The simulation of a simple $A{\rightarrow}B$ reaction scheme:

Values for the reaction variables must be entered, with a suitable time span,

$A := 0.0002 \quad k := 0.004 \quad H := 2{\cdot}10^{10} \quad n := 1.7 \quad Q := A{\cdot}H$

$t := 0, 200 .. 200000$

The rate of change of power with time is described by the reaction,

$a(t) := \left[k{\cdot}t{\cdot}H^{1-n}{\cdot}(n-1) + Q^{1-n}\right]^{\frac{n}{1-n}} {\cdot}k{\cdot}H^{1-n}$

Therefore, the observed calorimetric signal is,

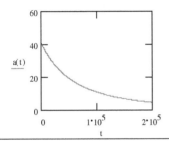

Figure 2   The use of a MathCad worksheet to construct power–time data that conform to a simple one-step process.

model returning the lowest χ^2 value being the better fit. When comparing different models, it should be remembered that as the number of variables used for fitting increases, the more likely it is that the model will fit the data, and better fits will be obtained for more complex models (cf. when fitting data to a polynomial equation, it becomes much easier to fit the data as the number of terms in the polynomial expression increases). When fitting calorimetric data, it is necessary to choose the model that best describes the reaction under study, such that the reaction parameters determined are accurate. If the mechanism of reaction is unknown, then the model chosen for fitting should be the one that gives the most reasonable fit with the fewest number of parameters.

It should also be noted that when simulating data care must be taken with the signs of the enthalpy values obtained from the fitting procedure. Some calorimeters [such as the Thermal Activity Monitor (TAM), Thermometric AB] record exothermic events with a positive sign. This runs counter to scientific convention and will result in the fitting program reporting a positive enthalpy value for an exothermic event. The operator must be aware of the way in which the calorimeter being used records data and ensure that values are properly reported. Here, positive enthalpy values are used to generate data that correspond to exothermic outputs from the TAM.

In principle, it is possible to fit the data to the model without prior knowledge of any of the reaction parameters; however, as already noted, it may be advantageous to obtain some initial values (e.g., the reaction order) such that the iterative burden on the fitting program is reduced. This can be highlighted through a consideration of simulated data. The data generated in the earlier sections were imported into the fitting program and values for n (1.7) and Q (4×10^6 μJ) were entered and set as constants (these could be calculated for real systems). Initial estimates for k and ΔH were also entered, the values being chosen randomly as 0.0001 and 4×10^{10} μJ/mol, respectively, and the fitting procedure was started. The iteration was repeated until the value of χ^2 remained constant, and appeared to be at a minimum, and a best-fit line was obtained (Fig. 3). As can be seen, the fit line is very good, and the value of χ^2 is very small, but the values obtained for k and ΔH are incorrect. This highlights both the power of simulated data and one of the drawbacks of the method of analysis. Without knowing the value of the actual reaction parameters, it would not have been possible to know that the values of the parameters obtained were incorrect, unless good literature values were obtainable. The reason that the correct values for the parameters were not obtained from the fitting procedure is that there is a requirement that the value of A_0, the quantity of material available for reaction, or the value of ΔH, the reaction enthalpy, is known, and is set as a constant. The equation used for fitting the data, Equation (7), does not contain the term A_0 directly, but contains Q (equivalent to $A_0 \Delta H$).

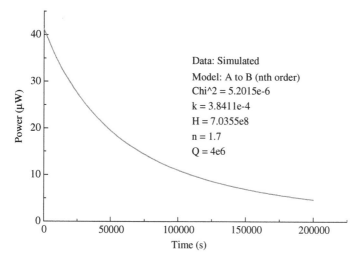

Figure 3 The result of fitting the simulated data to Equation (7), having set the values of n and Q constant. Note that the values of k and ΔH are incorrect, yet the fit line (*dotted*) overlays the raw data.

If the fitting procedure is repeated, replacing Q with $A_0 \Delta H$ in the fitting equation, and the value of A_0 is entered (which is usually known), then, if the same initial estimates as used already for the remaining parameters are entered, a different best fit line is obtained (Fig. 4). The correct reaction parameters have now been recovered.

This example illustrates the most fundamental requirement for the method to work; the number of moles of material available for reaction, or the value of the reaction enthalpy, must be known. The power of the fitting procedure is such that, as long as either the value of A_0 or the value of ΔH has been entered correctly, the correct reaction parameters will be recovered without the need for accurate initial estimates, and it is not usually necessary to determine the value of n or Q independently. In fact, it is observed that the fitting procedure will usually recover the correct values of n and k without the need for either the value of A_0 or ΔH to be known. Thus, as well as being powerful, the method is also extremely fast.

It follows, however, that requiring prior knowledge of such parameters may lead to problems for some reacting systems. For reactions that occur in solution, it is likely that the value of A_0 will equal the number of moles of reactant in solution, a number which is determinable, but the situation is considerably more complex for systems involving a solid-phase component. If a reaction occurs between a liquid and a solid component, then it may be the case that the solid component is dissolved as the reaction proceeds and, hence, the total number of moles of reactant available for reaction can be determined. If, however, the

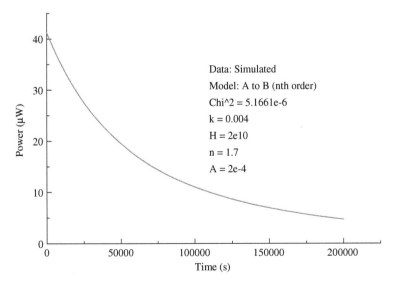

Figure 4 The result of fitting the simulated data to Equation (7), having substituted Q for $A_0 \Delta H$. The values of n and A_0 were held constant. Note that the correct values for the reaction parameters have been obtained, and that the value of χ^2 is slightly smaller than the previous fitting (Fig. 3). Again, the fit line (*dotted*) overlays the raw data.

solid component is not dissolved, or there is a reaction between a solid and a gas or a solid and a solid, then a considerable degree of complexity may be introduced to the system. For example, reaction may occur on the surface of the solid. After all the material on the surface has reacted, the reaction may stop. In that case, the number of moles of material available for reaction will not equal the number of moles of material loaded into the cell, and it may be difficult to determine, or even estimate, the value of A_0.

Similarly, the value of the reaction enthalpy may often be obtained from the literature; however, where this is not possible, an estimate may need to be made from knowledge of bond enthalpies. This requires that a reasonable idea of the mechanism of reaction (more specifically that the products, if not the actual pathway) is known, and this may be difficult for some systems. As will be discussed later, the situation becomes considerably more complex when the reaction being studied follows a consecutive or parallel path. In such cases, there will often be several reaction enthalpy values, and reasonable estimates will be required for each of them. In systems following parallel or consecutive pathways, where there is only one initial reactant, it is sufficient to know only the value of A_0 to recover the other reaction parameters.

There is, however, the possibility of obtaining mechanistic information from such a fitting procedure. If, for example, a solid-state reaction is being studied, and the value of ΔH is known, then it is possible to determine exactly the value of A_0. Such information may then be used to infer information on the type of reaction pathway being followed. For example, the value of A_0 may be significantly less than the number of moles of reactant being studied and it may be deduced that reaction only occurs on the surface of the reactant particles.

Simulation of Parallel Reaction Schemes

Parallel reaction schemes describe a large number of commonly encountered reactions, simply because most pharmaceuticals have many components, and may often be described satisfactorily by the simple addition of kinetic expressions. The following text shows how a range of parallel reaction schemes may be analyzed to give kinetic and thermodynamic parameters. A kinetic treatment is discussed for each scheme, and equations are derived that allow power–time data to be simulated. The principles are common to all parallel schemes and can hence be extended to other systems not described here.

Parallel First- and Second-Order Reactions with the Same Reactant

Many reactions may occur via more than one reaction pathway, and each pathway may have a different reaction order. Take, for example, the following case:

$$A \xrightarrow{k_1} P \text{ (first order, } \Delta H_1)$$

$$A \xrightarrow{k_2} P \text{ (second order, } \Delta H_1)$$

The reactant A may form product P via a first-order process of rate constant k_1 or via a second-order process of rate constant k_2. The rates of formation of P via each pathway are described by Equations (8) and (9) (where the subscripts 1 and 2 refer to the two pathways, respectively):

$$\frac{d[P]_1}{dt} = k_1 [A] \tag{8}$$

$$\frac{d[P]_2}{dt} = k_2 [A]^2 \tag{9}$$

The overall rate expression for the formation of P is then given by Equation (10):

$$\frac{d[P]}{dt} = \frac{-d[A]}{dt} = k_1 [A] + k_2 [A]^2 \tag{10}$$

Integration of Equation (10) allows $[A]$ to be determined with respect to time [Equation (11)]:

$$[A] = \frac{k_1 A_0}{e^{k_1 t}(k_1 + k_2 A_0) - k_2 A_0} \tag{11}$$

Substitution of this expression into Equations (8) and (9) gives:

$$\frac{d[P]_1}{dt} = k_1 \left[\frac{k_1 A_0}{e^{k_1 t}(k_1 + k_2 A_0) - k_2 A_0} \right] \tag{12}$$

$$\frac{d[P]_2}{dt} = k_2 \left[\frac{k_1 A_0}{e^{k_1 t}(k_1 + k_2 A_0) - k_2 A_0} \right]^2 \tag{13}$$

As $[P]_1$ equals $q_1/\Delta H_1$ and $[P]_2$ equals $q_2/\Delta H_1$, then:

$$\frac{dq_1}{dt} = k_1 \Delta H_1 \left[\frac{k_1 A_0}{e^{k_1 t}(k_1 + k_2 A_0) - k_2 A_0} \right] \tag{14}$$

$$\frac{dq_2}{dt} = k_2 \Delta H_1 \left[\frac{k_1 A_0}{e^{k_1 t}(k_1 + k_2 A_0) - k_2 A_0} \right]^2 \tag{15}$$

It follows that the power–time signal observed in the calorimeter for the overall reaction is given by:

$$\frac{dq_{obs}}{dt} = k_1 \Delta H_1 \left[\frac{k_1 A_0}{e^{k_1 t}(k_1 + k_2 A_0) - k_2 A_0} \right]$$
$$+ k_2 \Delta H_1 \left[\frac{k_1 A_0}{e^{k_1 t}(k_1 + k_2 A_0) - k_2 A_0} \right]^2 \tag{16}$$

Equations (14) and (15) may be used to construct power–time data for the two component pathways of the reaction, $A \rightarrow P$ (first order) and $A \rightarrow P$ (second order), respectively, using a mathematical worksheet. Equation (16) may be used to fit the overall power–time data that would be observed from such a reaction. Examples of the generation of data for parallel reactions are shown in Figures 5 and 6. The examples given show the effects of varying the values of the rate constants. The two individual power–time traces in each example can be summed, producing power–time data that are equivalent to those that would be obtained from a calorimeter used to study a real experiment.

It can be seen from these worksheets that, once the equations describing the rate of change of power with time have been derived for a particular reaction scheme, it is possible to explore the effects of varying the values of the reaction parameters. In this way, for example, the effects of systematically changing the magnitude of the rate constants for a reaction could be investigated, while maintaining the values of the other parameters constant. Such an investigation would be interesting, but outside the scope of the text presented here. Rather, this work

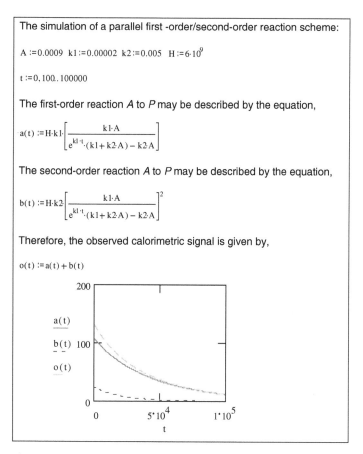

The simulation of a parallel first -order/second-order reaction scheme:

$A := 0.0009 \quad k1 := 0.00002 \quad k2 := 0.005 \quad H := 6 \cdot 10^9$

$t := 0, 100.. 100000$

The first-order reaction A to P may be described by the equation,

$$a(t) := H \cdot k1 \cdot \left[\frac{k1 \cdot A}{e^{k1 \cdot t} \cdot (k1 + k2 \cdot A) - k2 \cdot A} \right]$$

The second-order reaction A to P may be described by the equation,

$$b(t) := H \cdot k2 \cdot \left[\frac{k1 \cdot A}{e^{k1 \cdot t} \cdot (k1 + k2 \cdot A) - k2 \cdot A} \right]^2$$

Therefore, the observed calorimetric signal is given by,

$o(t) := a(t) + b(t)$

Figure 5   The simulation of a parallel first- and second-order reaction where $k_1 < k_2$.

The simulation of a parallel first-order/second-order reaction scheme:

$A := 0.0002 \quad k1 := 0.0002 \quad k2 := 0.000005 \quad H := 3 \cdot 10^9$

$t := 0, 100 .. 10000$

The first-order reaction A to P may be described by the equation,

$$a(t) := H \cdot k1 \cdot \left[\frac{k1 \cdot A}{e^{k1 \cdot t} \cdot (k1 + k2 \cdot A) - k2 \cdot A} \right]$$

The second-order reaction A to P may be described by the equation,

$$b(t) := H \cdot k2 \cdot \left[\frac{k1 \cdot A}{e^{k1 \cdot t} \cdot (k1 + k2 \cdot A) - k2 \cdot A} \right]^2$$

Therefore, the observed calorimetric signal is given by,

$o(t) := a(t) + b(t)$

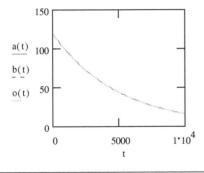

Figure 6 The simulation of a parallel first- and second-order reaction, where $k_1 > k_2$. The rate of the second reaction is so slow, compared with the first, that the observed power–time signal derives almost entirely from the first-order reaction and the data overlay.

is intended to show how different reaction schemes may be modeled, and it is left to the reader to use the equations presented for further investigations.

Parallel First- and Higher-Order Reactions with Different Reactants

A related parallel reaction scheme that may be encountered is the formation of a product from two different reactants:

$$A \xrightarrow{k_1} P \text{ (first-order, } \Delta H_1)$$

$$B \xrightarrow{k_2} P \text{ (first-order, } \Delta H_2)$$

The rates of disappearance of A and B are given by Equations (17) and (18):

$$\frac{d[A]}{dt} = -k_1 [A] \tag{17}$$

$$\frac{d[B]}{dt} = -k_2 [B] \tag{18}$$

Integration of Equations (17) and (18) yields:

$$[A] = A_0 e^{-k_1 t} \tag{19}$$

$$[B] = B_0 e^{-k_2 t} \tag{20}$$

It follows that:

$$\frac{d[A]}{dt} = -k_1 A_0 e^{-k_1 t} \tag{21}$$

$$\frac{d[B]}{dt} = -k_2 B_0 e^{-k_2 t} \tag{22}$$

As $[A]$ equals $q_A/\Delta H_1$ and $[B]$ equals $q_B/\Delta H_2$, Equations (21) and (22) become:

$$\frac{dq_A}{dt} = \Delta H_1 k_1 A_0 e^{-k_1 t} \tag{23}$$

$$\frac{dq_B}{dt} = \Delta H_2 k_2 B_0 e^{-k_2 t} \tag{24}$$

Equations (23) and (24) may be used to construct power–time data that represent the reactions $A \rightarrow P$ and $B \rightarrow P$, respectively. As the rate of formation of P is given by:

$$\frac{d[P]}{dt} = k_1 [A] + k_2 [B] \tag{25}$$

It follows that:

$$\frac{d[P]}{dt} = k_1 A_0 e^{-k_1 t} + k_2 B_0 e^{-k_2 t} \tag{26}$$

and, hence, the observed calorimetric signal for the overall reaction is given by:

$$\frac{dq_{obs}}{dt} = k_1 A_0 \Delta H_1 e^{-k_1 t} + k_2 B_0 \Delta H_2 e^{-k_2 t} \tag{27}$$

The use of these equations to generate data is illustrated in Figures 7 and 8.

It is possible to extend such an analysis to reactions that are second, third, or mixed order. Consider, for example, a reaction that proceeds via two

The simulation of a parallel first-order reaction scheme, with different starting reagents

$A := 0.0005$ $B := 0.0005$ $k1 := 0.00002$ $k2 := 0.00002$ $H1 := 3 \cdot 10^9$ $H2 := 5 \cdot 10^9$

$t := 0, 1000.. 100000$

The reaction A to P can be described by the equation,

$a(t) := k1 \cdot A \cdot H1 \cdot e^{-k1 \cdot t}$

The reaction B to P can be described by the equation,

$b(t) := k2 \cdot B \cdot H2 \cdot e^{-k2 \cdot t}$

Therefore, the observed calorimetric signal is given by,

$o(t) := a(t) + b(t)$

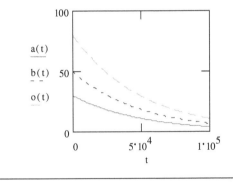

Figure 7 The simulation of a parallel first-order reaction scheme where $k_1 = k_2$ and $A_0 = B_0$. The differences in the traces arise as a result of the different reaction enthalpies for each step.

simultaneous reaction pathways:

$$A \xrightarrow{k_1} P \ (n\text{th order}, \Delta H_1)$$

$$B \xrightarrow{k_2} P \ (\text{first-order}, \Delta H_2)$$

The power change for the reaction $B \rightarrow P$ is described by Equation (24), while the reaction $A \rightarrow P$ requires a new expression. Starting with the premise:

$$\frac{d[A]}{dt} = -k_1 [A]^n \tag{28}$$

The simulation of a parallel first-order reaction scheme, with different starting reagents:

A := 0.0009 B := 0.0007 k1 := 0.00008 k2 := 0.00002 H1 := 3·10⁹ H2 := 8·10⁹

t := 0, 1000 . 100000

The reaction A to P can be described by the equation,

a(t) := k1·A·H1·e$^{-k1·t}$

The reaction B to P can be described by the equation,

b(t) := k2·B·H2·e$^{-k2·t}$

Therefore, the observed calorimetric signal is given by,

o(t) := a(t) + b(t)

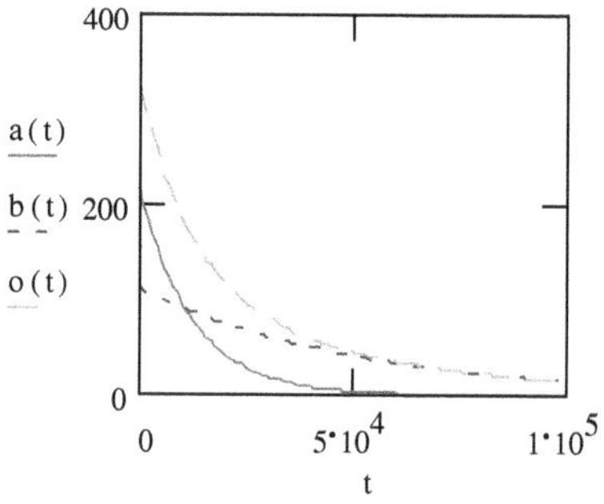

Figure 8 The simulation of a parallel first-order reaction where $k_1 \neq k_2$ and $A_0 \neq B_0$. The reaction $A \rightarrow P$ proceeds at a much faster rate than the reaction $B \rightarrow P$ and is completed before the end of the observation period.

Equation (28) may be readily integrated to yield:

$$[A] = (k_1 t (n-1) + A_0^{1-n})^{1/1-n} \tag{29}$$

Hence:

$$\frac{d[P]_A}{dt} = k_1 [(k_1 t (1-n) + A_0^{1-n})^{1/1-n}]^n \tag{30}$$

where $[P]_A$ represents the quantity of P formed from reactant A. Equation (30) may be expressed in calorimetric terms:

$$\frac{dq_A}{dt} = \Delta H_1 k_1 [(k_1 t (1-n) + A_0^{1-n})^{1/1-n}]^n \tag{31}$$

Equation (31) may be used to simulate power–time data for the reaction $A \to P$. Combining Equations (24) and (31) gives the equation that describes the power–time signal that would be observed calorimetrically for a parallel nth/first-order reaction [Equation (32)]:

$$\frac{dq_{obs}}{dt} = \Delta H_1 \, k_1 \, [(k_1 \, t \, (1-n) + A_0^{1-n})^{1/1-n}]^n + \Delta H_2 \, k_2 \, B_0 \, e^{-k_2 t} \qquad (32)$$

It follows that using these equations described, it is possible to simulate many parallel reaction schemes so long as each scheme does not involve a consecutive step (examples are given in Figs. 9 and 10). The treatment of

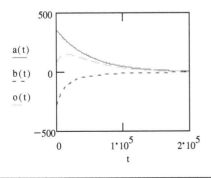

Figure 9 The simulation of a parallel first- and nth-order reaction scheme.

The simulation of a parallel mth- and nth-order reaction scheme, with different starting reagents:

$A := 0.0007 \quad B := 0.001 \quad k1 := 0.0003 \quad k2 := 0.003 \quad H1 := 7 \cdot 10^9 \quad H2 := 7 \cdot 10^9 \quad m := 1.5 \quad n := 1.8$

$t := 0, 1000 .. 400000$

The mth-order reaction A to P can be described by the equation,

$$a(t) := k1 \cdot H1 \cdot \left[\left[k1 \cdot t \cdot (m-1) + A^{1-m} \right]^{\frac{1}{1-m}} \right]^{m}$$

The nth-order reaction B to P can be described by the equation,

$$b(t) := k2 \cdot H2 \cdot \left[\left[k2 \cdot t \cdot (n-1) + B^{1-n} \right]^{\frac{1}{1-n}} \right]^{n}$$

The observed calorimetric signal is given by,

$$c(t) := a(t) + b(t)$$

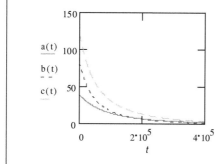

Figure 10   The simulation of a parallel mth- and nth-order reaction scheme.

parallel/consecutive reactions is discussed later. To simulate or fit data arising from parallel reaction schemes that comprise more than two simultaneous reaction pathways, all that is required is to sum the required number of equations together (Fig. 11).

Simulation of Consecutive Reactions

Series reaction mechanisms are commonly discussed, and form an important group of reactions, although they are more complex to analyze than the parallel schemes previously discussed. As before, the following text shows how a series

The simulation of a parallel first-, *m*th- and *n*th-order reaction scheme, with different starting reagents:

$A := 0.0007 \quad B := 0.0002 \quad C := 0.0005 \quad k1 := 0.0009 \quad k2 := 0.03 \quad k3 := 0.00006$

$t := 0, 1000 .. 100000 \quad H1 := 7 \cdot 10^9 \quad H2 := -6 \cdot 10^9 \quad H3 := 5 \cdot 10^9 \quad m := 1.5 \quad n := 1.8$

The *m*th-order reaction *A* to *P* can be described by the equation,

$$a(t) := k1 \cdot H1 \cdot \left[\left[k1 \cdot t \cdot (m-1) + A^{1-m} \right]^{\frac{1}{1-m}} \right]^m$$

The *n*th-order reaction *B* to *P* can be described by the equation,

$$b(t) := k2 \cdot H2 \cdot \left[\left[k2 \cdot t \cdot (n-1) + B^{1-n} \right]^{\frac{1}{1-n}} \right]^n$$

The first-order reaction *C* to *P* can be described by the equation,

$$c(t) := k3 \cdot H3 \cdot C \cdot e^{-k3 \cdot t}$$

The observed calorimetric signal is given by,

$$o(t) := a(t) + b(t) + c(t)$$

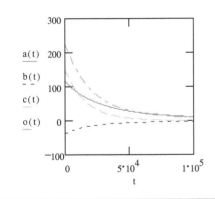

Figure 11 The simulation of a parallel first-, *m*th-, and *n*th-order reaction scheme.

of consecutive reaction schemes may be analyzed to yield thermodynamic and kinetic information. A kinetic treatment is discussed for each case, and equations are derived that allow power–time data to be simulated. Using such equations, the effects of changing the reaction parameters on each scheme are investigated. Similar principles can be used for reaction schemes not specifically discussed.

Consecutive, First-Order, $A \rightarrow B \rightarrow P$ Reactions

The simplest consecutive reaction schemes often describe a wide range of commonly encountered chemical reactions. Take, for example:

$$A \xrightarrow{k_1} B \xrightarrow{k_2} P$$

where both steps are first-order, with reaction enthalpies ΔH_1 and ΔH_2. (Examples of reactions that follow this type of mechanism include certain hydrolyses and the radioactive decay of ^{218}Po.) The rate equations that describe each reaction step are:

$$\frac{d[A]}{dt} = -k_1 [A] \tag{17}$$

$$\frac{d[B]}{dt} = k_1 [A] - k_2 [B] \tag{33}$$

$$\frac{d[P]}{dt} = k_2 [B] \tag{34}$$

As already noted, the integral of Equation (17) is:

$$[A] = A_0 e^{-k_1 t} \tag{19}$$

Equation (19) may be substituted into Equation (33) to yield:

$$\frac{d[B]}{dt} = k_1 A_0 e^{-k_1 t} - k_2 [B] \tag{35}$$

Equation (35) may be integrated using an integrating factor, if $B_0 = 0$, to give:

$$[B] = \frac{A_0 k_1}{k_2 - k_1} (e^{-k_1 t} - e^{-k_2 t}) \tag{36}$$

As the sum of the reactant concentrations must equal A_0, if neither B nor P is present initially, then:

$$[P] = A_0 - [A] - [B] \tag{37}$$

So:

$$[P] = A_0 \left[1 + \frac{1}{k_1 - k_2} (k_2 e^{-k_1 t} - k_1 e^{-k_2 t}) \right] \tag{38}$$

Differentiation of Equation (38) yields:

$$\frac{d[P]}{dt} = k_1 k_2 A_0 \left(\frac{e^{-k_1 t} - e^{-k_2 t}}{k_2 - k_1} \right) \tag{39}$$

and hence:

$$\frac{dq_P}{dt} = k_1 k_2 \, \Delta H_2 \, A_0 \left(\frac{e^{-k_1 t} - e^{-k_2 t}}{k_2 - k_1} \right) \tag{40}$$

where dq_P/dt represents the power associated with the reaction $B \to P$. Equation (40) may be used to simulate the power–time data for the reaction $B \to P$. The reaction $A \to B$ may, as noted earlier, be described by Equation (23). Combining Equations (23) and (40) gives an equation that describes the overall power–time data that would be observed for a reaction following a consecutive, two-step, first-order reaction scheme:

$$\frac{dq_{obs}}{dt} = k_1 \, \Delta H_1 \, A_0 \, e^{-k_1 t} + k_1 k_2 \, \Delta H_2 \, A_0 \, \frac{e^{-k_1 t} - e^{-k_2 t}}{k_2 - k_1} \tag{41}$$

The use of these equations is illustrated in Figures 12–14, showing the power–time traces that would be obtained for each of the component steps for the particular set of constants shown. The examples show the effects of changing the values of the rate constants and reaction enthalpies. In each example, the component power–time traces are shown summed, representing the power–time traces that would be observed calorimetrically.

Consecutive, First-Order, $A \to B \to C \to P$ Reactions

A more complicated example of a consecutive reaction scheme is represented by:

$$A \xrightarrow{k_1} B \xrightarrow{k_2} C \xrightarrow{k_3} P$$

where each step is first order, with reaction enthalpies of ΔH_1, ΔH_2, and ΔH_3, respectively.

The reactions describing the first two steps are the same as already derived in Equations (23) and (40) [remembering that in the case of Equation (40), P is replaced with B]. The term $[C]$ depends on the rates of the reactions $B \to C$ and $C \to P$, and may be described by a kinetic equation of the form:

$$\frac{d[C]}{dt} = k_2[B] - k_3[C] \tag{42}$$

Substitution of Equation (36) into Equation (42) yields:

$$\frac{d[C]}{dt} = k_2 \left[\frac{A_0 k_1}{k_2 - k_1} (e^{-k_1 t} - e^{-k_2 t}) \right] - k_3[C] \tag{43}$$

The simulation of an $A \rightarrow B \rightarrow P$ consecutive reaction scheme:

$A := 0.0002 \quad k1 := 0.00005 \quad k2 := 0.00008 \quad H1 := 3 \cdot 10^9 \quad H2 := 4.5 \cdot 10^9$

$t := 0, 1000.. \, 100000$

The reaction A to B can be described by the equation,

$a(t) := k1 \cdot \left[e^{-k1 \cdot t} \cdot (A \cdot H1) \right]$

The reaction B to P can be described by the equation,

$b(t) := H2 \cdot A \cdot k1 \cdot k2 \cdot \dfrac{e^{-k1 \cdot t} - e^{-k2 \cdot t}}{k2 - k1}$

Therefore, the observed calorimetric signal is given by,

$o(t) := a(t) + b(t)$

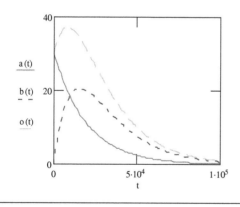

Figure 12 The simulation of a consecutive first-order $A \rightarrow B \rightarrow P$ reaction scheme with $k_1 \approx k_2$.

Equation (43) may be integrated using the integrating factor $e^{\int k_3 dt}$, that is, $e^{k_3 t}$:

$$e^{k_3 t} \frac{d[C]}{dt} + e^{k_3 t} k_3[C] = e^{k_3 t} \left[k_2 \left[\frac{A_0 k_1}{k_2 - k_1} (e^{-k_1 t} - e^{-k_2 t}) \right] \right] \tag{44}$$

$$\Rightarrow \frac{d}{dt}(e^{k_3 t}[C]) = e^{k_3 t} \left[k_2 \left[\frac{A_0 k_1}{k_2 - k_1} (e^{-k_1 t} - e^{-k_2 t}) \right] \right] \tag{45}$$

$$\Rightarrow [C] = \left(\frac{k_2^2 k_1 A_0 e^{-k_2 t}}{-k_1 k_2 - k_3 k_2 + k_2^2 + k_3 k_1} \right) + \left(\frac{k_2 k_1^2 A_0 e^{-k_1 t}}{-k_1 k_2 + k_3 k_2 - k_3 k_1 + k_1^2} \right)$$
$$- \left(\frac{k_2 k_1 A_0 e^{-k_3 t}}{-k_1 k_2 - k_3 k_2 + k_2^2 + k_3 k_1} \right) - \left(\frac{k_2 k_1 A_0 e^{-k_3 t}}{-k_1 k_2 + k_3 k_2 - k_3 k_1 + k_1^2} \right) \tag{46}$$

The simulation of an $A \rightarrow B \rightarrow P$ consecutive reaction scheme:

A := 0.0003   k1 := 0.00004   k2 := 0.000008   H1 := 7·10^9 H2 := 4.5 10^9

t := 0, 1000 . 150000

The reaction A to B can be described by the equation,

$$a(t) := k1 \cdot \left[e^{-k1 \cdot t} \cdot (A \cdot H1) \right]$$

The reaction B to P can be described by the equation,

$$b(t) := H2 \cdot A \cdot k1 \cdot k2 \cdot \frac{e^{-k1 \cdot t} - e^{-k2 \cdot t}}{k2 - k1}$$

Therefore, the observed calorimetric signal is given by,

$$o(t) := a(t) + b(t)$$

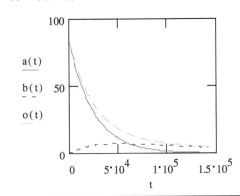

Figure 13   The simulation of a consecutive first-order $A \rightarrow B \rightarrow P$ reaction scheme with $k_1 > k_2$. The reaction $A \rightarrow B$ has nearly gone to completion before the reaction $B \rightarrow P$ reaches a maximum rate.

As the sum of the reactant concentrations must equal A_0, if B, C, and P are not present initially, then:

$$[P] = A_0 - [A] - [B] - [C] \tag{47}$$

Hence:

$$[P] = A_0 - (A_0 e^{-k_1 t}) - \left(\frac{A_0 k_1}{k_2 - k_1} (e^{-k_1 t} - e^{-k_2 t}) \right)$$

$$- \left(\frac{k_2^2 k_1 A_0 e^{-k_2 t}}{-k_1 k_2 - k_3 k_2 + k_2^2 + k_3 k_1} \right) - \left(\frac{k_2 k_1^2 A_0 e^{-k_1 t}}{-k_1 k_2 + k_3 k_2 - k_3 k_1 + k_1^2} \right)$$

$$+ \left(\frac{k_2 k_1 A_0 e^{-k_3 t}}{-k_1 k_2 - k_3 k_2 + k_2^2 + k_3 k_1} \right) + \left(\frac{k_2 k_1 A_0 e^{-k_3 t}}{-k_1 k_2 + k_3 k_2 - k_3 k_1 + k_1^2} \right) \tag{48}$$

The simulation of an $A{\to}B{\to}P$ consecutive reaction scheme:

$A := 0.0003 \quad k1 := 0.00004 \quad k2 := 0.00008 \quad H1 := -7 \cdot 10^9 \quad H2 := 4.5 \cdot 10^9$

$t := 0, 1000 . 150000$

The reaction A to B can be described by the equation,

$$a(t) := k1 \cdot \left[e^{-k1 \cdot t} \cdot (A \cdot H1) \right]$$

The reaction B to P can be described by the equation,

$$b(t) := H2 \cdot A \cdot k1 \cdot k2 \frac{e^{-k1 \cdot t} - e^{-k2 \cdot t}}{k2 - k1}$$

Therefore, the observed calorimetric signal is given by,

$$o(t) := a(t) + b(t)$$

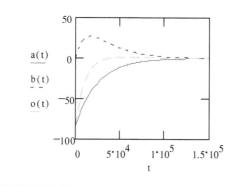

Figure 14   The simulation of a consecutive first-order $A \to B \to P$ reaction scheme with $k_2 > k_1$ and ΔH_1 negative.

Equation (48) may be differentiated to give Equation (49):

$$\frac{d[P]}{dt} = A_0 k_1 e^{-k_1 t} - \frac{A_0 k_1}{(k_2 - k_1)} \left(-k_1 e^{-k_1 t} + k_2 e^{-k_2 t} \right)$$

$$+ \frac{k_2^2 k_1 A_0 e^{-k_2 t}}{-k_1 k_2 - k_3 k_2 + k_2^2 + k_3 k_1} + \frac{k_2 k_1^2 A_0 e^{-k_1 t}}{-k_1 k_2 + k_3 k_2 - k_3 k_1 + k_1^2}$$

$$- \frac{k_2 k_1 A_0 k_3 e^{-k_3 t}}{-k_1 k_2 - k_3 k_2 + k_2^2 + k_3 k_1} - \frac{k_2 k_1 A_0 k_3 e^{-k_3 t}}{-k_1 k_2 + k_3 k_2 - k_3 k_1 + k_1^2} \tag{49}$$

and hence:

$$\frac{dq_P}{dt} = \Delta H_3 \left(A_0 k_1 e^{-k_1 t} - \frac{A_0 k_1}{(k_2 - k_1)} (-k_1 e^{-k_1 t} + k_2 e^{-k_2 t}) \right.$$

$$+ \frac{k_2^2 k_1 A_0 e^{-k_2 t}}{-k_1 k_2 - k_3 k_2 + k_2^2 + k_3 k_1} + \frac{k_2 k_1^2 A_0 e^{-k_1 t}}{-k_1 k_2 + k_3 k_2 - k_3 k_1 + k_1^2}$$

$$\left. - \frac{k_2 k_1 A_0 k_3 e^{-k_3 t}}{-k_1 k_2 - k_3 k_2 + k_2^2 + k_3 k_1} - \frac{k_2 k_1 A_0 k_3 e^{-k_3 t}}{-k_1 k_2 + k_3 k_2 - k_3 k_1 + k_1^2} \right) \quad (50)$$

where, dq_P/dt represents the power associated with the reaction $C \rightarrow P$. Equation (50) may be used to construct power–time data that describe the reaction $C \rightarrow P$. Summation, and simplification, of Equations (23), (40), and (50) therefore results in an equation that describes the overall power–time data for a consecutive, three-step, first-order reaction scheme:

$$\frac{dq_{obs}}{dt} = k_1 \Delta H_1 A_0 e^{-k_1 t} + k_1 k_2 \Delta H_2 A_0 \frac{e^{-k_1 t} - e^{-k_2 t}}{k_2 - k_1}$$

$$+ \Delta H_3 \left[A_0 k_1 k_2 k_3 \left(\frac{e^{-k_1 t}}{(k_2 - k_1)(k_3 - k_1)} + \frac{e^{-k_2 t}}{(k_1 - k_2)(k_3 - k_2)} \right. \right.$$

$$\left. \left. + \frac{e^{-k_3 t}}{(k_1 - k_3)(k_2 - k_3)} \right) \right] \quad (51)$$

The use of these equations is illustrated in Figures 15–17, showing the power--time traces that would be obtained for each of the component steps for the particular set of constants shown. The examples shown demonstrate the effects of changing the values of the rate constants and reaction enthalpies on the component pathways. In each example, the component power–time traces are shown summed, representing the power–time traces that would be observed if such a reaction were studied in a calorimeter. It should be noted that in the following figures, Equation (50) has been omitted for clarity, but should be included when using a MathCad worksheet.

Consecutive First-Order $A \rightarrow B \rightarrow C \rightarrow D \rightarrow P$ Reactions

It is possible to extend the analysis discussed already to a four-step, consecutive reaction scheme.

Consider, for example:

$$A \xrightarrow{k_1} B \xrightarrow{k_2} C \xrightarrow{k_3} D \xrightarrow{k_4} P$$

where each step is first-order, with reaction enthalpies of ΔH_1, ΔH_2, ΔH_3, and ΔH_4, respectively.

The first three steps are described by Equations (23), (40), and (50), respectively (remembering to change the term P for the terms B and C). The

The simulation of an $A{\to}B{\to}C{\to}P$ consecutive reaction scheme:

$A := 0.0009$ $k1 := 0.00008$ $k2 := 0.00007$ $k3 := 0.00001$

$t := 0, 100 .. 100000$ $H1 := 8 \cdot 10^9$ $H2 := 4.5 \cdot 10^9$ $H3 := 7 \cdot 10^9$

The reaction A to B can be described by the equation,

$a(t) := k1 \cdot \left[e^{-k1 \cdot t} \cdot (A \cdot H1) \right]$

The reaction B to C can be described by the equation,

$b(t) := H2 \cdot A \cdot k1 \cdot k2 \dfrac{e^{-k1 \cdot t} - e^{-k2 \cdot t}}{k2 - k1}$

The reaction C to P can be described by the equation

$c(t)$ = Equation 50 (see main text)

Therefore, the observed calorimetric signal is given by,

$o(t) := a(t) + b(t) + c(t)$

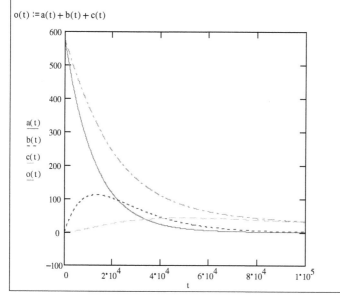

Figure 15 The simulation of a consecutive first-order $A \to B \to C \to P$ reaction scheme with $k_1 \approx k_2 > k_3$.

term $[D]$ depends of the rates of the reactions $C \to D$ and $D \to P$, and may be described by a kinetic expression of the form:

$$\frac{d[D]}{dt} = k_3[C] - k_4[D] \tag{52}$$

The simulation of an $A{\rightarrow}B{\rightarrow}C{\rightarrow}P$ consecutive reaction scheme:

$A := 0.0009 \quad k1 := 0.00008 \quad k2 := 0.00001 \quad k3 := 0.0002$

$t := 0, 100.. 100000 \quad H1 := 6 \cdot 10^9 \quad H2 := 4.5 \cdot 10^9 \quad H3 := 7 \cdot 10^9$

The reaction A to B can be described by the equation,

$a(t) := k1 \cdot \left[e^{-k1 \cdot t} \cdot (A \cdot H1) \right]$

The reaction B to C can be described by the equation,

$b(t) := H2 \cdot A \cdot k1 \cdot k2 \cdot \dfrac{e^{-k1 \cdot t} - e^{-k2 \cdot t}}{k2 - k1}$

The reaction C to P can be described by the equation,

$c(t) =$ Equation 50 (See main text)

Therefore, the observed calorimetric signal is given by,

$o(t) := a(t) + b(t) + c(t)$

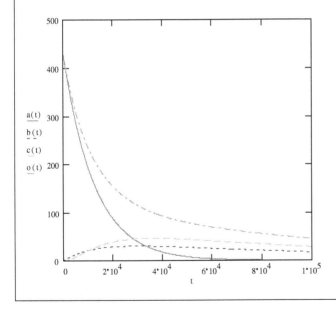

Figure 16 The simulation of a consecutive first-order $A \rightarrow B \rightarrow C \rightarrow P$ reaction scheme with $k_3 > k_1 > k_2$.

The simulation of an $A{\rightarrow}B{\rightarrow}C{\rightarrow}P$ consecutive reaction scheme:

$A := 0.0009$ $k1 := 0.00008$ $k2 := 0.00009$ $k3 := 0.00002$

$t := 0, 100.. 100000$ $H1 := -6 \cdot 10^{9}$ $H2 := 4.5 \cdot 10^{9}$ $H3 := 7 \cdot 10^{9}$

The reaction A to B can be described by the equation,

$$a(t) := k1 \cdot \left[e^{-k1 \cdot t} \cdot (A \cdot H1) \right]$$

The reaction B to C can be described by the equation,

$$b(t) := H2 \cdot A \cdot k1 \cdot k2 \cdot \frac{e^{-k1 \cdot t} - e^{-k2 \cdot t}}{k2 - k1}$$

The reaction C to P can be described by the equation,

$c(t)$ = Equation 50 (See main text)

Hence,

$o(t) := a(t) + b(t) + c(t)$

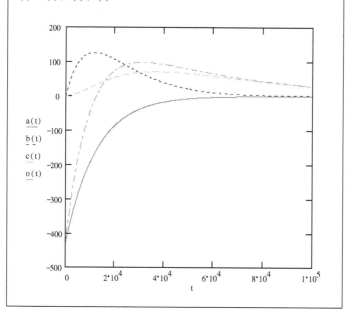

Figure 17   The simulation of a consecutive first-order $A \rightarrow B \rightarrow C \rightarrow P$ reaction scheme with $k_1 \approx k_2 > k_3$.

Substitution of Equation (46) into Equation (52) yields:

$$\frac{d[D]}{dt} = k_3 \left[\left(\frac{k_2^2 k_1 A_0 e^{-k_2 t}}{-k_1 k_2 - k_3 k_2 + k_2^2 + k_3 k_1} \right) + \left(\frac{k_2 k_1^2 A_0 e^{-k_1 t}}{-k_1 k_2 + k_3 k_2 - k_3 k_1 + k_1^2} \right) \right.$$
$$\left. - \left(\frac{k_2 k_1 A_0 e^{-k_3 t}}{-k_1 k_2 - k_3 k_2 + k_2^2 + k_3 k_1} \right) - \left(\frac{k_2 k_1 A_0 e^{-k_3 t}}{-k_1 k_2 + k_3 k_2 - k_3 k_1 + k_1^2} \right) \right]$$
$$- k_4 [D] \tag{53}$$

Equation (53) may be integrated using the integrating factor $e^{\int k_4 dt}$, that is, $e^{k_4 t}$:

$$e^{k_4 t} \frac{d[D]}{dt} + e^{k_4 t} k_4 [D] = e^{k_4 t} \left[k_3 \left[\left(\frac{k_2^2 k_1 A_0 e^{-k_2 t}}{-k_1 k_2 - k_3 k_2 + k_2^2 + k_3 k_1} \right) \right. \right.$$
$$+ \left(\frac{k_2 k_1^2 A_0 e^{-k_1 t}}{-k_1 k_2 + k_3 k_2 - k_3 k_1 + k_1^2} \right) - \left(\frac{k_2 k_1 A_0 e^{-k_3 t}}{-k_1 k_2 - k_3 k_2 + k_2^2 + k_3 k_1} \right)$$
$$\left. \left. - \left(\frac{k_2 k_1 A_0 e^{-k_3 t}}{-k_1 k_2 + k_3 k_2 - k_3 k_1 + k_1^2} \right) \right] \right] \tag{54}$$

$$\Rightarrow \frac{d}{dt} (e^{k_4 t} [D]) = e^{k_4 t} \left[k_3 \left[\left(\frac{k_2^2 k_1 A_0 e^{-k_2 t}}{-k_1 k_2 - k_3 k_2 + k_2^2 + k_3 k_1} \right) \right. \right.$$
$$+ \left(\frac{k_2 k_1^2 A_0 e^{-k_1 t}}{-k_1 k_2 + k_3 k_2 - k_3 k_1 + k_1^2} \right) - \left(\frac{k_2 k_1 A_0 e^{-k_3 t}}{-k_1 k_2 - k_3 k_2 + k_2^2 + k_3 k_1} \right)$$
$$\left. \left. - \left(\frac{k_2 k_1 A_0 e^{-k_3 t}}{-k_1 k_2 + k_3 k_2 - k_3 k_1 + k_1^2} \right) \right] \right] \tag{55}$$

$$\Rightarrow [D] = \frac{k_2 k_1 k_3 A_0 e^{-k_3 t} (k_1 - k_2)^2}{-k_1 k_4 k_2 + k_1 k_4 k_3 - k_1 k_3^2 + k_2 k_1 k_3 - k_3^2 k_2 + k_3^2 + k_4 k_3 k_2 - k_4 k_3^2}$$
$$+ \frac{k_2 k_1 k_3 A_0 e^{-k_2 t} (k_1 - k_2)}{-k_3 k_2 - k_4 k_2 + k_4 k_3 + k_2^2} - \frac{k_2 k_1 k_3 A_0 e^{-k_1 t} (k_1 - k_2)}{k_1^2 - k_3 k_1 - k_1 k_4 + k_4 k_3}$$
$$- \frac{k_2 k_1 k_3 A_0 e^{-k_4 t} (k_1 - k_2)}{-k_3 k_2 - k_4 k_2 + k_4 k_3 + k_2^2}$$
$$+ \frac{k_1 k_2 k_3 A_0 e^{-k_4 t} (k_1 - k_2)^2}{-k_1 k_4 k_2 + k_1 k_4 k_3 - k_1 k_3^2 + k_2 k_1 k_3 - k_3^2 k_2 + k_3^2 + k_4 k_3 k_2 - k_4 k_3^2}$$
$$+ \frac{k_1 k_2 k_3 A_0 e^{-k_4 t} (k_1 - k_2)}{k_1^2 - k_3 k_1 - k_1 k_4 + k_4 k_3} \tag{56}$$

As the sum of the reactant concentrations must equal A_0, if B, C, D, and P are not present initially, then:

$$[P] = A_0 - [A] - [B] - [C] - [D] \tag{57}$$

Hence:

$$[P] = A_0 - (A_0 e^{-k_1 t}) - \left(\frac{A_0 k_1}{k_2 - k_1} (e^{-k_1 t} - e^{-k_2 t}) \right)$$

$$- \left(\frac{k_2^2 k_1 A_0 e^{-k_2 t}}{-k_1 k_2 - k_3 k_2 + k_2^2 + k_3 k_1} \right) - \left(\frac{k_2 k_1^2 A_0 e^{-k_1 t}}{-k_1 k_2 + k_3 k_2 - k_3 k_1 + k_1^2} \right)$$

$$+ \left(\frac{k_2 k_1 A_0 e^{-k_3 t}}{-k_1 k_2 - k_3 k_2 + k_2^2 + k_3 k_1} \right) + \left(\frac{k_2 k_1 A_0 e^{-k_3 t}}{-k_1 k_2 + k_3 k_2 - k_3 k_1 + k_1^2} \right)$$

$$- \frac{k_2 k_1 k_3 A_0 e^{-k_3 t}(k_1 - k_2)^2}{-k_1 k_4 k_2 + k_1 k_4 k_3 - k_1 k_3^2 + k_2 k_1 k_3 - k_3^2 k_2 + k_3^2 + k_4 k_3 k_2 - k_4 k_3^2}$$

$$+ \frac{k_2 k_1 k_3 A_0 e^{-k_2 t}(k_1 - k_2)}{-k_3 k_2 - k_4 k_2 + k_4 k_3 + k_2^2}$$

$$- \frac{k_2 k_1 k_3 A_0 e^{-k_1 t}(k_1 - k_2)}{k_1^2 - k_3 k_1 - k_1 k_4 + k_4 k_3} - \frac{k_2 k_1 k_3 A_0 e^{-k_4 t}(k_1 - k_2)}{-k_3 k_2 - k_4 k_2 + k_4 k_3 + k_2^2}$$

$$+ \frac{k_1 k_2 k_3 A_0 e^{-k_4 t}(k_1 - k_2)^2}{-k_1 k_4 k_2 + k_1 k_4 k_3 - k_1 k_3^2 + k_2 k_1 k_3 - k_3^2 k_2 + k_3^2 + k_4 k_3 k_2 - k_4 k_3^2}$$

$$+ \frac{k_1 k_2 k_3 A_0 e^{-k_4 t}(k_1 - k_2)}{k_1^2 - k_3 k_1 - k_1 k_4 + k_4 k_3} \tag{58}$$

Equation (58) may be differentiated:

$$\frac{d[P]}{dt} = A_0 k_1 e^{-k_1 t} - A_0 \frac{k_1}{k_2 - k_1} (-k_1 e^{-k_1 t} + k_2 e^{-k_2 t})$$

$$+ \frac{k_2^2 k_1 A_0 e^{-k_2 t}}{-k_2 k_1 - k_3 k_2 + k_2^2 + k_3 k_1} + \frac{k_2 k_1^2 A_0 e^{-k_1 t}}{-k_2 k_1 + k_3 k_2 - k_3 k_1 + k_1^2}$$

$$- \frac{k_2 k_1 A_0 k_3 e^{-k_3 t}}{-k_2 k_1 - k_3 k_2 + k_2^2 + k_3 k_1} - \frac{k_2 k_1 A_0 k_3 e^{-k_3 t}}{-k_2 k_1 + k_3 k_2 + k_1^2 - k_3 k_1}$$

$$- \frac{k_2 k_1 k_3^2 A_0 e^{-k_3 t}(k_1 - k_2)^2}{-k_1 k_4 k_2 + k_1 k_4 k_3 - k_1 k_3^2 + k_2 k_1 k_3 - k_3^2 k_2 + k_3^2 + k_4 k_3 k_2 - k_4 k_3^2}$$

$$+ \frac{k_2^2 k_1 k_3 A_0 e^{-k_2 t}(k_1 - k_2)}{-k_3 k_2 - k_4 k_2 + k_4 k_3 + k_2^2} - \frac{k_2 k_1^2 k_3 A_0 e^{-k_1 t}(k_1 - k_2)}{k_1^2 - k_3 k_1 - k_1 k_4 + k_4 k_3}$$

$$- \frac{k_2 k_1 k_3 A_0 k_4 e^{-k_4 t}(k_1 - k_2)}{-k_3 k_2 - k_4 k_2 + k_4 k_3 + k_2^2}$$

$$+ \frac{k_1 k_2 k_3 k_4 A_0 e^{-k_4 t}(k_1 - k_2)^2}{-k_1 k_4 k_2 + k_1 k_4 k_3 - k_1 k_3^2 + k_2 k_1 k_3 - k_3^2 k_2 + k_3^2 + k_4 k_3 k_2 - k_4 k_3^2}$$

$$+ \frac{k_1 k_2 k_3 k_4 A_0 e^{-k_4 t}(k_1 - k_2)}{k_1^2 - k_3 k_1 - k_1 k_4 + k_4 k_3} \tag{59}$$

and hence:

$$\frac{dq_P}{dt} = H_4\left[A_0k_1e^{-k_1t} - A_0\frac{k_1}{k_2-k_1}(-k_1e^{-k_1t}+k_2e^{-k_2t})\right.$$

$$+\frac{k_2^2k_1A_0e^{-k_2t}}{-k_2k_1-k_3k_2+k_2^2+k_3k_1} + \frac{k_2k_1^2A_0e^{-k_1t}}{-k_2k_1+k_3k_2-k_3k_1+k_1^2}$$

$$-\frac{k_2k_1A_0k_3pe^{-k_3t}}{-k_2k_1-k_3k_2+k_2^2+k_3k_1} - \frac{k_2k_1A_0k_3e^{-k_3t}}{-k_2k_1+k_3k_2+k_1^2-k_3k_1}$$

$$-\frac{k_2k_1k_3^2A_0e^{-k_3t}(k_1-k_2)^2}{-k_1k_4k_2+k_1k_4k_3-k_1k_3^2+k_2k_1k_3-k_3^2k_2+k_3^2+k_4k_3k_2-k_4k_3^2}$$

$$+\frac{k_2^2k_1k_3A_0e^{-k_2t}(k_1-k_2)}{-k_3k_2-k_4k_2+k_4k_3+k_2^2} - \frac{k_2k_1^2k_3A_0e^{-k_1t}(k_1-k_2)}{k_1^2-k_3k_1-k_1k_4+k_4k_3}$$

$$-\frac{k_2k_1k_3A_0k_4e^{-k_4t}(k_1-k_2)}{-k_3k_2-k_4k_2+k_4k_3+k_2^2}$$

$$+\frac{k_1k_2k_3k_4A_0e^{-k_4t}(k_1-k_2)^2}{-k_1k_4k_2+k_1k_4k_3-k_1k_3^2+k_2k_1k_3-k_3^2k_2+k_3^2+k_4k_3k_2-k_4k_3^2}$$

$$\left.+\frac{k_1k_2k_3k_4A_0e^{-k_4t}(k_1-k_2)}{k_1^2-k_3k_1-k_1k_4+k_4k_3}\right] \tag{60}$$

where, dq_P/dt represents the power associated with the reaction $D \rightarrow P$. Equation (60) may be used to construct power–time data that describe the reaction $D \rightarrow P$. Summation of Equations (23), (40), (50), and (60) gives an equation that may be used to fit observed power–time data for a reaction following a consecutive, four step, first-order reaction scheme [Equation (61)]:

$$\frac{dq_{obs}}{dt} = k_1\Delta H_1A_0e^{-k_1t} + k_1k_2\Delta H_2A_0\frac{e^{-k_1t}-e^{-k_2t}}{k_2-k_1}$$

$$+\Delta H_3\left[A_0k_1k_2k_3\left(\frac{e^{-k_1t}}{(k_2-k_1)(k_3-k_1)} + \frac{e^{-k_2t}}{(k_1-k_2)(k_3-k_2)}\right.\right.$$

$$\left.\left.+\frac{e^{-k_3t}}{(k_1-k_3)(k_2-k_3)}\right)\right] + H_4\left[A_0k_1e^{-k_1t} - A_0\frac{k_1}{k_2-k_1}\right.$$

$$\times(-k_1e^{-k_1t}+k_2e^{-k_2t}) + \frac{k_2^2k_1A_0e^{-k_2t}}{-k_2k_1-k_3k_2+k_2^2+k_3k_1}$$

$$+\frac{k_2k_1^2A_0e^{-k_1t}}{-k_2k_1+k_3k_2-k_3k_1+k_1^2} - \frac{k_2k_1A_0k_3e^{-k_3t}}{-k_2k_1-k_3k_2+k_2^2+k_3k_1}$$

$$-\frac{k_2 k_1 A_0 k_3 e^{-k_3 t}}{-k_2 k_1 + k_3 k_2 + k_1^2 - k_3 k_1}$$

$$-\frac{k_2 k_1 k_3^2 A_0 e^{-k_3 t}(k_1 - k_2)^2}{-k_1 k_4 k_2 + k_1 k_4 k_3 - k_1 k_3^2 + k_2 k_1 k_3 - k_3^2 k_2 + k_3^2 + k_4 k_3 k_2 - k_4 k_3^2}$$

$$+\frac{k_2^2 k_1 k_3 A_0 e^{-k_2 t}(k_1 - k_2)}{-k_3 k_2 - k_4 k_2 + k_4 k_3 + k_2^2} - \frac{k_2 k_1^2 k_3 A_0 e^{-k_1 t}(k_1 - k_2)}{k_1^2 - k_3 k_1 - k_1 k_4 + k_4 k_3}$$

$$-\frac{k_2 k_1 k_3 A_0 k_4 e^{-k_4 t}(k_1 - k_2)}{-k_3 k_2 - k_4 k_2 + k_4 k_3 + k_2^2}$$

$$+\frac{k_1 k_2 k_3 k_4 A_0 e^{-k_4 t}(k_1 - k_2)^2}{-k_1 k_4 k_2 + k_1 k_4 k_3 - k_1 k_3^2 + k_2 k_1 k_3 - k_3^2 k_2 + k_3^2 + k_4 k_3 k_2 - k_4 k_3^2}$$

$$+\left.\frac{k_1 k_2 k_3 k_4 A_0 e^{-k_4 t}(k_1 - k_2)}{k_1^2 - k_3 k_1 - k_1 k_4 + k_4 k_3}\right] \tag{61}$$

Note here that the complexity of the equations is increasing significantly with the number of reaction steps. The derivations for further steps are not considered here, for reasons of brevity, but follow the same principles.

The use of these equations is illustrated in Figures 18–20, showing the power–time traces that would be obtained for each of the component steps for the particular set of constants shown. The examples show the effects of changing the rate constants and reaction enthalpies. In each example, the component traces are summed, showing the power–time trace that would be observed in a calorimeter. It should be noted that in the following figures, Equations (50) and (60) have been omitted for clarity, but should be included when using a MathCad worksheet.

Parallel-Consecutive Reactions

Sometimes, a material may degrade via two or more parallel pathways, one or more of which involves consecutive steps. The following examples illustrate some of the possible combinations of pathways; as for the previous examples, the same principles can be applied to any other reaction schemes not discussed here.

First-Order, Parallel-Consecutive $A \rightarrow B \rightarrow P, A \rightarrow Q$ Reactions

From the previous discussion, it can be seen that it is possible to derive equations that allow relatively simple parallel and consecutive reactions to be simulated. It is possible to combine the parallel and consecutive reaction schemes already discussed to allow the simulation of a reactant that degrades via simultaneous parallel

The simulation of an $A{\rightarrow}B{\rightarrow}C{\rightarrow}D{\rightarrow}P$ consecutive reaction scheme:

$A := 0.0009$ $k1 := 0.00008$ $k2 := 0.00005$ $k3 := 0.00004$ $k4 := 0.00005$

$t := 0, 100..\,100000$ $H1 := 8 \cdot 10^9$ $H2 := 4.5 \cdot 10^9$ $H3 := 7 \cdot 10^9$ $H4 := 5 \cdot 10^9$

The reaction A to B can be described by the equation,

$$a(t) := k1 \cdot \left[e^{-k1 \cdot t} \cdot (A \cdot H1) \right]$$

The reaction B to C can be described by the equation,

$$b(t) := H2 \cdot A \cdot k1 \cdot k2 \cdot \frac{e^{-k1 \cdot t} - e^{-k2 \cdot t}}{k2 - k1}$$

The reaction C to D can be described by the equation,

$c(t)$ = Equation 50 (See main text)

The reaction D to P can be described by the equation,

$d(t)$ = Equation 60 (See main text)

Therefore, the observed calorimetric signal is given by,

$o(t) := a(t) + b(t) + c(t) + d(t)$

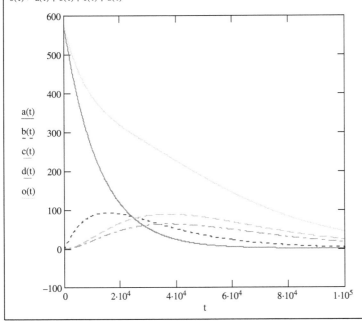

Figure 18 The simulation of a consecutive first-order $A \rightarrow B \rightarrow C \rightarrow D \rightarrow P$ reaction scheme with $k_1 > k_2 \approx k_3 \approx k_4$.

The simulation of an $A{\rightarrow}B{\rightarrow}C{\rightarrow}D{\rightarrow}P$ consecutive reaction scheme:

$A := 0.0009 \quad k1 := 0.00008 \quad k2 := 0.00009 \quad k3 := 0.00001 \quad k4 := 0.00007$

$t := 0, 100..\, 100000 \quad H1 := 8 \cdot 10^9 \quad H2 := 4.5 \cdot 10^9 \quad H3 := 7 \cdot 10^9 \quad H4 := 5 \cdot 10^9$

The reaction A to B can be described by the equation,

$a(t) := k1 \cdot \left[e^{-k1 \cdot t} \cdot (A \cdot H1) \right]$

The reaction B to C can be described by the equation,

$b(t) := H2 \cdot A \cdot k1 \cdot k2 \cdot \dfrac{e^{-k1 \cdot t} - e^{-k2 \cdot t}}{k2 - k1}$

The reaction C to D can be described by the equation,

$c(t) =$ Equation 50 (See main text)

The reaction D to P can be described by the equation,

$d(t) =$ Equation 60 (See main text)

Therefore, the observed calorimetric signal is given by,

$o(t) := a(t) + b(t) + c(t) + d(t)$

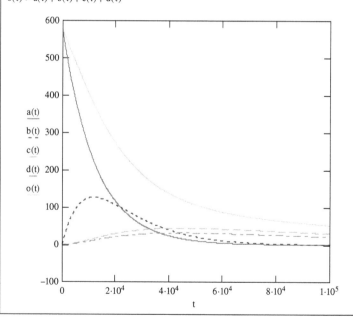

Figure 19 The simulation of a consecutive first-order $A \rightarrow B \rightarrow C \rightarrow D \rightarrow P$ reaction scheme with $k_1 \approx k_2 \approx k_4 > k_3$.

The simulation of an $A \rightarrow B \rightarrow C \rightarrow D \rightarrow P$ consecutive reaction scheme:

$A := 0.0009 \quad k1 := 0.00008 \quad k2 := 0.00005 \quad k3 := 0.00004 \quad k4 := 0.00005$

$t := 0, 100 .. 100000 \quad H1 := -8 \cdot 10^9 \quad H2 := 4.5 \cdot 10^9 \quad H3 := 7 \cdot 10^9 \quad H4 := -5 \cdot 10^9$

The reaction A to B can be described by the equation,

$a(t) := k1 \cdot \left[e^{-k1 \cdot t} \cdot (A \cdot H1) \right]$

The reaction B to C can be described by the equation,

$b(t) := H2 \cdot A \cdot k1 \cdot k2 \cdot \dfrac{e^{-k1 \cdot t} - e^{-k2 \cdot t}}{k2 - k1}$

The reaction C to D can be described by the equation,

$c(t) =$ Equation 50 (See main text)

The reaction D to P can be described by the equation,

$d(t) =$ Equation 60 (See main text)

Therefore, the observed calorimetric signal is given by,

$o(t) := a(t) + b(t) + c(t) + d(t)$

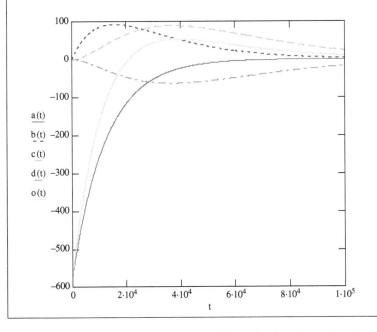

Figure 20   The simulation of a consecutive first-order $A \rightarrow B \rightarrow C \rightarrow D \rightarrow P$ reaction scheme with $k_1 > k_2 \approx k_3 \approx k_4$.

and consecutive pathways. Consider, for example, the following reaction scheme:

$$Q \xleftarrow{k_2} A \xrightarrow{k_1} B \xrightarrow{k_3} P$$

where all steps are first-order, with reaction enthalpies ΔH_2, ΔH_1, and ΔH_3, respectively. The reactant A may form product Q via a first-order process of rate constant k_2 or may form product P via a two-step, consecutive, first-order process of rate constants k_1 and k_3. The terms $[A]$ and $[B]$ depend on the rates of the reactions $A \rightarrow B$ and $A \rightarrow Q$. The rates of the steps in each pathway may be expressed by the following kinetic expressions:

$$\frac{d[A]}{dt} = -k_1 [A] - k_2 [A] \tag{62}$$

$$\frac{d[B]}{dt} = k_1 [A] - k_3 [B] \tag{63}$$

$$\frac{d[Q]}{dt} = k_2 [A] \tag{64}$$

Equation (62) may be readily integrated to give:

$$[A] = A_0 e^{(-k_1 - k_2)t} \tag{65}$$

Equation (65) may be substituted into Equation (63) to yield:

$$\frac{d[B]}{dt} = k_1 (A_0 e^{(-k_1 - k_2)t}) - k_3 [B] \tag{66}$$

Equation (66) may be integrated using the integrating factor $e^{\int k_3\, dt}$, that is, $e^{k_3 t}$:

$$e^{k_3 t} \frac{d[B]}{dt} + e^{k_3 t} k_3 [B] = e^{k_3 t} k_1 A_0 e^{(-k_1 - k_2)t} \tag{67}$$

$$\Rightarrow \frac{d}{dt}(e^{k_3 t} [B]) = e^{k_3 t} k_1 A_0 e^{(-k_1 - k_2)t} \tag{68}$$

$$\Rightarrow [B] = \frac{-k_1 A_0 e^{-(k_1 + k_2)t}}{k_1 + k_2 - k_3} + \frac{k_1 A_0 e^{-k_3 t}}{k_1 + k_2 - k_3} \tag{69}$$

The term $[Q]$ may be determined by substitution of Equation (65) into Equation (64). Hence:

$$\frac{d[Q]}{dt} = k_2 A_0 e^{(-k_1 - k_2)t} \tag{70}$$

Equation (70) integrates readily to:

$$[Q] = \frac{k_2 A_0}{k_1 + k_2} (1 - e^{-(k_1 + k_2)t}) \tag{71}$$

As the sum of the reactant concentrations must equal A_0, if B, P, and Q are not present initially, then:

$$[P] = A_0 - [A] - [B] - [Q] \tag{72}$$

Hence:

$$[P] = A_0 - A_0 e^{(-k_1-k_2)t} - \frac{-k_1 A_0 e^{-(k_1+k_2)t}}{k_1 + k_2 - k_3}$$
$$+ \frac{k_1 A_0 e^{-k_3 t}}{k_1 + k_2 - k_3} - \frac{k_2 A_0}{k_1 + k_2}\left(1 - e^{-(k_1+k_2)t}\right) \tag{73}$$

Equation (73) may be differentiated:

$$\frac{d[P]}{dt} = -A_0(-k_1 - k_2)e^{(-k_1-k_2)t} + \frac{k_1 A_0\,(-k_1 - k_2)\,e^{(-k_1-k_2)t}}{k_1 + k_2 - k_3}$$
$$+ \frac{k_1 A_0 k_3 e^{-k_3 t}}{k_1 + k_2 - k_3} + \frac{k_2 A_0\,(-k_1 - k_2)\,e^{(-k_1-k_2)t}}{k_1 + k_2} \tag{74}$$

Hence:

$$\frac{dq_P}{dt} = \Delta H_3\left[-A_0(-k_1 - k_2)e^{(-k_1-k_2)t} + \frac{k_1 A_0\,(-k_1 - k_2)\,e^{(-k_1-k_2)t}}{k_1 + k_2 - k_3} \right.$$
$$\left. + \frac{k_1 A_0 k_3 e^{-k_3 t}}{k_1 + k_2 - k_3} + \frac{k_2 A_0\,(-k_1 - k_2)\,e^{(-k_1-k_2)t}}{k_1 + k_2} \right] \tag{75}$$

where dq_P/dt represents the power associated with the reaction $B \to P$. Equation (75) may be used to construct power–time data that describe the reaction $B \to P$. The reaction $A \to B$ can be described by the kinetic expression:

$$\frac{d[B]}{dt} = k_1\,[A] \tag{76}$$

Hence:

$$\frac{d[B]}{dt} = k_1 A_0\, e^{(-k_1-k_2)t} \tag{77}$$

Or, calorimetrically:

$$\frac{dq_B}{dt} = \Delta H_1\, k_1 A_0 e^{(-k_1-k_2)t} \tag{78}$$

where dq_B/dt represents the power associated with the reaction $A \to B$. A similar equation may be derived for the reaction $A \to Q$:

$$\frac{dq_Q}{dt} = \Delta H_2\, k_2 A_0\, e^{(-k_1-k_2)t} \tag{79}$$

where dq_Q/dt represents the power associated with the reaction $A \rightarrow Q$. Hence, simulated data may be constructed for the reaction steps $A \rightarrow B$, $A \rightarrow Q$, and $B \rightarrow P$ using Equations (75), (78), and (79). These equations may be combined to give an equation that describes the overall power–time data that would be measured for a reaction following the mechanistic scheme given already:

$$
\begin{aligned}
\frac{dq_{obs}}{dt} = {}& \Delta H_1 \, k_1 \, A_0 \, e^{(-k_1 - k_2)t} + \Delta H_2 \, k_2 \, A_0 \, e^{(-k_1 - k_2)t} \\
&+ \Delta H_3 \left[-A_0 \, (-k_1 - k_2) \, e^{(-k_1 - k_2)t} + \frac{k_1 \, A_0 \, (-k_1 - k_2) \, e^{(-k_1 - k_2)t}}{k_1 + k_2 - k_3} \right. \\
&\left. + \frac{k_1 \, A_0 \, k_3 \, e^{-k_3 t}}{k_1 + k_2 - k_3} + \frac{k_2 A_0 \, (-k_1 - k_2) \, e^{(-k_1 - k_2)t}}{k_1 + k_2} \right]
\end{aligned}
\tag{80}
$$

The use of these equations is illustrated in Figures 21–23, showing the power–time traces that would be obtained for each of the component steps for the particular set of constants shown. The examples show the effects of changing the rate constants and reaction enthalpies. Each example shows the result of summing the component traces, giving the power–time trace that would be observed calorimetrically.

Concurrent Reaction Schemes

Having derived equations that allow the modeling of individual reaction schemes, it is now possible to model concurrent reaction schemes. Such schemes would occur in any system where more than one process was occurring, but where the processes are independent of each other. These may be common in multicomponent pharmaceuticals, for instance. The examples given in Figures 24 and 25 are represented by the following schemes.

The reaction scheme modeled in Figure 24 is represented by the concurrent scheme:

$$
A \xrightarrow{k_1} B \xrightarrow{k_2} C \xrightarrow{k_3} D \xrightarrow{k_4} P
$$

(first-order, ΔH_1, ΔH_2, ΔH_3, and ΔH_4, respectively) and:

$$
Q \xleftarrow{k_5} F \xrightarrow{k_6} G \xrightarrow{k_7} R
$$

(first-order, ΔH_5, ΔH_6, and ΔH_7, respectively).

The reaction modeled in Figure 25 is represented by the concurrent schemes:

$$
D \xleftarrow{k_1} A \xrightarrow{k_2} B \xrightarrow{k_3} C
$$

(first-order, ΔH_1, ΔH_2, and ΔH_3, respectively) and:

$$
Q \xleftarrow{k_5} F \xrightarrow{k_6} G \xrightarrow{k_7} R
$$

(first-order, ΔH_5, ΔH_6, and ΔH_7, respectively).

The simulation of a parallel $A{\rightarrow}B{\rightarrow}P$, $A{\rightarrow}Q$ reaction scheme :

$A := 0.0005$ $k1 := 0.00004$ $k2 := 0.00005$ $k3 := 0.00003$

$t := 0, 100.. 100000$ $H1 := 5 \cdot 10^9$ $H2 := 6 \cdot 10^9$ $H3 := 7 \cdot 10^9$

The reaction A to B can be described by the equation,

$b(t) := H1 \cdot k1 \cdot \left[A \cdot e^{-(k1+k2) \cdot t} \right]$

The reaction A to Q can be described by the equation,

$q(t) := H2 \cdot k2 \cdot \left[A \cdot e^{-(k1+k2) \cdot t} \right]$

The reaction B to P can be described by the equation,

$p(t) =$ Equation 75 (See main text)

Therefore, the observed calorimetric signal is given by,

$o(t) := b(t) + p(t) + q(t)$

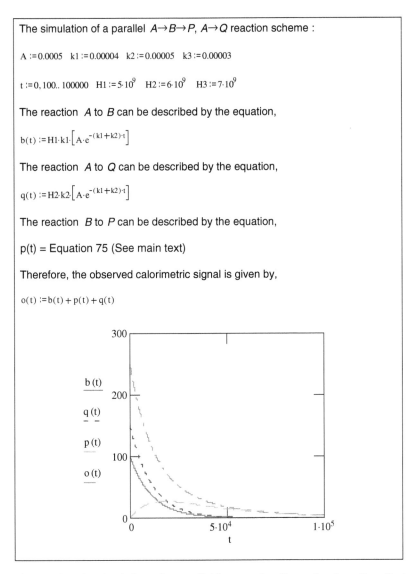

Figure 21   The simulation of a parallel/consecutive first-order $A \rightarrow B \rightarrow P$, $A \rightarrow Q$ reaction scheme with $k_1 \approx k_2 \approx k_3$.

Flow Calorimetry

This chapter has thus far dealt with the calorimetric equations that describe the calorimetric output for static ampoule experiments. There is the option in some calorimeters of flowing the reagents (if they are in the solution phase or are suspensions) through the calorimetric cell. The calorimetric data returned from flow calorimetric experiments still contain the same kinetic and thermodynamic

The simulation of a parallel $A{\rightarrow}B{\rightarrow}P$, $A{\rightarrow}Q$ reaction scheme:

$A := 0.0005 \quad k1 := 0.00008 \quad k2 := 0.000005 \quad k3 := 0.00007$

$t := 0, 100.. \, 100000 \quad H1 := 6 \cdot 10^9 \quad H2 := 5 \cdot 10^9 \quad H3 := 7 \cdot 10^9$

The reaction A to B can be described by the equation,

$b(t) := H1 \cdot k1 \cdot \left[A \cdot e^{-(k1+k2) \cdot t} \right]$

The reaction A to Q can be described by the equation,

$q(t) := H2 \cdot k2 \cdot \left[A \cdot e^{-(k1+k2) \cdot t} \right]$

The reaction B to P can be described by the equation,

$p(t) =$ Equation 75 (See main text)

Therefore, the observed calorimetric signal is given by,

$o(t) := b(t) + p(t) + q(t)$

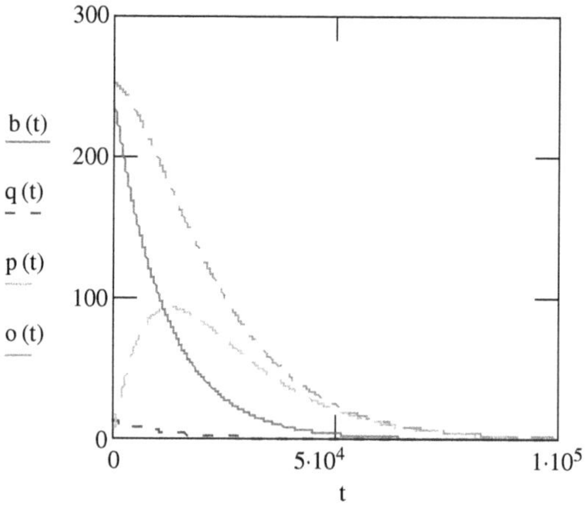

Figure 22   The simulation of a parallel/consecutive first-order $A \rightarrow B \rightarrow P$, $A \rightarrow Q$ reaction scheme with $k_1 \approx k_3 > k_2$.

information as that already described for static calorimetric data. However, because the reagents are flowing, the calorimetric signal is dependent on the characteristics of the flowing medium, particularly the flow rate, and therefore the equations which describe flow calorimetric data must take this dependence into account. This then places a further burden on the analysis of flow calorimetric data.

The simulation of a parallel $A \rightarrow B \rightarrow P$, $A \rightarrow Q$ reaction scheme:

A := 0.0005   k1 := 0.00009   k2 := 0.00003   k3 := 0.00007

t := 0, 100 .. 40000   H1 := 6·10^9 H2 := -7·10^9 H3 := 7·10^9

The reaction A to B can be described by the equation,

$$b(t) := H1 \cdot k1 \cdot \left[A \cdot e^{-(k1 + k2) \cdot t} \right]$$

The reaction A to Q can be described by the equation,

$$q(t) := H2 \cdot k2 \cdot \left[A \cdot e^{-(k1 + k2) \cdot t} \right]$$

The reaction B to P can be described by the equation,

p(t) = Equation 75 (See main text)

Therefore, the observed calorimetric signal is given by,

$$o(t) := b(t) + p(t) + q(t)$$

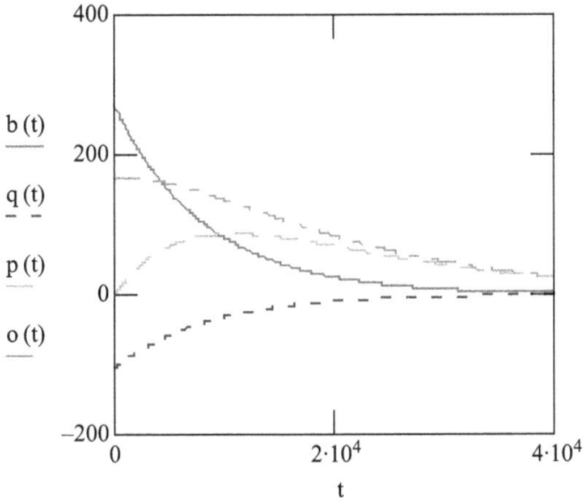

Figure 23 The simulation of a parallel/consecutive first-order $A \rightarrow B \rightarrow P$, $A \rightarrow Q$ reaction scheme with $k_1 > k_3 > k_2$.

Chapter 2 briefly highlighted the two modes of operation of the flow calorimeter, flow-mix and flow-through. The flow-mix mode of operation allows the investigation of fast reactions (i.e., rapid with respect to the time constant of the instrument). The flow-through mode of operation is used for relatively slow reactions with long half-lives (i.e., the reaction is slow compared with the time constant of the instrument). Usually, flow calorimeters are operated

The simulation of an $A{\to}B{\to}C{\to}D{\to}P$ consecutive reaction scheme with a concurrent $F{\to}Q$ and $F{\to}G{\to}R$ reaction scheme:

$A := 0.0009 \quad B := 0.0006 \quad t := 0, 1000..\,200000 \quad k1 := 0.00008 \quad k2 := 0.00005$

$k3 := 0.00004 \quad k4 := 0.00005 \quad k5 := 0.00005 \quad k6 := 0.00003 \quad k7 := 0.00007$

$H1 := 8 \cdot 10^9 \quad H2 := 4.5 \cdot 10^9 \quad H3 := -7 \cdot 10^9 \quad H4 := -5 \cdot 10^9 \quad H5 := 7 \cdot 10^9 \quad H6 := -3 \cdot 10^9 \quad H7 := -9 \cdot 10^9$

The reactions A to B and B to C can be described by the equations,

$$a(t) := k1 \cdot \left[e^{-k1 \cdot t} \cdot (A \cdot H1) \right] \qquad b(t) := H2 \cdot A \cdot k1 \cdot k2 \cdot \frac{e^{-k1 \cdot t} - e^{-k2 \cdot t}}{k2 - k1}$$

respectively. The reactions C to D and D to P can be described by,

c(t) = Equation 50 (See main text)   d(t) = Equation 60 (See main text)

respectively. The reactions F to G and F to Q can be described by the equations,

$$g(t) := H5 \cdot k5 \cdot \left[B \cdot e^{-(k5+k6) \cdot t} \right] \qquad i(t) := H6 \cdot k6 \cdot \left[B \cdot e^{-(k5+k6) \cdot t} \right]$$

respectively. The reaction G to R can be described by the equation,

h(t) = Equation 75 (See main text)

Therefore, the observed calorimetric signal is given by,

$$o(t) := a(t) + b(t) + c(t) + d(t) + g(t) + h(t) + i(t)$$

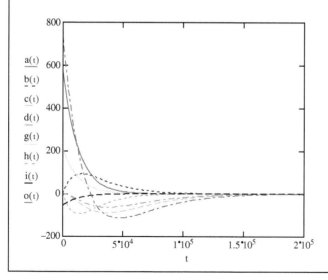

Figure 24 The simulation of two concurrent reaction schemes. The two reactions occur independently of one another.

The simulation of an $A{\to}D$, $A{\to}B{\to}C$ parallel reaction scheme with a concurrent $F{\to}Q$, $F{\to}G{\to}R$ parallel reaction scheme:

$A := 0.001 \quad B := 0.0006 \quad t := 0, 1000.. 100000$

$k1 := 0.00007 \quad k2 := 0.00005 \quad k3 := 0.00004 \quad k5 := 0.00006 \quad k6 := 0.00003 \quad k7 := 0.00008$

$H1 := 9 \cdot 10^{9} \quad H2 := 5 \cdot 10^{9} \quad H3 := 7 \cdot 10^{9} \quad H5 := 7 \cdot 10^{9} \quad H6 := 3 \cdot 10^{9} \quad H7 := 9 \cdot 10^{9}$

The reactions A to B and A to D can be described by the equations,

$$b(t) := H1 \cdot k1 \cdot \left[A \cdot e^{-(k1+k2) \cdot t} \right] \quad d(t) := H2 \cdot k2 \cdot \left[A \cdot e^{-(k1+k2) \cdot t} \right]$$

respectively. The reaction B to C can be described by the equation,

$c(t)$ = Equation 75 (See main text)

The reactions F to G and F to Q can be described by the equations,

$$g(t) := H5 \cdot k5 \cdot \left[B \cdot e^{-(k5+k6) \cdot t} \right] \quad i(t) := H6 \cdot k6 \cdot \left[B \cdot e^{-(k5+k6) \cdot t} \right]$$

respectively. The reaction G to R can be described by the equation,

$h(t)$ = Equation 75 (See main text)

Therefore, the observed calorimetric signal is given by,

$$o(t) := b(t) + c(t) + d(t) + g(t) + h(t) + i(t)$$

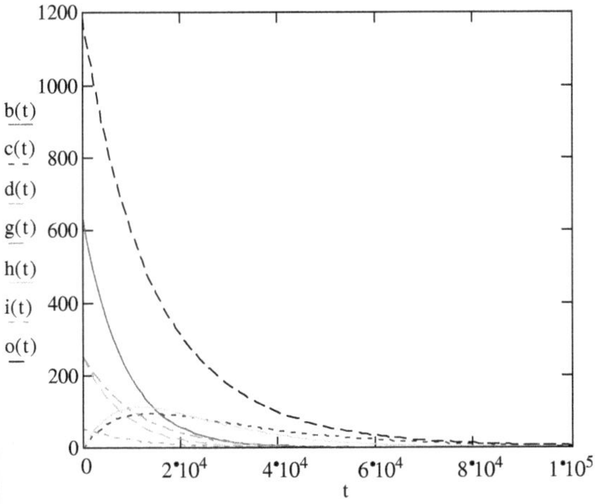

Figure 25 The simulation of two concurrent independent parallel/consecutive reaction schemes.

in the flow-through mode and hence, in the following discussion, a greater emphasis will be placed on the calculation of the desired parameters from flow-through data.

Flow-Mix Calorimeters

The flow rates of two streams in a flow-mix experiment have flow rates F_1 and F_2 (dm^3/sec), respectively (4). Stream 1 contains reagent with initial concentration $[C]_0^{(1)}$ and stream 2 contains reagent with initial concentration $[C]_0^{(2)}$ (both mol/dm^3). If reagent 1 is in excess such that $(F_1[C]_0^{(1)} > F_2[C]_0^{(2)})$, and the reaction is instantaneous, the rate of formation of product, P (mol dm^3/sec), is given by:

$$\frac{d[P]}{dt} = F_2[C_0]^{(2)} \tag{81}$$

The power output for the product, Φ (J/sec) is thus given by:

$$\Phi = -F_2[C_0]^{(2)}\Delta H_R \tag{82}$$

As the reagents are mixed in the cell, both will be diluted by some factor, D, as they enter the cell:

$$D_m = \frac{F_m}{(F_1 + F_2)} \tag{83}$$

where $m = 1$ or 2 for reagents 1 and 2, respectively.

Hence, there is an associated enthalpy of dilution ΔH_{D1} and ΔH_{D2} for reagents 1 and 2, respectively. The thermal power from dilution is then:

$$\Phi_D = -\left(F_1[C_0]^{(1)}\Delta H_{D1} + F_2[C_0]^{(2)}\Delta H_{D2}\right) \tag{84}$$

The overall calorimetric signal is therefore given by:

$$\Phi = -\left(F_1[C_0]^{(1)}\Delta H_{D1}\right) + \beta - \left(F_2[C_0]^{(2)}(\Delta H_{D2} + \Delta H_R)\right) \tag{85}$$

Here, β is a term that encompasses all other heat-generating processes within the cell, such as heat derived from turbulent flow.

In flow-mix experiments a baseline is usually generated by mixing reagent 1 with solvent (with all flow rates being the same) and recording the output. All subsequent measurements are taken relative to this baseline. The output, relative to this baseline, is given by:

$$\Phi = -F_2[C_0]^{(2)}(\Delta H_{D2} + \Delta H_R) \tag{86}$$

In practice, the enthalpy of dilution, ΔH_{D2}, is generally small compared with ΔH_R and hence the calorimetric output is dependent effectively only on the value of $F_2[C]_0^{(2)}$. Consequently, the concentration of reagent 2 can be calculated.

Slow reactions: For slow reactions, the measured power is the average output over the residence time, τ (τ is the time spent by the reacting solution in the calorimetric cell). The residence time is dependent on the flow rate of the solution and the thermal volume of the cell, V_c [Equation (87)]. The concept of thermal volume results from heat being carried from the cell by the flowing solution; it is discussed in more detail in Chapter 2:

$$\tau = \frac{V_c}{(F_1 + F_2)} \tag{87}$$

In this instance, the calorimetric signal is equal to the instantaneous calorimetric signal only when the reaction is independent of time, that is, zero-order kinetics.

Zero-order reactions: For a reaction which conforms to zero-order kinetics, the rate of formation of product is constant with respect to time (here the residence time) and is given by:

$$\frac{d[P]}{dt} = k_0 V_c \tag{88}$$

The resulting power output is therefore:

$$\Phi = -k_0 V_c \Delta H_R \tag{89}$$

The overall calorimetric signal is given by:

$$\Phi = -\left(k_0 V_c \Delta H_R + F_2 [C]_0^{(2)} \Delta H_{D2} \right) \tag{90}$$

If $k_0 V_c \Delta H_R > F_2 [C]_0^{(2)} \Delta H_{D2}$ (which is often the case), then the overall calorimetric signal is simply described by Equation (89).

First-order reaction in reagent 2 and zero-order in reagent 1: For any given time, the rate of formation of product is given by:

$$\frac{d[P]}{dt} = k_1 [C]^{(2)} \tag{91}$$

Here, k_1 is the apparent first-order rate constant (sec^{-1}) and $[C]^{(2)}$ is the instantaneous concentration of reagent 2 in a given volume fraction. If the rate of mixing is fast compared with the residence time, it follows that Equation (92) is true:

$$k_1 [C]^{(2)} = \left(\frac{k_1 [C]^{(2)}}{(F_1 + F_2)} \right) e^{-k_1 t} \tag{92}$$

Here, t (seconds) is the time elapsed since the volume fraction entered the cell.

As there is now a time element, the rate of reaction must be averaged over that time spent in the calorimetric cell, τ, and hence integration of Equation (92) over that residence time yields:

$$\frac{d[P]}{dt} = \left(\frac{F_2[C]_0^{(2)}}{(F_1 + F_2)\tau}\right)\left(1 - e^{-k_1\tau}\right) \tag{93}$$

To obtain the amount of product formed per unit time (mol/sec), it is necessary to multiply this by the thermal volume V_c. The thermal volume is shown by Equation (87) to be equal to $(F_1 + F_2)\tau$ and hence Equation (93) becomes:

$$\frac{d[P]}{dt} = \left(F_2[C]_0^{(2)}\right)\left(1 - e^{-k_1\tau}\right) \tag{94}$$

Therefore, the power that would be observed experimentally (average output for the lifetime in the cell), relative to the baseline, is given by:

$$\Phi = -F_2[C]_0^{(2)}\left[\Delta H_{D2} + \Delta H_R\left(1 - e^{-k_1\tau}\right)\right] \tag{95}$$

The average concentration of reagent 2, $\overline{[C]}_0^{(2)}$, over the residence time, is given by:

$$\overline{[C]}_0^{(2)} = \left(\frac{F_2[C]_0^{(2)}}{(k_1(F_1 + F_2)\tau)}\right)\left(1 - e^{-k_1\tau}\right) \tag{96}$$

If the flow is stopped and reagent 1 is in excess, then this concentration will decay exponentially. The heat output in a stopped flow experiment will vary with time according to Equation (97):

$$\Phi = -F_2[C]_0^{(2)}\Delta H_r(1 - e^{-k_1\tau})(e^{-k_1 t}) \tag{97}$$

Flow-Through Calorimeters

Zero-order reactions: Zero-order systems, as they have no time dependence, can be described in a manner identical to that for the flow-mix case, and the calorimetric output is again given by Equation (97).

First order with respect to reagent 2:   If a reaction is first order with respect to reagent 2, then the reacting mixture enters the cell t seconds after initiation and the concentration is given by Equation (98):

$$[C]_t^2 = [C]_0^2(e^{-k_1\tau}) \tag{98}$$

As the reaction medium is flowing, the accompanying power must be averaged between t and $t + \tau$. Using the same method described for the flow-mix system, it can be shown that the thermal output for a first-order flow reaction can also be described by Equation (97).

Second-order reaction in reagent 2 and zero-order in reagent 1: It was seen earlier that the rate of reaction can be expressed by Equation (2):

$$\frac{dx}{dt} = k(A_0 - x)^n \tag{2}$$

The rate constant for a second-order reaction can be written as:

$$k = \frac{1}{t}\left(\frac{x}{[A_0]([A_0] - x)}\right) \tag{99}$$

This can be rearranged to yield an expression in terms of x for any time, t:

$$x_t = \frac{kt[A_0]^2}{1 + kt[A_0]} \tag{100}$$

Therefore, the amount of material reacted during the residence time in the cell is given by subtracting Equation (100) from Equation (101):

$$x_{t+\tau} = \frac{k(t + \tau)[A_0]^2}{1 + k(t + \tau)[A_0]} \tag{101}$$

$$x_{t+\tau} - x_t = \frac{k(t + \tau)[A_0]^2}{1 + k(t + \tau)[A_0]} - \frac{kt[A_0]^2}{1 + kt[A_0]}$$

$$= \frac{k\tau[A_0]^2}{(1 + k(t + \tau)[A_0])(1 + kt[A_0])} \tag{102}$$

The average extent of reaction over the residence time is given by:

$$\overline{x_\tau} = \frac{k[A_0]^2}{(1 + k(t + \tau)[A_0])(1 + kt[A_0])} \tag{103}$$

A dimensional analysis of Equation (103) reveals that the units of $\overline{x_\tau}$ are mol/dm^3 sec, and therefore to derive the equation which describes the calorimetric output it is necessary to convert these units to J/sec. This is readily achieved by incorporation of the enthalpy and thermal volume term:

$$\Phi = \frac{k\Delta H V_c[A_0]^2}{(1 + k(t + \tau)[A_0])(1 + kt[A_0])} \tag{104}$$

As described in the derivation of equations for ampoule experiments, the output is described by equations derived from a kinetic equation that describes the reaction mechanism. If a suitable fit of the data is achieved then, in the absence of other information regarding the reaction, these equations can still only be viewed as a model of the calorimetric output. If the mechanism for reaction is known, then analysis of the calorimetric data using the appropriate equations can yield values for the reaction enthalpy and rate constant for each individual step within a more complex scheme.

CALCULATION VS. ITERATION

Thus far, it has been seen that calorimetric data can, in the main, be successfully analyzed via iterative procedures. Indeed, until recently (5–8), it was required that the data for complex reaction mechanisms be iterated (9–11) because calculation methods were not available. However, although iterative procedures appear robust they are unsatisfactory for several reasons. The first is that they require a reaction model to be known. This is less of a challenge if the reaction mechanism is known or if the process consists of a single step. The problem arises when the system becomes complex (multiple steps) and/or the reaction mechanism is unknown. If there are several parameters to be iterated for, the burden on the iteration software to return the correct values increases as there are increasingly more combinations of values which will describe the calorimetric data. These problems can be mitigated somewhat if initial estimates can be made for some of the desired parameters. Models in which desired parameters appear as product functions also pose problems. For example, consider Equation (105):

$$\frac{dq}{dt} = k\Delta H \left(A_0 - \frac{q}{\Delta H} \right)^n \tag{105}$$

If the values of A_0 and V are known prior to iteration, then the derived values of k and ΔH can be assumed to be accurate. It has previously been indicated (11) that, in principle, all the parameters in a calorimetric equation can be iterated for. However, this may not be the case; in the following model, the parameters A_0 and ΔH are mutually dependent (12).

The total area under the calorimetric curve (for a reaction that has gone to completion), Q, is a constant value and is equal, in this case, to the product of the number of moles of material reacting and the reaction enthalpy [Equation (3)]:

$$Q = A_0 \Delta H \tag{3}$$

These values are, in reality, also fixed. However, if they are not known prior to the iterative procedure, they are essentially variables. It is apparent therefore, that if A_0 is halved then it would require ΔH to be doubled in order to maintain a constant value for Q. This is clearly unsatisfactory, and hence it is essential to have prior knowledge of A_0 if procedures of this type are to be conducted. In practice, it is difficult to know the initial value of A_0; issues such as the purity of the sample, the limitations in weighing, and the reaction mechanism (especially in the solid-state) will have a bearing on the accuracy of A_0. Quantification of A_0 in complex heterogeneous systems is even more difficult. Generally, it is impossible to know A_0 in such situations, unless ancillary data are obtained.

Prior knowledge of A_0 is also essential for the derivation of accurate values for the rate constant, k. In the case of the model described by Equation (105), the initial condition where $t = 0$ (hence $q = 0$) can be described by:

$$\frac{dq}{dt} = k\Delta H A_0^n \tag{106}$$

If the initial concentration of material is now multiplied by some factor, a, the reaction enthalpy will be altered by $1/a$ in order to compensate for the effect. It is also seen that the derived rate constant is affected. If the factor, a, is now entered into Equation (105), Equation (107) can be derived:

$$\frac{dq}{dt} = \frac{1}{a}k\Delta H \left(aA_0 - \frac{q}{\frac{1}{a}\Delta H}\right)^n \tag{107}$$

If the initial conditions are considered, then Equation (107) becomes:

$$\frac{dq}{dt} = k\Delta H a^n A_0^n \tag{108}$$

Combining Equations (106) and (108) yields:

$$\Delta H A_0^n k_{real} = \frac{1}{a}k_{apparent}\Delta H a^n A_0^n \tag{109}$$

Hence:

$$k_{real} = \frac{k_{apparent}a^n}{a} \tag{110}$$

$$k_{real} = k_{apparent}a^{n-1} \tag{111}$$

$$k_{apparent} = k_{real}a^{1-n} \tag{112}$$

It is clear, from Equation (112), that using an inaccurate value for A_0 alters the apparent rate constant by the function of a^{1-n}. It is essential, therefore, when employing iterative procedures that A_0 is known.

In certain instances, then, iteration for the desired parameters can be inadequate and misleading. Therefore, iterative techniques, although useful, must be used with caution.

CALCULATION TECHNIQUES

An alternative to iteration, and one which sidesteps many of the limitations of iteration, is direct calculation (5–8). Here, data are examined using the same kinetic models but the equations which describe the model have been transformed to allow calculation of each of the desired parameters without the requirement of any prior knowledge of any associated parameter.

This methodology allows each parameter to be calculated individually; that is, there is no requirement to separate product functions. It is necessary, however, to follow a defined protocol.

For the purposes of this demonstration, the analyses will be presented through consideration of the calorimetric equations derived earlier; however, the principles employed are equally valid for the analysis of flow calorimetric data.

Calculation for the Initial Calorimetric Signal Φ_0

In Chapter 2, it was noted that for most calorimeters there is a period of time during which the calorimeter is not recording meaningful data; for instance, during sample loading, equilibration time, frictional heat dissipation, and so on. Figure 26 shows the effect of loading an ampoule into the calorimeter and the heat that can be associated with frictional effects as the ampoules are lowered.

The spike caused by frictional heat is clearly seen at the start of the plot. For this experiment, it is nearly 1.5 hours before the frictional heat has dissipated, allowing data to be properly recorded. This period of time necessarily incorporates the initial calorimetric signal. Therefore, in order to obtain an appropriate value of Φ_0, it is required that the value be inferred in some way from the recorded data set. A convenient strategy is to apply a polynomial series to the first 10 hours of any experimental data set and to extrapolate to a value for Φ_0 at $t = 0$ (Fig. 27).

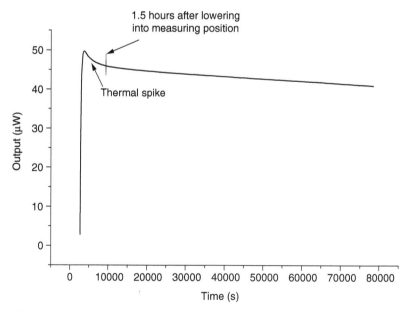

Figure 26 Calorimetric output showing spike caused by thermal shock.

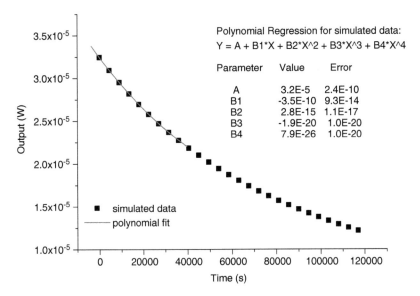

Figure 27 Fourth-order polynomial fit to a simulated calorimetric output.

Tables 1 and 2 show values of Φ_0 calculated for relatively fast- and medium-term reactions, respectively. It can be seen that the correct value for Φ_0 is found from a fourth-order polynomial equation for reactions that are relatively fast and that the value for Φ_0 for a slower reaction is determined much more readily from a first-order polynomial equation. In all instances, the value of Φ_0 is found from the limiting value of the polynomial equation, that is, when $x(t) = 0$.

Calculation for the Reaction Order

The shape of a calorimetric curve is governed by the value of n. The reaction order is independent from all other reaction parameters. Knowledge of its

Table 1 Results from Extrapolation to Φ_0 for a Relatively Fast Reaction

k (sec^{-1})				0.00271				
Φ_0 (W)				0.00229				

Polynomial order	1	2	3	4	5	6	7	8	9
Φ_0 (calculated)	0.00201	0.00223	0.00228	0.00229	0.0023	0.0023	0.0023	0.0023	0.0023
Goodness of fit	0.94864	0.99797	0.99993	1	1	1	1	1	1

Table 2   Results from Extrapolation to Φ_0 for a Medium-Term Reaction

k (sec^{-1})		2.71×10^{-6}		
Φ_0 (W)		2.296×10^{-6}		
Polynomial order	1	2	3	4
Φ_0 (calculated)	2.30×10^{-6}	2.30×10^{-6}	2.30×10^{-6}	2.30×10^{-6}
Goodness of fit	1	1	1	1

value is not only vital for some of the following analyses but it also can give some mechanistic insight into a data set which contains no molecular information.

By selecting two values of Φ from the calorimetric data, Φ_1 and Φ_2, it has been shown that the ratio of the two associated times, t_1 and t_2 for Φ_1 and Φ_2 is dependent only on the order of reaction (2). Rearrangement of Equation (7) for the two-time points gives:

$$t_1 = \frac{\left(\frac{\Phi_1}{k\,\Delta H^{1-n}}\right)^{1-n/n} - Q^{1-n}}{k\,\Delta H^{1-n}\,(n-1)} \tag{113}$$

and:

$$t_2 = \frac{\left(\frac{\Phi_2}{k\,\Delta H^{1-n}}\right)^{1-n/n} - Q^{1-n}}{k\,\Delta H^{1-n}\,(n-1)} \tag{114}$$

and therefore:

$$\frac{t_2}{t_1} = \frac{(\Phi_2)^{1-n/n}}{(\Phi_1)^{1-n/n}} \tag{115}$$

It follows that the order of reaction may be determined from knowledge of the value of t_2/t_1. This is most easily achieved through the use of a suitable mathematical worksheet (Fig. 28). The values of Φ_1 and Φ_2 are converted into percentages of the initial calorimetric signal (Φ_0) and, by using the worksheet, a table of values of t_2/t_1, calculated from Equation (115), can be constructed as a function of the rate constant for a particular pair of (Φ_1/Φ_0) and (Φ_2/Φ_0) ratios. The experimental t_2/t_1 constant may then be compared with the table of t_2/t_1 values, and the order of reaction can be determined. This calculation method relies on the quality of the data. For instance, if the data are particularly noisy, errors will be introduced into the ratio of Φ values. Hence, the derived value of n may also be uncertain. More important is the requirement that significant curvature is present in the data; that is, the percentage difference in magnitude between the initial and final data point is large. As a rough guide, it appears experimentally that the first percentage value should be between 60% and 80% of the initial power signal and the second less than 30% of the initial power signal. If this curvature is not present, the analysis is somewhat compromised. In order to

The determination of reaction order using two time points.

The two chosen time intervals, calculated as percentages, are entered into the spreadsheet;

$$\%1(n) := \frac{d(n)}{100} \cdot 63.07 \qquad\qquad \%2(n) := \frac{d(n)}{100} \cdot 24.27$$

Where,

$$d(n) := k \cdot H^{1-n} \cdot q(n)^n$$

The equations describing the two chosen time intervals are;

$$t1(n) := \frac{\left(\dfrac{\%1(n)}{k \cdot H^{1-n}}\right)^{\frac{1-n}{n}} - Q^{1-n}}{k \cdot H^{1-n} \cdot (n-1)} \qquad t2(n) := \frac{\left(\dfrac{\%2(n)}{k \cdot H^{1-n}}\right)^{\frac{1-n}{n}} - Q^{1-n}}{k \cdot H^{1-n} \cdot (n-1)}$$

Hence, a table of t_2/t_1 values may be constructed, with the corresponding values of n. Here, t_2/t_1 equals 3.786. Therefore, n equals 1.7.

$\dfrac{t2(n)}{t1(n)}$	n
3.206	1.1
3.33	1.2
3.44	1.3
3.54	1.4
3.629	1.5
3.711	1.6
3.785	1.7
3.853	1.8
3.915	1.9
3.972	2

Figure 28 The use of a mathematical worksheet to determine the order of a reaction using two time points, selected from the power–time data constructed in Figure 2.

determine accurately the desired parameters, it is preferred that all data are used in the calculation such that an average can be taken, thus minimizing any errors that may be introduced.

Consider the logarithmic form of Equation (106):

$$\ln(\Phi_0) = \ln k\Delta H + n \ln(A_T) \tag{116}$$

If a series of experiments are carried out with varying values of A_T, it becomes possible to make a plot of $\ln(\Phi_0)$ versus $\ln(A_T)$ (Fig. 29). It is then apparent from Equation (116) that this plot will be linear and the reaction order can be calculated from the slope of the line. This is clearly a more satisfactory method as it takes all data points into consideration allowing an averaged value to be calculated.

The sample quantity, A_T (i.e., that loaded into the calorimeter), is used because if there is incomplete reaction (i.e., an equilibrium exists between reactants and products) then only A_T is known. It is also true that if the reaction mechanism is constant then the same fraction of any sample quantity loaded into the calorimeter will react for any given set of conditions (temperature, pH, relative humidity, pO_2, and so on). As a result, the slope of the $\ln(\Phi_0)$ versus $\ln(A_T)$ plot will always be equal to the reaction order, n, while the y-intercept ($\ln k\Delta H$) will vary. Note that zero-order reactions are quantity/concentration independent and therefore the value of Φ_0 will not vary as a function of A_T. This, of course, will immediately identify the reaction as being zero-order and hence $n = 0$.

Calculation for the Total Heat Released for Complete Reaction

If the hypothetical reaction considered here is allowed to proceed to completion, the total heat, Q, given out by that reaction can be calculated by integrating the area under the calorimetric curve from $t = 0$ to t_{end}. This approach is only valid if the reaction rate is rapid enough to permit complete reaction within a

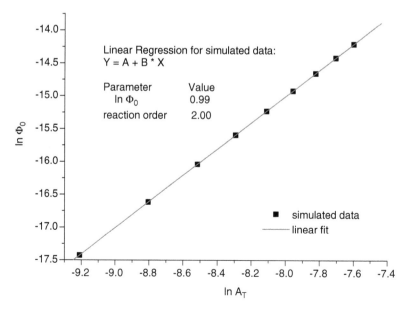

Figure 29   A plot of $\ln \Phi_0$ versus $\ln A_T$.

measurable time frame. It is more common, however, that the reaction under observation will not be complete within many months or even years; this is especially the case with pharmaceutical materials as they are usually chosen, in part, for their stability characteristics. It is necessary, therefore, to calculate a value for Q from the experimental data.

Recall Equation (5):

$$\Phi = k\Delta H^{1-n}(Q - q)^n \tag{5}$$

If two values of Φ, at different points along the calorimetric curve, are taken and their associated values of q are noted such that:

$$\Phi_1 = \Delta H^{1-n}k(Q - q_1)^n \tag{117}$$

$$\Phi_2 = \Delta H^{1-n}k(Q - q_2)^n \tag{118}$$

then:

$$\frac{\Phi_1}{\Phi_2} = \frac{(Q - q_1)^n}{(Q - q_2)^n} \tag{119}$$

$$\left(\frac{\Phi_1}{\Phi_2}\right)^{1/n} = \frac{(Q - q_1)}{(Q - q_2)} \tag{120}$$

setting:

$$\left(\frac{\Phi_1}{\Phi_2}\right)^{1/n} = R \tag{121}$$

The value for Q can then be derived from Equation (122):

$$Q = \frac{(q_1 - Rq_2)}{1 - R} \tag{122}$$

At this point, it should be noted that the values of q_1 and q_2 must include the area under the curve for the missing data always encountered at the start of any calorimetric experiment. It is clear that this value cannot be measured directly but it can be approximated (with reasonable accuracy) from the calculated value of Φ_0 and the accompanying polynomial equation used to derive it, described earlier.

Calculation for Reaction Half-Life

The reaction half-life, $t_{1/2}$, is defined as the time taken for half the reactable material to be consumed. Its value can be calculated via two methods. The first method is to integrate systematically the area under the power−time curve until $q = Q/2$; this is clearly an extremely laborious and time-consuming process and, more importantly, it requires that the reaction is observed to completion. A better alternative is to derive the value by calculation.

Equation (123) describes $t_{1/2}$, for the simple reaction $A \rightarrow B$, for any reaction order $n > 1$:

$$t_{1/2} = \frac{(2^{n-1} - 1)}{[(n-1)kA_0^{n-1}]} \tag{123}$$

Hence, if k, A_0, and n are known, then $t_{1/2}$ can be calculated.

Calculation of $t_{1/2}$ is possible even if the values for k and A_0 are unknown. The fact that the derived value for k is dependent on the value of A_0 is crucial to this treatment. Starting from:

$$\Phi = -kH\left(\frac{(A_0^{n-1})}{((n-1)A_0^{n-1}kt) + 1}\right)^{n/n-1} \tag{124}$$

If A_0 is varied, the derived values of k will vary accordingly [Equation (109)]. If incorrect values of k and A_0 are entered into Equation (123), it is found that the correct value for $t_{1/2}$ is returned. This is because the error in A_0 is compensated for in the derived value of k and hence the correct value of $t_{1/2}$ is always returned.

Calculation for Rate Constant

As the reaction order, n, is known, a kinetic equation that describes the reaction can easily be written. This equation can then be manipulated to reveal the rate constant. Taking a second-order reaction as an example:

$$\Phi = -k\Delta H\left(\frac{A_0}{1 + kA_0 t}\right)^2 \tag{125}$$

If the ratio of two data points Φ_1 and Φ_2 at times t_1 and t_2 are taken, Equation (126) is obtained:

$$\frac{\Phi_1}{\Phi_2} = \frac{(1 + kA_0 t_2)}{(1 + kA_0 t_1)} \tag{126}$$

This equation can be expanded and rearranged to yield a quadratic expression in terms of k:

$$k^2(RA_0^2 t_1^2 - A_0^2 t_2^2) + k(2RA_0 t_1 - 2A_0 t_2) + (R - 1) = 0 \tag{127}$$

This quadratic function can then be solved in the normal way where here $R = \dfrac{\Phi_1}{\Phi_2}$.

Calculation for a Rate Constant from Flow Calorimetric Data

The process for calculating the rate constant from flow calorimetric data is identical to that already described for the ampoule/batch equation. However, the process is slightly complicated by the appearance of the residence time in the equation. The quadratic equation obtained (for a second-order reaction) is derived in exactly the same way as done already with the result being:

$$k^2(RA_0^2 t_1^2 + Rt_1 \tau A_0^2 - t_2^2 A_0^2 - t_2 A_0^2 \tau) + k(2Rt_1 A_0 - 2t_t A_0 + R\tau A_0 - \tau A_0)$$
$$+ (R - 1) = 0 \tag{128}$$

Calculation for Reaction Enthalpy

The calculation for the enthalpy of reaction can be approached in two ways. Earlier in this chapter, it was shown that the total heat output, Q, is given by Equation (3):

$$Q = A_0 H \tag{3}$$

Consequently, knowledge of A_0 and Q allows ΔH to be calculated. If the number of moles of material involved in the reaction is known (i.e., the sample is pure and well characterized), then the value of the returned enthalpy will be expressed in J/mol. However, if the sample is not pure, or if it is not well characterized (a heterogeneous mix, for example), then it is only possible to know the mass A_T of sample and, hence, only values for the specific enthalpy (J/unit mass) can be calculated. If this is the case, then this second method should be employed whereby the reaction enthalpy can be found by substituting the parameters with known values into the appropriate equation and rearranging as required.

Reactions Which Proceed to a Point of Equilibrium

It was shown earlier that the area under the calorimetric curve is representative of the amount of material that has reacted to any given time, t (q for incomplete reaction and Q for complete reaction). In some instances, it is possible that not all the material loaded into the calorimeter will react. One important example is if the reaction reaches equilibrium. In these instances, the amount of the sample that reacts is dependent upon the equilibrium constant. If the reaction is suspected to reach equilibrium, then it is imperative that this is determined. If the reaction is assumed to go to completion and the enthalpy is calculated using A_0 and Q, then that value of ΔH will be incorrect. The first challenge then is to determine whether the reaction proceeds to completion or not.

Note that in the following treatment, Q represents the amount of material that will react and is distinct from Q_T, which is the value of Q if all the sample (i.e., A_0) content had reacted.

Test for Complete Reaction

A test for complete reaction is simply a study of the reaction (for identical loads) over a range of temperatures. Noting that because the equilibrium constant (if one exists), K, will change as a function of temperature, Q must also vary as a function of temperature. It is clear that if the reaction proceeds to completion then the value of Q will remain constant across the temperature range and will be equal to Q_T for all temperatures.

Determination of K

If a simple equilibrium reaction $A \Leftrightarrow B$ is considered, then for the reaction $A \Leftrightarrow B$ the equilibrium constant is simply given by:

$$K = \frac{[A]}{[B]} \tag{129}$$

Note that the presence of square brackets, in this case, does not imply a concentration. They are merely the general form of the expression for K. For this reaction, B is representative of the number of moles of product formed at time t, as described. It is clear, therefore, that equilibrium at $[B]$ is given by Equation (130) and $[A]$ is described by Equation (131):

$$[B] = \frac{Q}{H} \tag{130}$$

$$[A] = A_T - [B] \tag{131}$$

Hence:

$$[A] = \frac{Q_T}{H} - \frac{Q}{H} \tag{132}$$

Thus:

$$K = \frac{(Q_T/H - Q/H)}{(Q/H)} \tag{133}$$

and therefore:

$$K = \frac{Q}{(Q_T - Q)} \tag{134}$$

Equation (134) can be written for studies at different temperatures, T_m ($m = 1, 2, 3, \ldots$). Note that Q_T will remain constant for all values of T, providing that the reaction mechanism does not change and that there is no dependence of change in heat capacity, ΔC_p, over the chosen temperature range.

Equation (134) then permits K to be calculated for any chosen temperature. However, a value for Q_T is required in order to effect this calculation. The value of Q_T is not directly available from the calorimetric data and must be calculated separately.

Calculation of Q_T

The calculation of Q_T is effected through consideration of the effect of change in temperature on the equilibrium constant.

The enthalpy change of a reaction, at constant pressure, can be cast in terms of the Gibbs, G, energy of that system [Equation (135)]:

$$\left[\frac{\delta(\Delta G/T)}{\delta(1/T)} \right]_p = \Delta H \tag{135}$$

The Gibbs energy can also be written in terms of the equilibrium constant:

$$\Delta G = -RT \ln K \tag{136}$$

Substituting Equation (136) into Equation (135) yields:

$$\frac{\delta(\ln K)}{\delta(1/T)} = -\frac{\Delta H}{R} \tag{137}$$

This equation is known as the van't Hoff isochore and relates K and T through $\Delta H/R$. If it is assumed that ΔH is independent of temperature, then integration of Equation (137) gives:

$$\frac{K_1}{K_2} = -\frac{H}{R}\left(\frac{1}{T_1} - \frac{1}{T_2}\right) \tag{138}$$

The ratio K_1/K_2 will be equal to K_2/K_3, for temperatures T_2 and T_3, if the temperatures T_1, T_2, and T_3 are such that Equation (139) is true:

$$\frac{T_1 T_2}{(T_2 - T_1)} = \frac{T_2 T_3}{(T_3 - T_2)} \tag{139}$$

(for instance, if $T_1 = 298$ K and $T_2 = 303$ K, then $T_3 = 308.5$ K). If this condition is met, then Equation (140) can be written as:

$$\frac{K_1}{K_2} = \frac{(Q_1/Q_T - Q_1)}{(Q_2/Q_T - Q_2)} = \frac{K_2}{K_3} = \frac{(Q_2/Q_T - Q_2)}{(Q_3/Q_T - Q_3)} \tag{140}$$

This can be solved for Q_T (for temperatures $m = 1$, 2, and 3):

$$Q_T = \left(\frac{Q_2^2 Q_1 + Q_2^2 Q_3 - 2Q_3 Q_2 Q_1}{Q_2^2 - Q_3 Q_1}\right) \tag{141}$$

The enthalpy is now accessible because Q_T and A_T are known:

$$\frac{Q_T}{A_T} = \Delta H \tag{142}$$

Once ΔH is known, A can be calculated:

$$\frac{Q_m}{\Delta H} = A \tag{143}$$

Once these parameters are known, it is a simple matter to calculate the rate constant, k, using the methods described earlier.

As Q_T and Q are accessible, the equilibrium constant, K, can be derived from Equation (134) and hence the values for the Gibbs function and entropy are readily obtained from Equations (136) and (144), respectively:

$$\frac{(\Delta H - \Delta G)}{T} = \Delta S \tag{144}$$

The calculated values of K will allow an "internal" check on the validity of the procedure through the derived value of ΔH. A plot of ln K versus $1/T$ will yield a straight line with slope $-\Delta H/R$. If the procedure is invalid, the mechanism changes with T, or there is a significant temperature dependency of heat capacity (over the temperature range employed), the van't Hoff plot will not be linear and/or the value of ΔH will not match that derived from Equation (142).

As Q and Q_T are known and if ΔH is accurately recovered then it is now possible quantitatively to determine the reactable material content in a heterogeneous sample.

Extension to More Complex Reaction Schemes

The analysis already described is not applicable only to simple reaction mechanisms; it can be developed further to incorporate more complicated reaction systems (6).

Consider the equilibrium $A \Leftrightarrow B + C$. The equilibrium constant for this system is given by:

$$\frac{[B][C]}{[A]} = K \tag{145}$$

Again an expression can be derived for K based upon Q_m and Q_T values:

$$K = \frac{(Q_m/\Delta H)(Q_m/\Delta H)}{(Q_T/\Delta H - Q_m/\Delta H)} \tag{146}$$

As the value for ΔH in this expression is a constant it can be taken over to yield a modified equilibrium constant:

$$K\Delta H = K' = \frac{Q_m^2}{(Q_T - Q_m)} \tag{147}$$

Again the equilibrium constant, K', is dependent only on the values of Q_T and Q_m. Hence, a plot of ln K' versus $1/T$ will still yield a straight line with slope equal to $-\Delta H/R$, allowing ΔH to be determined. Consequently, the values A_T and A can be calculated (i.e., the total amount of reactable material present in an undefined sample and the amount of that material that will react, under a given set of conditions).

This analysis can be applied to the study of an uncharacterized complex sample for which there is a determinable reaction order. The value of Q_m can be determined for standardized loads studied in the calorimeter over the appropriate temperature range. This then allows determination of Q_T for that complex sample. There is then the possibility of applying various equilibrium models until an appropriate model is found; that is, one in which Q_T remains constant. The necessary requirement is that the van't Hoff isochore holds for the reaction under study.

SOLID-STATE SYSTEMS

This chapter has thus far illustrated the derivation and development of a number of kinetic models that describe both simple and complex reaction schemes to determine values for reaction enthalpy change, rate constant, order of reaction, activation energy, etc. However, each of the models discussed has been specifically written for reactions which proceed in the solution phase. As the majority of pharmaceuticals are formulated in solid dosage forms it is clear that, certainly for pharmaceutical applications, similar equations must be derived for reactions which proceed in the solid phase. A general equation, Equation (148), which is said to describe all solid-state reaction mechanisms (in which one of the products is a gas), was published in 1972 (13). It is known as the Ng equation:

$$\frac{d\alpha}{dt} = k(1 - \alpha)^n (\alpha)^m \tag{148}$$

Here, α is the fractional extent of reaction at time t (it hence has values that range between zero and one) and k, the rate coefficient has, necessarily, the dimensions of s^{-1}. The superscripts m and n are fitting constants (related to the reaction mechanism), but are not reaction orders.

Theoretical Development

In a manner identical to that described for the solution phase models, the Ng equation can be converted to a form which describes calorimetric data obtained from a solid-state reaction (one that conforms to the Ng model). The fractional extent of reaction, α, can be cast in terms of the number of joules of heat released by the reaction process. The total number of joules involved in the reaction to time, $t = \infty$ (i.e., reaction completion) is denoted Q, and q is the number of joules evolved up to any time t. Thus, α can be set equal to q/Q and Equation (148) becomes:

$$\frac{dq}{dt} = \phi = kQ\left[1 - \left(\frac{q}{Q}\right)\right]^n \left(\frac{q}{Q}\right)^m \tag{149}$$

As was seen earlier in this chapter, it is possible to manipulate the kinetic equations which describe solution phase systems in such a way as to be able to obtain values for the target parameters, from the calorimetric data, by an iterative process. The same is also true for data obtained from a solid-state reaction process. However, the problem associated with solid-state reactions, and not present in the analysis of both simple and complex solution phase systems where the constituent reactions have integral orders, is that the fitting (not order) parameters expressed in Equation (149) are not usually integral and that the equations are complex. Moreover, the majority of calorimetric studies [usually differential scanning calorimetry (DSC) and other techniques], of solid-state systems consider reactions which can be monitored over relatively short times to significant fractional (α) extents of completion. This is not always applicable to pharmaceutical systems because generally a pharmaceutically active/excipient and/or formulation is chosen, in part, for its stability and hence, any associated reactions are likely to have long half-lives. This then imposes a greater burden on the software chosen to perform the iterative analysis and may reduce the confidence in any returned values. Consequently, it is recommended that data of this type are analyzed for the target parameters by direct calculation methods. The calculation method is explained here through its application to a simulated solid-state reaction data set.

Determination of *m* and *n*

The first step in the analysis of calorimetric data is the determination of the reaction order or, in the case of solid-state reactions, the fitting parameters m and n. The values for m and n lie between zero and one, and the particular combination of values describes the mechanism of the process under study. The values which m and n can take and the mechanisms they describe can be found in Willson (2).

The values of m and n can be calculated using a technique of data pairing (8). The procedure permits values to be determined from a method which relies only on the knowledge of values of Φ and q, for paired time points, throughout the power–time curve ($\Phi - t$) recorded during the observation period. This method is robust but does require that the percentage difference between data points is greater than some defined value, below which the analysis becomes increasingly unreliable.

Determination of *Q*

As pharmaceutically related solid-state reactions generally have long reaction half-lives (i.e., slow rates of reaction), it is not possible to determine experimentally the value of Q. Hence, some way of inferring that value from the available data is required. This again is achieved through analysis of paired data points. Writing Equation (149) for two data points and forming the ratio between them yields:

$$\frac{\phi_1}{\phi_2} = \frac{(1 - q_1/Q)^n (q_1/Q)^m}{(1 - q_2/Q)^n (q_2/Q)^m} \tag{150}$$

If the values of q_1 and q_2 are selected such that q_2 is a known factor of q_1, that example, q_2 is equal to cq_1 and hence $q_2/q_1 = c$ and setting R as:

$$R = \left(\frac{\phi_1}{\phi_2} (c)^m \right)^{1/n} = \frac{(1 - q_1/Q)}{(1 - cq_1/Q)} \tag{151}$$

then Equation (151) is solvable for Q:

$$Q = \frac{q_1(cR - 1)}{(R - 1)} \tag{152}$$

If Q is calculated for all values of q, then it is possible to obtain an averaged value for Q over the lifetime of the reaction.

Once a value for Q is found, it then becomes a simple algebraic rearrangement of Equation (149) to find values of k:

$$k = \frac{\phi}{Q[1 - (q/Q)]^n (q/Q)^m} \tag{153}$$

Limitations of the Method

There are, of course, as with any data analysis, limitations to this calculation method. The main issue for this analysis is what minimum fraction of reaction is required for a successful determination of Q, and hence the remaining target parameters. The calculation method, already described, has been used to analyze simulated data in order to explore these limitations (14,15).

Data for given values of m, n, Q, and k, have been simulated using MathCad for hypothetical solid-state reactions where Q ranged from 10 to 10 000 J; the rate coefficient, k, ranged from 10^{-4} to 10^{-8}/sec, and values for m and n were between zero and one. The data were produced in the form of Φ versus q for a range of values of α up to a maximum of $\alpha = 1$. The data were then analyzed for values of m and n using Willson's (8) algorithm written in MathCad. The returned values for m and n were then applied in the analysis for Q using an algorithm written in Microsoft Excel constructed from the equations already outlined.

Through the simulated data sets (14–15), the minimum value for α [i.e., (q/Q)] for given values of n and m that allow characterization of the model system can be established, the test being that the set values compare favorably with the calculated values. It can be shown that from an α value as small as ~ 0.01 (when Q is assigned a value of 100 J, k is 3×10^{-4} and m and n are 1 and 0.5, respectively), it is possible to recover the correct values for the target parameters. Figure 30 illustrates the overall data plot of Φ versus q and Figure 31 the fraction of the data set required for successful recovery of the target parameters. It should be noted that for successful analysis the ratio of ϕ_2/ϕ_1 should be as large as possible.

As the value of Φ_1 approaches Φ_2 (hence Φ_2/Φ_1 approaches 1), the analysis becomes more difficult. The separation required between Φ_1 and Φ_2 depends on the values of Q and α. If Q is small then α must be large enough to allow sufficient separation between Φ_1 and Φ_2.

The maximum required value of α depends on the value of Q; the maximum value of α required for satisfactory analysis is 0.1 for solid-state reactions with values of Q as low as 2 J. That is, long, slow solid-state reactions are amenable

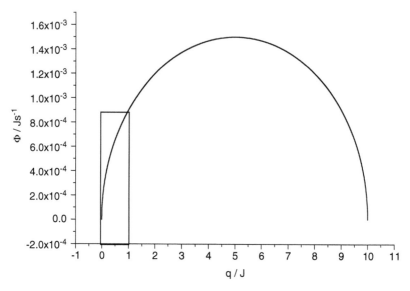

Figure 30 Simulation for the α range 0–1 for $m = 0.5$, $n = 0.5$, $k = 3 \times 10^{-4}/\text{sec}$, $Q = 10$ J. The portion of data enclosed in the box is expanded in Figure 31. The whole data set is a representation of 1.7 million data points of which only 17,000 are required for successful analysis. For example, 17,000 data points is equivalent to recording one data point every 10 seconds over a 48-hour period.

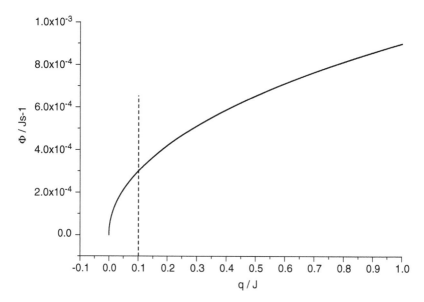

Figure 31 The portion of data from Figure 30 up to $q = 1$ J, that is, $\alpha = 0.1$. The required data set for successful analysis is that prior to the vertical line at $q = 0.1$ J, that is, $\alpha = 0.01$.

to study. It is important to remember that, currently, there is no method available to integrate the calorimetric form of the Ng equation such that it can be cast in terms of time. Therefore, the analyses performed here give no indication as to how long it will take to reach an α value of 0.1 (for the given rate parameters).

COMPLEX REACTION SYSTEMS

The equations and techniques described in this chapter thus far, although robust, all suffer one flaw. They all require that a reaction mechanism be known (i.e., it is required that the order of reaction can be found). This is not necessarily a major difficulty where only one reaction step is occurring; however, it does become problematic and severely limiting when reactions comprising multiple steps are studied. The analysis for complex systems already described required that the reaction mechanism be known prior to the iterative analysis in order for it to be successful at all. The alternative would be systematically to fit the data to increasingly complex kinetic models until a fit was obtained which was viewed to be satisfactory (i.e., fits of theoretical and experimental data would be sought from increasingly complex sequential/parallel reaction mechanisms spanning the range of integral orders 0, 1, 2, and 3, until a satisfactory fit is obtained). Such an approach is unsatisfactory for two reasons. First, it is extremely time consuming and requires the derivation of many complex equations and secondly because it is necessary to impose a model on the system.

Calorimetry is now being used for a wide range of applications, particularly for the investigation of stability of a variety of pharmaceutical formulations (creams, lotions, tablets, and so on). Such formulations are complex, multicomponent systems, and their stability may not just be dependent on the degradation kinetics of the active pharmaceutical ingredient; indeed, it is more likely that there may be several independently degrading components. Moreover, the kinetic behavior of these degradation processes may be known individually but in the presence of other components could be significantly different. As calorimeters record all thermal events, without discrimination, the resultant calorimetric data sets will reflect the complexity of the system (the data now being a summation of all thermal processes from each mechanistic step). As has already been seen, this results in the defining equations becoming more complex and unwieldy.

Chemometric Analysis

The problem of complexity has been recognized and extensively debated in the literature for nonisothermal (scanning) calorimetric studies by Roduit (16) and references therein, and Galwey (17,18) for spectroscopic studies (19). One proposed approach to the problem of complexity in multivariate data sets is the application of chemometric procedures via techniques such as PCA (20). Such approaches are well established for spectroscopic data and there are many commercially available software packages for this purpose (for instance, DiKnow,

Infometrix, Netzsch) (21–23). These approaches are also well known for the analysis of nonisothermal (scanning) calorimetric data (in particular that of solid-state systems), where the complexity arises from the underlying (applied) change in temperature over the course of the experiment as well as any multistep chemical/physical process. The challenge, then, is to elucidate the true kinetic character [so-called "kinetic triplet" consisting of the conversion function (or kinetic model) and the Arrhenius parameters E_a and A], of the system, free from the effect of the change in temperature. Much debate exists as to the best chemometric approach for the analysis of scanning calorimetric data. This has been extensively covered in the literature with a study dedicated to the comparison of the different methods available; for example, the International Confederation for Thermal Analysis and Calorimetry (ICTAC) Kinetics Project (16,24–27), and the reader is directed toward these reviews for a more complete review.

While the concept of "model-free" chemometric kinetic analysis is well known for nonisothermal systems, there appears to be little work published for the analysis of IC data. This, in part, can be ascribed to the fact that, until recently, the use of IC as a quantitative tool was limited and hence, there was no real need for the development of such techniques. Recent years have seen a growth in the use of isothermal calorimetric instruments for the study of complex systems, particularly in the pharmaceutical and food industries, and its potential for quantitative analysis is now beginning to be realized. In order for IC to fulfill its true potential as a quantitative tool, new techniques need to be developed to deal with complex data sets. One approach is to adapt chemometric analysis (commonly used for analysis of spectroscopic data) to the analysis of calorimetric data.

Principles of Chemometrics

Chemometric techniques involve pattern searching from a matrix of data in a model-free manner. Conventionally, a matrix of data containing at least three variables (e.g., power, time, and temperature for scanning calorimetry) is required in order to facilitate multivariate (pattern searching) type analysis. IC data consist only of two variables, power and time; however, the fact that each component of the system exhibits a unique maximum for any given temperature suggests that the data could be analyzed via multivariate techniques providing that a suitable matrix can be generated.

The deconvolution of the data matrix, without prior knowledge, will at least provide the scientist with the basic underlying factors or species in the reaction over a time period, t. For the example provided here, a modified target factor analysis (TFA) technique was used to analyze the data. TFA is used to determine whether or not a hypothetical vector, gleaned from chemical principles or heuristic intuition, lies inside the factor space and thus contributes to the phenomenon (28,29). The strength of this analysis lies in the fact that each hypothetical vector can be tested individually for significance in the presence of a host of other unknown factors. When a data matrix has been decomposed into abstract factors (also known as principal components) in the row and column space of the data, P, and the numbers of significant factors have been determined, these significant factors can now be subjected to various forms of mathematical scrutiny to determine if they have real chemical or physical meaning. For this example, singular value decomposition (30) was used for the decomposition process involved in the TFA process.

Equations (154) to (159) summarize the mathematical principles used to deconvolute the data, P. The principal factor matrices, $\mathbf{R}$ and $\mathbf{C}$, describe the row and column information, respectively. E is the error or noise added to the data from the instrument to data P. Recombination of the significant row and columns $\mathbf{R}$ and $\mathbf{C}$ produces a new matrix $\mathbf{P_x}$ [Equation (155)]. $\mathbf{X_t}$ are hypothetical j sets of vectors that are considered fully to describe the data $\mathbf{P_x}$. Various schemes have been proposed to determine $\mathbf{X_t}$. In this analysis, the so-called needle search method, fully described elsewhere (28-29), was used. Essentially, initial hypothetical test vectors are generated by setting all time points to zero except those points that show significant power intensity for each of the j factors identified. From these test vectors, prototype profiles are generated after target testing. An iterative or interactive method is then used to select the unique set of $\mathbf{X_t}$ that best describes the data. Operating the pseudoinverse of $\mathbf{R}$, that is, $\mathbf{R}^+$ with $\mathbf{X_t}$ yields $\mathbf{T}$, the transformation matrix indicated in Equation (156). $\mathbf{T}$ is used to obtain $\mathbf{X_j}$ and $\mathbf{Y}$. $\mathbf{X_n}$ represents the predicted test matrix based on the $\mathbf{X_t}$ [Equation (157)] and it describes the row domain real information, which is the predicted evolution profile for each of the significant j factors identified in the data matrix $\mathbf{P}$. $\mathbf{Y}$ is the column domain real information matrix, which is the predicted power output profile obtained from Equation (158). Recombination of $\mathbf{X_n}$ and $\mathbf{Y}$ produces $\mathbf{P_n}$, the significant part of the data matrix $\mathbf{P}$ [Equation (159)]. The differences between $\mathbf{P_n}$ and $\mathbf{P_x}$ are minimized:

$$\mathbf{P} = [\mathbf{R}\,\mathbf{C}] + E \tag{154}$$

$$\mathbf{P_x} = \mathbf{R}\,\mathbf{C} \tag{155}$$

$$\mathbf{T} = \mathbf{R}^+\,\mathbf{X_t} \tag{156}$$

$$\mathbf{X_n} = \mathbf{R}\,\mathbf{T} \tag{157}$$

$$\mathbf{Y} = \mathbf{T}^+\,\mathbf{C} \tag{158}$$

$$\mathbf{P_n} = \mathbf{X_n}\,\mathbf{Y} \tag{159}$$

Chemometric analysis has successfully been applied to a number of systems and to spectroscopic systems in particular. Such systems allow the generation of vast quantities of data (intensity and wave number, for example) as a function of other variables such as pH, time, and temperature. This is not readily achieved using currently available calorimetric instruments and so it is necessary to engineer a third variable from the system. Although it is not immediately obvious what form this third variable should take, one option is to record a large number of replicate runs of the system under study. The natural variations between these runs (small differences in the load sample mass, for example) are sufficient to permit successful chemometric analysis of simulated data.

Application to Calorimetric Data

The power–time data generated from the calorimetric instrument are the sum of power output for each component reaction at defined times. From the perspective of pattern analysis, each component in the system will exhibit a unique maximum during the course of the reaction at a set temperature. This unique maximum for each component suggests that the data may be analyzed by a multivariate method if a suitable matrix can be generated. The crucial difference between spectroscopic and calorimetric data is that the former captures copious data as a function of another variable such as time, pH, and temperature. This type of data matrix has x, y, z properties, where x, y, and z describe row, column, and intensity information, respectively. Figure 32 shows a diagrammatic layout of the data.

For calorimetric measurements at a fixed temperature, T, there are t time points, and P powers. Power–time data are univariate in nature and are the typical data obtained from the calorimeter. There is, therefore, the need to engineer S, here the intensity information already noted, from the same sample system. S can be obtained by replicated runs of the same sample. These runs can be designed such that each one varies from the other in the total power output. Variations in sample loading or weight for repeated runs provides S sets of runs. A system in which X species evolve after time t will require at least $2X + 2$ replicate runs (i.e., $S = 2X + 2$) to create a suitable matrix [**P**] which can then be analyzed to determine the rank of the data through multivariate data decomposition. For calorimetric data sets, factor analysis can be used to decompose the data matrix on the assumption that (*i*) the power signal at time t is a linear sum of all species present at that moment and (*ii*) the power signal is proportional to the sample load in the calorimeter.

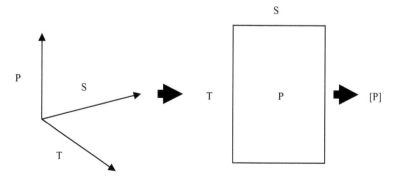

Figure 32 Schematic data format suitable for deconvolution. *Abbreviations*: P, power; S, sample mass; T, time.

Chemometric analysis, at the very least, will allow determination of the number of reaction steps, which contribute, to the overall mechanism. Because of the unique maximum of each step, it can aid identification of products and intermediates as well as the time at which the maximum concentration of each species is present. If the reaction is quenched at these times, then the species can be more easily identified from ancillary sources, such as high performance liquid chromatography, for example. More important, in some respects, is the fact that the chemometric analysis returns the component heat outputs for each reaction step. If the deconvolution is successful, then the individual "patterns" must conform to a single-step kinetic equation such as that described earlier and therefore, in principle, can be analyzed for rate constant, enthalpy, order of reaction, etc. It should also be noted that this approach is truly model-free in nature, the only information presented to the software is the overall calorimetric output for each of the repeats.

Limitations of Chemometric Analysis

The analysis does suffer some limitations, however. The returned form of the data after deconvolution is expressed as intensity (cf. power) versus time. The relationship between power and intensity is not clear and therefore accurate thermodynamic information cannot be found directly from the deconvoluted data. (It has been shown that it is possible to calculate the reaction enthalpies, for complex reaction mechanisms, from the calorimetric data if the mechanism is known and the appropriate calorimetric equations can be written; this is described in more detail in the following section.) However, because the shape of the data is not altered (i.e., the rate of change in the data is not affected by the overall magnitude), it is possible, in principle, to derive kinetic information from the deconvoluted data using the methods outlined earlier. It should also be noted that calorimetric data do not contain any molecular information, and therefore, identification of products, and hence confirmation of the mechanism, must be achieved through ancillary studies. One severe constraint on the applicability of this method is the requirement for multiple repeats to be performed [a minimum of $(2n + 1)$ repeats must be performed for a system containing n reaction steps] in order for successful analysis. This may be a problem for studies arising from the discovery laboratory, for example, where the quantity of available material is limited and, of course, there is the time constraint of performing multiple experiments, necessary for successful chemometric analysis of systems in which many thermal events are being recorded. This issue may be overcome with future developments in calorimetric arrays.

Calculation of Reaction Enthalpies for Complex Data Sets

Reaction enthalpies can be calculated from either two power–time data or two heat–time data. Using two power–time data is simpler, but the approach is subject to random noise error. The heat–time data approach is more complex, but essentially averages the data over the chosen time periods, thus minimizing errors.

Determination of ΔH from Two Power–Time Data

The following equation was derived earlier, which describes the power output for a reaction that proceeds through two consecutive, first-order steps:

$$\frac{dq}{dt} = k_1 \, \Delta H_1 \, A_0 \, e^{-k_1 t} + k_1 \, k_2 \, \Delta H_2 \, A_0 \, \frac{e^{-k_1 t} - e^{-k_2 t}}{k_2 - k_1} \tag{41}$$

It is, therefore, possible to select from the experimental data the instantaneous power outputs (P_1 and P_2) at two respective time points (t_1 and t_2) and write:

$$P_1 = k_1 \, \Delta H_1 \, A_0 \, e^{-k_1 t_1} + k_1 \, k_2 \, \Delta H_2 \, A_0 \, \frac{e^{-k_1 t_1} - e^{-k_2 t_1}}{k_2 - k_1} \tag{160}$$

$$P_2 = k_1 \, \Delta H_1 \, A_0 \, e^{-k_1 t_2} + k_1 \, k_2 \, \Delta H_2 \, A_0 \, \frac{e^{-k_1 t_2} - e^{-k_2 t_2}}{k_2 - k_1} \tag{161}$$

Equation (160) can be rewritten to make ΔH_1 the subject:

$$\Delta H_1 = \frac{P_1 \, (k_2 - k_1) - k_1 \, k_2 \, A_0 \, \Delta H_2 \, (e^{-k_1 t_1} - e^{-k_2 t_1})}{k_1 \, A_0 \, (k_2 - k_1) \, e^{-k_1 t_1}} \tag{162}$$

Substituting Equation (162) into Equation (161) and rearranging for ΔH_2 gives:

$$\Delta H_2 = \frac{(k_2 - k_1) . (P_1 \, e^{-k_1 t_2} - P_2 \, e^{-k_1 t_1})}{k_1 \, k_2 \, A_0 \, (e^{-k_1 t_1 - k_2 t_2} - e^{-k_1 t_2 - k_2 t_1})} \tag{163}$$

Thus, Equation (163) can be used to calculate the value of ΔH_2 if A_0, k_1, k_2, P_1, P_2, t_1, and t_2 are known. Knowledge of ΔH_2 then allows the value of ΔH_1 to be calculated by Equation (41).

Determination of ΔH from Two Heat–Time Data

In this case, the heat outputs (q_1 and q_2) at two time points (t_1 and t_2) are required. The heats correspond to the area under the curve up to the two time points and can easily be determined using a mathematical analysis package. However, in order to show the derivation of this approach, it is necessary to consider the mathematical equations that give these values; these are the integrals of Equation (43) between t_0 and t_1, and t_0 and t_2 (where t_0 is some initial time which does not have to be zero) and are represented by:

$$
q_1 = \frac{\begin{array}{l} A_0\,\Delta H_1\,(k_1\,e^{-t_1\,k_1} - k_2\,e^{-t_1\,k_1} + k_2\,e^{-t_0\,k_1} - k_1\,e^{-t_0\,k_1}) \\ + A_0\,\Delta H_2\,(k_1\,e^{-t_1\,k_2} - k_1\,e^{-t_0\,k_2} + k_2\,e^{-t_0\,k_1} - k_2\,e^{-t_1\,k_1}) \end{array}}{(k_2 - k_1)}
\tag{164}
$$

$$
q_2 = \frac{\begin{array}{l} A_0\,\Delta H_1\,(k_1\,e^{-t_2\,k_1} - k_2\,e^{-t_2\,k_1} + k_2\,e^{-t_0\,k_1} - k_1\,e^{-t_0\,k_1}) \\ + A_0\,\Delta H_2\,(k_1\,e^{-t_2\,k_2} - k_1\,e^{-t_0\,k_2} + k_2\,e^{-t_0\,k_1} - k_2\,e^{-t_2\,k_1}) \end{array}}{(k_2 - k_1)}
\tag{165}
$$

Equation (164) can be rewritten to make ΔH_1 the subject:

$$
\Delta H_1 = \frac{\left[\begin{array}{l} \frac{q_1\,(k_2 - k_1)}{A_0} - \Delta H_2\,k_2\,e^{-t_0\,k_1} + \Delta H_2\,k_1\,e^{-t_0\,k_2} \\ + \Delta H_2\,k_2\,e^{-t_1\,k_1} - \Delta H_2\,k_1 e^{-t_1\,k_2} \end{array}\right]}{(-k_2\,e^{-t_1\,k_1} + k_1\,e^{-t_1\,k_1} + k_2\,e^{-t_0\,k_1} - k_1\,e^{-t_0\,k_1})}
\tag{166}
$$

Substituting Equation (166) into Equation (165) and rearranging for ΔH_2 gives:

$$
\Delta H_2 = \frac{\begin{array}{l} -q_2\,[(-e^{-t_0\,k_1} + e^{-t_1\,k_1})(k_2 - k_1)] \\ -q_1\,[((k_2 - k_1)\,e^{-2t_0\,k_1}) + ((k_1 - k_2)\,e^{-(t_0+t_2)\,k_1}] \end{array}}{A_0\,(k_2\,\alpha + k_1\,\beta)}
\tag{167}
$$

where:

$$
\alpha = e^{-t_2\,k_1} - e^{-k_1\,(2t_0+t_2)} - e^{-t_0.k_1} + e^{-3t_0.k_1}
\tag{168}
$$

and:

$$
\begin{aligned}
\beta = {} & e^{-t_0\,k_1 - t_0\,k_2 - t_1\,k_1} + e^{-t_1\,k_2 - t_0\,k_1 - t_2\,k_1} + e^{-2t_0\,k_1 - t_2\,k_2} \\
& - e^{-t_2\,k_2 - t_0\,k_1 - t_1\,k_1} - e^{-t_1\,k_2 - 2t_0\,k_1} - e^{-t_0\,k_1 - t_0\,k_2 - t_2\,k_1}
\end{aligned}
\tag{169}
$$

Thus, Equation (167) can be used to calculate the value of ΔH_2 if A_0, k_1, k_2, q_1, q_2, t_0, t_1, and t_2 are known. Knowledge of ΔH_2, as before, allows ΔH_1 to be calculated by Equation (41). In the special case where t_0 is taken to represent the time at which the reaction was initiated, Equation (167) reduces to:

$$
\Delta H_2 = \frac{-q_2\,[(e^{-t_1\,k_1} - 1)\,(k_2 - k_1)] - q_1\,[(k_2 - k_1) + (k_1 - k_2)\,e^{-t_2\,k_1}]}{A_0\,k_1\,(e^{-t_1\,k_1} + e^{-t_1\,k_2 - t_2\,k_1} + e^{-t_2\,k_2} - e^{-t_2\,k_2 - t_1\,k_1} - e^{-t_1\,k_2} - e^{-t_2\,k_1})}
\tag{170}
$$

In most calorimetric experiments, there is a time delay between the initiation of reaction and is commencement of data measurement is commensurate with the fact that ampoules are usually prepared externally from the calorimeter. However, it is also usually the case that the percentage of data lost is negligible (for long-term reactions) or that the data can be extrapolated back to the initiation time (for short-term reactions). In either case, Equation (170) can then be used to determine ΔH_2.

The two calculational methods are based on the same principle and, when applied to simulated data, give the same result. However, when applied to real data, it is likely that the method based on two heats, rather than two powers, will be more robust, because it essentially averages the data up to the two selected time points and is therefore less affected by random data fluctuation. Analysis of calorimetric data in this way for a given system at more than one temperature allows recovery of both the Gibbs function and, hence, the entropy change associated with the reaction.

SUMMARY

There can be little doubt that the difficultly in extracting quantitative reaction parameters from complex calorimetric data is the biggest obstacle to IC becoming a routine analysis technique, in the way DSC is. The complexity of the data is a direct consequence of the versatility of the technique; while clever experimental design and sample preparation may be one way to simplify the data, the increasing desire to study whole systems means that the development of analysis strategies is the best way forward. In this chapter, four main strategies have been introduced and discussed. These range from simple empirical fits through model-based approaches to model-free chemometrics. Each has merits and drawbacks and, at least in the case of the latter two, requires further research and development, although these approaches are beginning to reverberate through the field. Analysis though iteration or direct calculation currently offers the best chance of recovering quantitative data, although these approaches clearly require the reaction mechanism to be known in advance. The complexity of the equations also increases dramatically with an increase in the number of reactants, intermediates, or pathways, although in an early stage, chemometric analysis appears to be capable of determining the number of reaction steps, which may assist the analyst in determining the exact reaction mechanism. Once this is known, analysis using the model-based approaches becomes simpler.

In any case, the information provided demonstrate that analysis of complex data is possible and should enable the reader to begin to analyze their own data; in addition, the text should provide a framework for developing models for other, more complex, reaction mechanisms.

REFERENCES

1. Bakri A, Janssen LHM, Wilting J. Determination of reaction rate parameters using heat conduction microcalorimetry. J Therm Anal 1988; 33:185–190.
2. Willson RJ. Isothermal microcalorimetry: Theoretical development and experimental studies. Ph.D. dissertation, University of Kent, Canterbury, UK, 1995.
3. Willson RJ, Beezer AE, Mitchell JC, Loh W. Determination of thermodynamic and kinetic parameters from isothermal heat conduction microcalorimetry: applications to long-term reaction studies. J Phys Chem 1995; 99:7108–7113.

4. Beezer AE, Steenson TI, Tyrrell HJV. Application of flow-microcalorimetry to analytical problems 2. Urea-urease system. Talanta 1974; 21:467–474.

5. Beezer AE. An outline of new calculation methods for the determination of both thermodynamic and kinetic parameters from isothermal heat conduction microcalorimetry. Thermochimica Acta 2001; 380:205–208.

6. Beezer AE, Morris AC, O'Neill MAA, et al. Direct determination of equilibrium thermodynamic and kinetic parameters from isothermal heat conduction microcalorimetry. J Phys Chem B 2001; 105:1212–1215.

7. Willson RJ, Beezer AE. The determination of equilibrium constants, delta G, delta H and delta S for vapour interaction with a pharmaceutical drug, using gravimetric vapour sorption. Int J Pharm 2003; 258:77–83.

8. Willson RJ, Beezer AE. A mathematical approach for the calculation of reaction order for common solution phase reactions. Thermochimica Acta 2003; 402:75–80.

9. Beezer AE, Gaisford S, Hills AK, Willson RJ, Mitchell JC. Pharmaceutical microcalorimetry: applications to long-term stability studies. Int J Pharm 1999; 179:159–165.

10. Beezer AE, Willson RJ, Mitchell JC, et al. Thermodynamic and kinetic parameters from isothermal heat conduction microcalorimetry. Pure App Chem 1998; 70:633–638.

11. Willson RJ, Beezer AE, Mitchell JC, Loh W. Determination of thermodynamic and kinetic parameters from isothermal heat-conduction microcalorimetry—applications to long-term reaction studies. J Phys Chem 1995; 99:7108–7113.

12. Hills AK. Theoretical and experimental studies in isothermal microcalorimetry. Ph.D. dissertation, University of Kent, Canterbury, UK, 2001.

13. Ng WL. Thermal decomposition in solid-state. Aust J Chem 1975; 28:1169–1178.

14. Beezer AE, O'Neill MAA, Urakami K, Connor JA, Tetteh J. Pharmaceutical microcalorimetry: recent advances in the study of solid-state materials. Thermochimica Acta 2004; 420:19–22.

15. O'Neill MAA, Beezer AE, Morris AC, Urakami K, Willson RJ. Solid-state reactions from isothermal heat conduction microcalorimetry—theoretical approach and evaluation via simulated data. J Therm Anal Cal 2003; 73:709–714.

16. Roduit B. Computational aspects of kinetic analysis: Part E: The ICTAC kinetics project—numerical techniques and kinetics of solid-state processes. Thermochimica Acta 2000; 355:171–180.

17. Galwey AK. Is the science of thermal analysis kinetics based on solid foundations? A literature appraisal. Thermochimica Acta 2004; 413:139–183.

18. Galwey AK, Brown ME. Isothermal kinetic analysis of solid-state reactions using plots of rate against derivative function of the rate equation. Thermochimica Acta 1995; 269–270:1–25.

19. Dias M, Hadgraft J, Raghavan SL, Tetteh J. The effect of solvent on permeant diffusion through membranes studied using ATR-FTIR and chemometric data analysis. J Pharm Sci 2004; 93:186–196.

20. Metcalfe E, Tetteh J. Combustion toxicity and chemometrics. Abs Papers Am Chem Soc 2000; 220:125-PMSE Part 2.

21. www.infometrix.com.

22. www.DiKnow.com.

23. www.ngb.netzsch.com.

24. Brown ME, Maciejewski M, Vyazovkin S, et al. Computational aspects of kinetic analysis: Part A: The ICTAC kinetics project-data, methods and results. Thermochimica Acta 2000; 355:125–143.

25. Maciejewski M. Computational aspects of kinetic analysis: Part B: The ICTAC kinetics project—the decomposition kinetics of calcium carbonate revisited, or some tips on survival in the kinetic minefield. Thermochimica Acta 2000; 355:145–154.
26. Vyazovkin S. Computational aspects of kinetic analysis: Part C. The ICTAC kinetics project—the light at the end of the tunnel? Thermochimica Acta 2000; 355:155–163.
27. Burnham AK. Computational aspects of kinetic analysis: Part D: The ICTAC kinetics project—multi-thermal history model-fitting methods and their relation to Iso-conversional methods. Thermochimica Acta 2000; 355:165–170.
28. Malinowski E. Factor Analysis in Chemistry. 2nd ed. New York: John Wiley and Sons, Inc., 1991.
29. Tetteh J. Enhanced target factor analysis and radial basis function neural networks. Ph.D. dissertation, University of Greenwich, UK, 1997.
30. Johnson RM. On a theorem stated by Eckart and Young. Pschometrika 1963; 28:259–263.

Role of Calorimetry in Preformulation Studies

INTRODUCTION

There are four phases through which all pharmaceuticals pass before their release onto the market. Although the phases are not rigidly defined and activities in some phases overlap with others, broadly speaking, they can be described as lead compound identification and optimization, formulation, process optimization and scale up, and regulatory review/market release. Typically, several candidate compounds result from the lead identification step. Once these active pharmaceutical ingredients (APIs) have been identified, they undergo a battery of tests to identify their fundamental physicochemical properties. Based upon the results of these tests, the number of APIs is whittled down to two or three candidates before a final selection process. Collectively, these tests are usually known as preformulation (1). The preformulation stage is therefore a very complex, time consuming but vital step in the process of getting a new product to market.

Preformulation is a dynamic phase in the new product process, involving a raft of analytical techniques. It is difficult, therefore, to define absolutely the preformulation process but Akers (2) provides a general definition:

> Preformulation testing encompasses all studies on a new drug compound in order to produce useful information for subsequent formulation of a stable and biopharmaceutically suitable drug form.

The analytical techniques available during preformulation are many and usually incorporate a spectroscopic finish. High-performance liquid chromatography is a good example. As has been discussed in Chapter 1, these techniques, although valuable, do suffer some distinct disadvantages. Isothermal calorimetric methods, in some cases, may offer a viable alternative to more traditional

Table 1 List of Some of the Important Topics in Preformulation Studies and the Calorimetric Techniques Used to Measure Them

Property	Technique
Melting point	DSC
Purity	DSC/IC
Presence of polymorphs	DSC/IC
API/excipient compatibility	IC/DSC
Solubility	DSC/IC
Glass transition temperature	DSC
Dissolution characteristics	IC
Hygroscopicity	IC

Abbreviations: API, active pharmaceutical ingredient; DSC, differential scanning calorimetry; IC, isothermal calorimetry.

methods for the quantification of some physicochemical parameters (3) [clearly, some cannot be investigated e.g., powder flow properties (if the drug is in the solid phase), compression properties, and so on.]. Data obtained from isothermal calorimetry (IC) do suffer the major limitation that they are not currently accepted by pharmaceutical regulatory authorities. However, they can be used as supporting evidence for data obtained from accepted techniques and can lend considerable strength to a new proposal. Differential scanning calorimetric data are accepted in some instances and the technique is widely used in the pre-formulation stage (4–8). IC is not as extensively used in the preformulation stage of product development, in part, because of the reluctance of regulatory authorities to accept isothermal calorimetric data in new drug proposals and partly because of the perception that a trained operator is required for experimental set up and data analysis. It is, however, as will be seen in Chapter 7, becoming more popular for the assessment of stability/compatibility (9–12) and for quantification of small amounts of amorphous material (13–15). This chapter will highlight those phys-icochemical parameters that are amenable to study by calorimetry [both differen-tial scanning calorimetry (DSC) and isothermal techniques] and how these parameters can be obtained from the returned calorimetric data. Where appropri-ate, examples from the literature will be given to emphasize the methodology.

Some of the physicochemical properties investigated in the preformulation stage and the calorimetric methods used for their detection are listed in Table 1.

USE OF DIFFERENTIAL SCANNING CALORIMETRY IN PREFORMULATION

DSC is perhaps the mostly widely used of all calorimetric techniques in the pre-formulation stage of product development. While the scope of this book is not intended to cover the use of DSC and its application, its use in the preformulation stage, is so widespread that a discussion of its capabilities is appropriate, if simply to provide comparison with isothermal techniques. DSC is used to characterize a variety of physicochemical parameters, knowledge of which is vital for the successful development of any pharmaceutical product.

Melting-Point Determination

Perhaps, the most common parameter obtained from DSC is the melting point of a compound, usually defined as the intersection between two tangential lines drawn through the premelting baseline and the melting endotherm. Determination of the melting point of a compound is of obvious importance as, for example, it allows rough identification of a compound's identity and it can be used to distinguish between isomers and salt forms. DSC is ideally suited to the measurement of the melting point as it allows a direct observation via the enthalpy of fusion of the substance. It confers the further advantage that it requires only small quantities of material (typically, a few milligrams), a particular boon when the material is scarce or expensive to produce.

Quantifying the Melting Point

Because of the nature of the measurement, there will always be some broadening of the peak associated with any phase change (for a perfectly pure material at infinitely slow scan rate, the peak associated with a phase change is essentially infinitely narrow). As a consequence, there are a number of ways in which the melting temperature can be reported. The most common method is to quote the linear onset temperature (Fig. 1) defined, as noted above, as the intersection between two tangential lines drawn through the premelting baseline and the melting endotherm.

Alternatively, the temperature at maximum peak height can be taken as the measure of the change (Fig. 2). This value is susceptible to greater variation from the effects of scan rate and from asymmetric peaks, and hence is not often used.

The final method for determining the temperature of a phase change is the temperature at half peak area (Fig. 3). This is often time consuming (especially, if the peak is not perfectly symmetrical), but is perhaps the most accurate method as it represents the temperature at which the process is 50% complete.

Purity Determination

In the early stages of any drug development, before synthesis routes are optimized, it is highly likely that the material obtained from the discovery laboratory will contain impurities at various levels. For example, it is possible that residual solvent exists or unreacted material remains. If a compound is observed to melt without degradation, then DSC analysis can be used to determine the absolute purity of that compound. DSC has been used since 1960s (16–20) to determine the purity of organic materials and its use is widely accepted within the pharmaceutical arena (21).

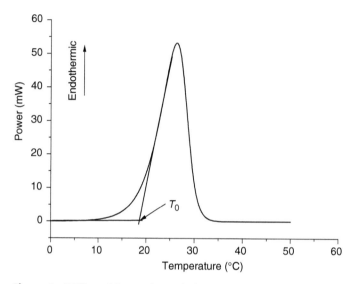

Figure 1 Differential scanning calorimetry response for a melting endotherm and the derivation of T_0.

Theory

The premise upon which the purity of a material can be quantified from DSC is the fact that even miniscule amounts of impurity in a material will affect the colligative properties of that material (22). Since the melting point of a material is

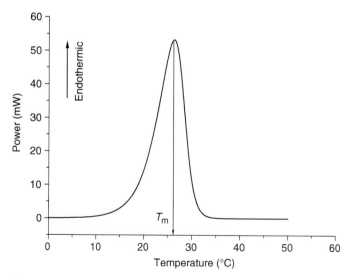

Figure 2 Differential scanning calorimetry response for a melting endotherm and the derivation of T_m.

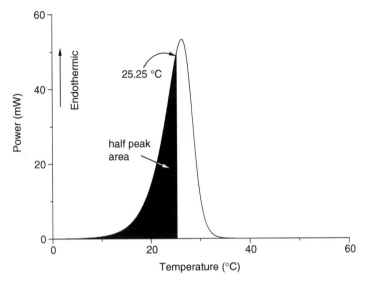

Figure 3 Differential scanning calorimetry response for a melting endotherm, showing the derivation of the melting temperature by the temperature at half peak area method.

one particular colligative property, the impurity will suppress the melting point (it is for this reason that salt is spread on the roads during cold weather, as the added impurity lowers the freezing point of water to a temperature below 0°C and thus helps to prevent ice from forming). The presence of an impurity will also result in the broadening of its melting range. Figure 4 demonstrates the effect of small amounts of impurity on the melting peak of a hypothetical organic compound. It is clear that as the quantity of impurity increases, the melting endotherm broadens and the peak lowers. Note that even relatively small levels of impurity result in significant changes to the shape and magnitude of the melting endotherm.

There exist two possible approaches for the analysis of the returned DSC data. The first is a qualitative approach. The peak shapes and melting points of unknown materials can be compared with known standards. This qualitative view is useful if the only information required is that the sample has a predefined minimum purity. If a quantitative analysis of purity is required, then the percent purity can be calculated using an approach, first described by Gray (17) and reviewed by Brennan (18) based on the van't Hoff isochore:

$$\frac{dq}{dT_s} = \frac{\Delta q(T_0 - T_m)}{(T_0 - T_s)^2} \tag{1}$$

$$(T_0 - T_m) = \frac{RT_0^2 X_2}{\Delta_f H} \tag{2}$$

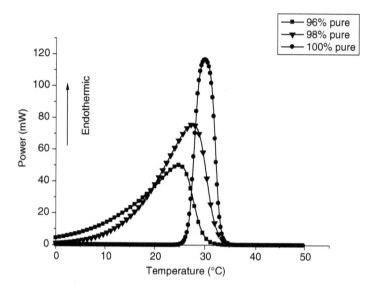

Figure 4 Schematic representation of the effect of impurity on the melting endotherm of a substance.

where dq/dT_s is a term describing the heat capacity of the sample (J K^{-1}), Δq the heat change associated with fusion (J), T_0 the melting temperature of a pure material (K), T_m the melting point of the impure material (K), T_s the sample temperature (K), X_2 the mol fraction of the impurity, $\Delta_f H$ the enthalpy of fusion (J mol^{-1}), and R the molar gas constant (8.314 J mol^{-1} K^{-1}). Integration of Equation (1) yields and the expression for the fraction of material melted, Equation (3):

$$F = \frac{T_0 - T_m}{T_0 - T_s} \tag{3}$$

where F is the fraction of material melted.

Rearranging Equation (3) to make T_s the subject and by substituting Equation (2) into this new expression yields Equation (4):

$$T_0 = T_s + \frac{RT_0^2 X_2}{\Delta_f H}\frac{1}{F} \tag{4}$$

Inspection of Equation (4) reveals that it is a linear function and hence a plot of T_0 versus $1/F$ should yield a straight-line plot of slope $RT_0^2 X_2/\Delta H_f$. The value of F, at a given T_s, is directly proportional to the area under the DSC curve up to that sample temperature and can be determined directly from the calorimetric data. Thus, the mol fraction impurity is easily determined.

Practical Considerations

There are several factors to consider when preparing samples for DSC purity analysis (18,23).

Temperature scan rate: The van't Hoff isochore assumes that equilibrium conditions exist during the melting of the sample. Therefore, the scan rate must be sufficiently slow to ensure that this equilibrium exists. Typically, scan rates of 2.5 K min^{-1} or lower are used.

Sample size: Careful choice of sample size is essential to ensure good thermal contact and that no thermal gradients exist. Normally, quantities between 1 mg and 5 mg are used.

Sample containment: As a solid material is heated, it can produce a significant vapor pressure at, or close to, its melting temperature. It is therefore necessary to seal the sample hermetically inside the DSC pan in order to minimize any erroneous powers associated with this phenomenon.

Thermal resistance: There is often a small temperature difference between the sample temperature and the temperature of the DSC, resulting from the thermal resistance of the sample, which is proportional to the thermal resistance constant R_0. This then causes a temperature lag while scanning, which must be corrected for. Methods for determining this constant and how it is used for temperature correction can be found in Gray (24).

Heat capacity contributions: The determination of the partial heats of fusion at different sample temperatures must incorporate a consideration of the heat capacity contributions from the sample and the pan if precise results are to be obtained. Again, this is dealt with in detail by Gray (24).

Materials amenable to study by differential scanning calorimetry: It has been estimated that the purity of approximately 75% of all organic compounds could be analyzed by the DSC method, if sufficiently pure (25). The DSC method does require that any impurity be insoluble in the solid phase but be soluble in the liquid phase. Consequently, materials that form solid solutions are not suitable for study using this technique.

Determination of Polymorphs

In principle, all organic molecules can exist in more than one distinct crystal form. This tendency to interconvert between different crystalline forms is known as polymorphism and different polymorphs may exhibit radically different properties. This applies not only to their mechanical properties (friability, density, and so on) but also to their physicochemical properties such that one polymorph may be readily bio-available, whereas another may be essentially useless. A discussion of polymorphism and its impact on the pharmaceutical process can be found in the excellent review article by Giron (26). It is clear therefore that an essential part of any preformulation study is to determine the polymorphic behavior of a candidate compound, and to assess the impact of the presence of any such polymorphs.

Different polymorphs of a compound often exhibit different melting temperatures, heat capacity, density, crystal hardness, dissolution rates, as well as different physical and chemical stabilities (27–30). For any given compound, only one polymorph is thermodynamically stable. All other polymorphs must, therefore, be unstable and will eventually convert to the stable form. A startling example of polymorphism is that of carbon. Carbon can exist as graphite (its equilibrium form) but it has a polymorph, diamond. The properties of diamond are strikingly different from those of graphite even at the level of physical appearance. On a more practical level, diamond is one of the known hardest natural materials (it is routinely used as a cutting material), whereas graphite is relatively soft (it can be used as a lubricant). These properties relate directly to the molecular arrangement (or the polymorphic character). Diamond is formed from a tetragonal arrangement of atoms, which gives it its great strength, whereas graphite is composed of layers of hexagonally arranged atoms with relatively weak interplanar bonds. These bonds are easily broken, allowing the planes of atoms to slide over one another, giving graphite its lubricating properties. Diamond is the metastable form of graphite and hence is not at thermodynamic equilibrium. It is an irrefutable law of thermodynamics that if a change can occur, it will occur. Because the Gibbs energy for the conversion of diamond to graphite is a negative value, the change occurs spontaneously, although the rate of change is vanishingly small. Nevertheless, diamond will, given enough time, eventually revert to graphite. De Beers is wrong: a diamond is not forever!

As for the physicochemical properties discussed earlier, DSC remains the most common calorimetric method for the investigation of polymorphs of a compound. It is, however, primarily used as a qualitative rather than quantitative tool.

Many examples exist in the literature of studies, using DSC as a method for predicting the presence of polymorphs of a compound (31–36). A good example is that of carbamazepine, a drug used in the treatment of epilepsy. It has been shown successfully (31) using DSC that three distinct polymorphs of carbamazepine exist. These findings were complemented by results obtained from Fourier-Transform Infra-red Spectroscopy (FT-IR) and X-ray Powder Diffraction (XRPD) experiments (31).

Although DSC has proved its worth in identifying the presence of multiple polymorphs, it is not always possible thermally to characterize the lower melting polymorphs of a compound. At the heating rates employed in standard DSC experiments there are often multiple thermal events, which are associated with the concurrent recrystallization to a new form and the subsequent melting of that new form. This behavior at such relatively low scan rates impacts on the capacity to determine the polymorphic purity of a given sample. Moreover, the presence of these multiple thermal events makes the quantification of thermodynamic parameters, such as the enthalpy of fusion, impossible to measure. Returning to the carbamazepine example, in the initial study, quantification of

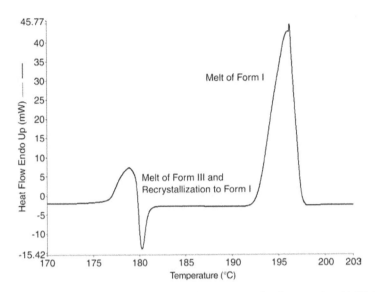

Figure 5 Differential scanning calorimetry scan of carbamazepine, highlighting the concurrent melt and recrystallization of Form III to Form I. *Source*: From Ref. 37.

the enthalpy of fusion of Form III was not possible because of a concurrent recrystallization of Form III to the higher melting Form I under slow heating rates $(5–10°C \text{ min}^{-1})$ (Fig. 5).

Recent advances in technology have permitted the construction of fast-scan DSC instruments capable of scan rates of up to $500°C \text{ min}^{-1}$. At higher scan rates, the kinetic nature of the melting transition is altered such that the recrystallization to the higher melting form does not occur from the melt and thus a quantitative measure can be made of the enthalpy of fusion for the lower melting form. Fast-scan DSC methods have been used to study carbamazepine (37). It was shown that, as the scan rate was increased, the recrystallization to Form I was inhibited (Fig. 6A) until it was completely retarded at $500°C \text{ min}^{-1}$ (Fig. 6B).

With the inhibition of the recrystallization of Form I, it was possible to measure directly the enthalpy of fusion from the calorimetric data, 109.5 J g^{-1}, a value that can only be estimated from normal DSC experiments.

Measurement of Polymorphic Composition

McGregor et al. (37) also describe a method by which application of high scan rates permits the determination of polymorphic composition. Here, binary mixtures of carbamazepine Form III and Form I were prepared in various ratios between 95:5 and 1:99 w/w Form III:Form I, in order to determine the limit of detection of Form III in a mixture. They were able to detect levels of 1% w/w of Form III present in a mixture with Form I (Fig. 7). They note, however, that crystal seeding of Form III by Form I in the mixture results in a partial

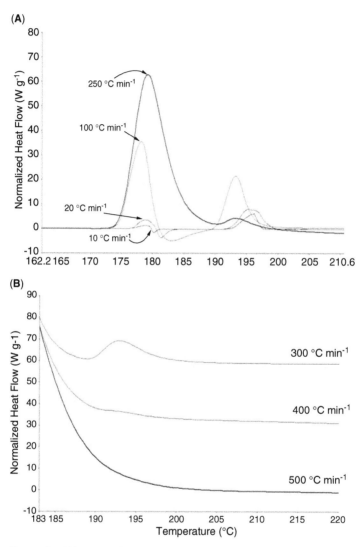

Figure 6 (**A**) Effect of scan rate on the differential scanning calorimetry (DSC) trace for carbamezipine showing that, as scan rate increases, the concurrent melt and recrystallization of Form III is retarded. (**B**) Effect of scan rate on the DSC trace for carbamezipine showing that, as scan rate reaches $500°C$ min^{-1}, the concurrent melts and recrystallization of Form III is completely inhibited. *Source*: From Ref. 37.

recrystallization of Form III, preventing accurate quantitative determination of the quantity of low levels of Form III.

Although it is not possible to quantify accurately the polymorphic purity of a system, the ability to discriminate for the presence of 1% w/w polymorphic

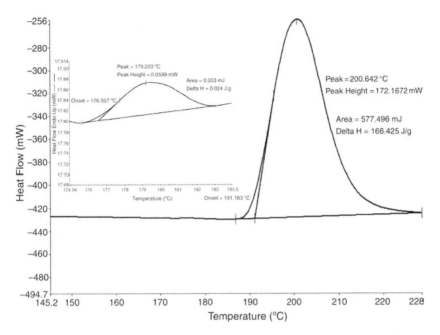

Figure 7 Differential scanning calorimetry trace for the detection of 1% w/w poly-morphic content of Carbamezipine. *Source*: From Ref. 37.

content is extremely desirable. Many techniques have a limit of detection of approximately 5% w/w, which often is outside the desired detection limit for a pharmaceutical compound. The rapidity of the measurement also offers the significant advantage of time and cost, particularly, as it allows a rapid screen of a product without the requirement for lengthy and expensive analysis to determine its polymorphic composition.

Stability/Compatibility

All pharmaceutical products are composed of a multitude of components [such as binders, disintegrants, preservatives as well as the APIs], all of which are slowly changing (chemically and/or physically) and potentially interacting in some way. Some interactions may be detrimental, such as accelerating the rate of degra-dation of the active, binding to the active, destabilization of the formulation, and so on by the excipient(s) and some beneficial, such as stabilization of the formulation or the synergistic antimicrobial action of combined parabens. Regardless of their potential advantages or disadvantages, these interactions must, at the very least, be identified; that is, preliminary screens to identify any stability issues and/or potential interactions, and more often than not a quantitat-ive assay, are required if the behavior of the product is to be predicted accurately. Such data help to ensure that the performance of a product will not fall outside its pharmacopoeial specifications within the manufacturer's specified use-by date.

Stability and compatibility testing are generally recognized as essential steps in the preformulation process (38,39), but as they are not always successful, some researchers question the validity of such studies (40). In addition, there are associated problems with, for example, the availability of raw material with which to assess stability and compatibility. A list of some of the factors that may affect stability/compatibility is given in Table 2 (41).

The most common method for assessing stability/compatibility is to leave the formulation(s) on long-term (months to years) storage under specified conditions and then analyzing for any physical or chemical change. Because most formulations are designed to be stable, it is often necessary to stress the system in order to maximize the potential of observing an interaction. There are obvious problems with this conventional approach to stability/compatibility testing. There is the clear disadvantage of having to leave samples on long-term storage and the associated cost and there is also the drawback that, by stressing the samples, they may be induced to change or interact in ways that are not likely to be observed in normal use.

Many studies utilizing DSC for the prediction of compatibility have been reported (42–48). The original system studied using this methodology was the known incompatibility between aspirin and magnesium stearate (49). There are, however, two principal objections to using DSC as a tool for the assessment of compatibility. The first is that any differences in the DSC scans of binary mixtures compared with individual components may arise for reasons other than incompatibility between the materials. While it is proper to assume that any change in the DSC curve may not be caused by an incompatibility, it is accepted that any change in the curve must be caused by an interaction between materials. Such changes in DSC curve have been demonstrated for drug in polyethylene glycol (PEG) mixtures (50). It is shown that the melting behavior is altered by a slow dissolution of the drug into the molten PEG phase and not an

Table 2   Some Factors Which May Affect Stability/Compatibility

Drug properties	Excipient properties	Formulation properties	Environmental properties
Moisture content	Moisture content	Drug:excipient ratio	Temperature
Chemical structure	Chemical structure	Granulation method	Relative humidity
Impurity profile	Impurity profile	Mixing/milling	Packaging
Physical form	Physical form	Powder packing	Light
Morphology	Morphology	—	—
Particle size/ surface area	Particle size/surface area	—	—

incompatibility between the drug and PEG. The second objection to the use of DSC is that, in the original aspirin/magnesium stearate study, the incompatibility was intrinsically linked to the melting behavior and hence this method may not be applicable to all systems (49).

The determination of compatibility is better suited to IC and will be dealt with briefly later in this chapter and in detail in Chapter 7.

Determination of Solubility for Ideal Solutions

In order for a drug to become bioavailable and to exert its intended therapeutic effect, it must be soluble in a fluid medium. It has been stated that if the drug compound does not possess a solubility value in excess of approximately $10 \, mg \, mL^{-1}$, then problems are likely to be encountered during absorption (51). If a compound has a solubility value of less than $1 \, mg \, mL^{-1}$, then the drug is usually derivatized to a salt form in order to increase its intrinsic solubility. Evaluation of the solubility of a drug therefore is vital to the drug development process. It is usually determined very early in the preformulation stage as it can impact severely on the formulation development stage and perhaps more importantly on the choice of dosage form.

Solubility data are required for efficient design of processes such as crystallization-based separations and chromatographic resolutions (43). In addition to ensuring efficient processing, knowledge of the solubility of a compound is important if a system has the potential to form a polymorph or if purity is an issue.

Solubility can be estimated using a calculation-based approach (52), but the success of such an approach is limited by the availability of the necessary thermodynamic data for the pure material and by the lack of knowledge of activity coefficient values.

The preferred method (and usually the more accurate method) is to quantify the solubility parameters by experimental measurement. The usual approach for the quantification of solubility is to measure the equilibrium concentration of a drug within a particular solvent (the shake-flask method). The concentration of drug dissolved is quantified using spectroscopic analyses. Although widely used, this approach suffers some drawbacks. Firstly, it requires that the drug has a chromophore. Secondly, that sufficient quantity of drug can dissolve in order to be detected. Thirdly, that there exists sufficient quantity of material in order to construct a calibration curve. As solubility is an equilibrium process, it is also required that the rate of dissolution is sufficiently high in order for sufficient material to dissolve in a reasonable time frame. Alternatives are available; for example, if no chromophore exists, then a dry weight determination can be made, in which a dry residue from a known volume of saturated solution is weighed. This approach is also not without its problems. For example, it requires that the residue is completely dry and that it does not absorb water during the measurement.

Because of the limitations described earlier, there are always new methods being sought for the measurement of solubility in order to reduce the time frame or increase the accuracy of the measurement. Calorimetric determination of solubility sidesteps many of these problems and provides a viable alternative to traditional techniques. As a dissolution event is a physical process, it means therefore that solubility, in principle, is accessible by calorimetry. Calorimetric methods do not require a spectroscopic finish; they are highly sensitive (modern calorimeters only require microgram quantities for successful analysis) and, providing that the drug completely dissolves in the solvent, they do not necessarily require a specific calibration curve to determine the solubility and hence large quantities of material are not required for the analysis.

DSC offers several advantages over conventional methods for solubility determination. These advantages include low detection limits (i.e., low sample mass requirements), controlled heating rates, and rapidity of measurement. These advantages have been recognized for some time (53–55) and DSC is now widely used as a tool for predicting solubility (56).

The ideal solubility, x, of a solute (i.e., the solubility of a solute that forms an ideal solution) in any solvent can be determined readily from knowledge of the enthalpy of fusion, $\Delta_f H$ of the solid; the melting temperature, T_f of the solid; and the universal gas constant, R, Equation (5). Both terms are readily accessible from DSC experimental data:

$$\ln x = \frac{\Delta_f H}{R}\left(\frac{1}{T_f} - \frac{1}{T}\right) \tag{5}$$

This is of limited use in that it is often the case that systems deviate significantly from ideality and hence alternative methods are required with which to calculate the true solubility in nonideal solutions.

Determination of Solubility for Nonideal Solutions

Dissolution of a solid is always accompanied by a change in the heat content of the system (usually, endothermic). This endothermic event can be used to determine the equilibrium solubility temperature by measuring the calorimetric response as the solid dissolves in the solvent. The general methodology is to heat the sample to a temperature at which all solid is dissolved and then cooled to a predefined temperature [e.g., 298 K (standard reporting temperature) or 310 K (biological reporting temperature)] and held for a period of time to ensure that the solute crystallizes (55). The sample is then heated at a variety of heating rates again to a predefined temperature ensuring that all solid dissolves. The peak completion temperature, T_c (the temperature at which the heat of dissolution peak returns to baseline) is then determined from the heating scan by extrapolation of the linear portion of the descending section of the peak to the baseline (Fig. 8).

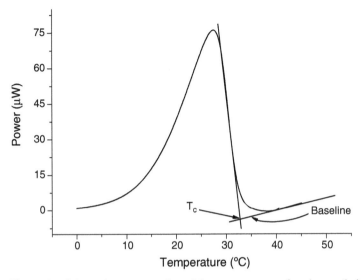

Figure 8 Schematic representation of the measurement of peak completion temperature for the analysis of solubility by differential scanning calorimetry.

The values of T_c are then plotted as a function of heating rate and a value for T_E (equilibrium solubility temperature) is obtained by extrapolation to zero heating rate (Fig. 9). If the experiment is then performed as a function of mole fraction, a solubility curve can be constructed (Fig. 10), and the equilibrium solubility at any temperature can be predicted. This methodology has been extended by Sapoundjiev et al. (43) to allow the determination of a solubility curve for a wide range of temperatures from a single calorimetric experiment. It is noted that this single calorimetric experiment method is not as accurate as traditional techniques for determining solubility. It is reported that this single experiment technique allows quantification of the equilibrium solubility to within 12%.

Limitations for the Measurement of Solubility

As with any analytical technique, there are limitations to the DSC method for determining solubility. Two key problems exist. The first is that if the crystallization kinetics are slow, it is difficult to determine whether the solution has attained equilibrium before the heating cycle commences and hence the accuracy of any measurement of solubility is limited. The second is associated with those systems where the increase in solubility with temperature is minimal. It must be reported (43) that a minimum increase of 3 wt% must be attained across a temperature range of 10 K for this method to be applicable. Further difficulties are encountered if the dissolution rate is slow.

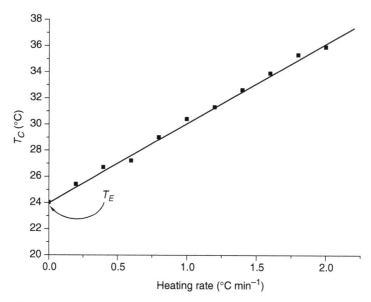

Figure 9   Schematic representation of the extrapolation for the equilibrium solubility temperature, T_E.

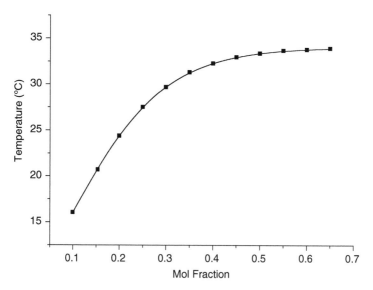

Figure 10   Hypothetical solubility curve derived from the measurement of equilibrium solubility temperature as a function of mol fraction.

Glass Transition Temperature

Many formulations utilize amorphous materials in order to improve dissolution characteristics and, consequently, bioavailability, but amorphicity is also introduced during processing (e.g., lyophilization, spray-drying, milling) and can be undesirable in terms of the formulation stability (57). (A detailed discussion of amorphicity and its quantification is deferred until Chap. 6.).

Vital to the understanding of how a system, which contains amorphous material, will behave is knowledge of the glass transition temperature, T_g. The glass transition temperature can be loosely defined as the point at which molecules within a glassy solid go from having little kinetic mobility (kinetically frozen) to having enough mobility to undergo a solid–solid transformation to become plastic; that is, at temperatures above the glass transition, an amorphous material can undergo spontaneous crystallization. This phenomenon can drastically affect the behavior and efficacy of a formulation. If the substance has a high T_g then crystallization is less likely to occur unless the T_g is lowered by a plasticizer such as a solvent or an impurity.

T_g is most commonly determined using DSC and is identified from a step change (Fig. 11) in the baseline of the calorimetric curve, which corresponds to a change in heat capacity between the amorphous and crystalline forms of the material under study.

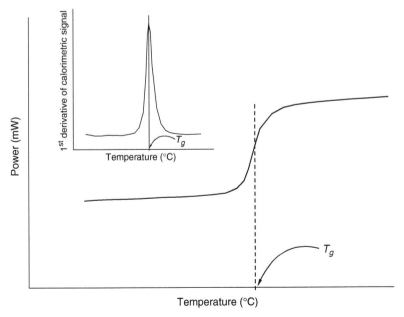

Figure 11 Schematic representation of the measurement of T_g directly from the calorimetric data (main figure) and by the first derivative of the signal (inset).

USE OF ISOTHERMAL CALORIMETRY IN PREFORMULATION

The exploitation of isothermal calorimetric methods is not as widespread in the preformulation stage as scanning calorimetry. IC is, however, beginning to find greater use as a preformulation tool as greater sensitivity and accuracy is sought in order to quantify better the physicochemical parameters important to preformulation scientists. It is, of course, unable to determine those parameters that require a change in temperature in order to facilitate the measurement (e.g., melting point or T_g). It will always be the case that no one technique will be suitable for all measurements. It is also important to remember that any successful characterization of a substance will require a range of techniques.

Purity Determination

Isothermal calorimetric measurements, in certain instances, can be used to estimate the level of impurity in a sample. The treatment requires that the sample undergoes some measurable chemical change and that there is no other contribution to the calorimetric signal other than that of the chemical reaction process.

The calculation to establish the quantity of impurity present is predicated upon the fact that a value for the reaction enthalpy of the chemical change for the pure material is known and that the sample quantity can be accurately known. The purity of the material can be calculated from the observed calorimetric data and knowledge of the reaction enthalpy for the pure material. Chapter 3 described a method for the calculation of Q, the area under the power–time curve when all reactant is consumed. It was also noted that Q is equal to the product of the quantity of reactant and the enthalpy for reaction, Equation (3). Therefore, if the derived value for Q, obtained from the suspected impure material, is divided by the value for Q obtained from the pure material, then that ratio will reflect the percentage purity of the suspect material. This method has used to estimate the purity of triacetin (a compound used for validation of isothermal calorimeters) by O'Neill (58). The presence of an impurity was first suspected through a poor correlation with the accepted reaction enthalpy for the hydrolysis reaction (see Chap. 2) and was confirmed from the results of a thin layer chromatography experiment. As the quantity of material and the reaction enthalpy for the pure material are known, this permits the estimation of the percentage impurity present. Using the methodology described in Chapter 2, the values for $Q_{suspect}$ and Q_{pure} were used to calculate the percent purity using Equation (6):

$$\frac{Q_{suspect}}{Q_{pure}} 100\% = \%\text{purity} \qquad (6)$$

The purity was calculated as being approximately 93%.

Limitations for the Assessment of Purity

Once any impurity has been quantified, it may become necessary to identify the impurity(ies) present (e.g., if the toxicity of any possible degradant is an issue). Calorimetric data do not yield any molecular information and, hence, neither method described earlier in this chapter will permit the identification of any impurity present in a sample. Hence it is always necessary to perform some ancillary measurement to identify the impurity. This is usually achieved using spectroscopic analytic methods. This topic is outside the remit of this text but an excellent overview of analytical methods can be found in (59).

Determination of Polymorphs

As was discussed in section "Determination of Polymorphs," DSC is perhaps the main thermal analysis method for the determination of polymorphs of a compound. The utilization of DSC, although very useful, can be problematic because data are required at many time points as fractions of conversion over the period of reaction. Providing the rate of conversion is reasonably fast [if the extent of conversion, α is >0.15 over the lifetime of the experiment (hours)], there is the possibility to study polymorphic behavior using isothermal calorimetric methods.

Seratrodast (an anti-asthmatic drug) has been studied via an isothermal technique (60). Here, the drug was studied under controlled atmospheres of defined relative humidity (RH) and allowed to interconvert over a period of approximately 20 hours. Elevated temperatures (50–65°C) were used to ensure that a sufficient fraction of the material had converted in order to perform the analysis. Not only was the enthalpy of the transition $(-5.70 \pm 1.13 \times 10^{-1}\, \text{kJ mol}^{-1})$ amenable to measurement, the quantification the kinetic nature of the transition was also accessible using the Hancock–Sharp equation (61), Equation (7):

$$\ln[\ln(1 - \alpha)] = \ln B + m \ln t \qquad (0.15 \le \alpha \le 0.5) \qquad (7)$$

where B is a constant.

The Hancock–Sharp method (61) consists of plotting $\ln[\ln(1 - \alpha)]$ versus $\ln t$ to obtain a series of straight lines, the slopes of which yield a parameter m. The value of this parameter then allows the prediction of a kinetic model that describes the mechanism of transformation from Form II to Form I. The theoretical equations of solid-state kinetic models and their corresponding values of m are given in Table 3.

From the values derived ($m = 2.00$, 2.09, and 2.1), the authors conclude that for each of the three sieved fractions, the mechanism of transformation of seratrodast is best described by the two dimensional growth of nuclei model. Using the Avrami–Erofeev equation, Equation (8), they show it is also possible to calculate the rate constant for the transformation for each sieve fraction and it

Table 3 Various Kinetic Models for Some Mechanisms of Solid-State Reactions

Kinetic model	m	Mechanism
$\alpha^2 = kt$	0.62	One-dimensional diffusion
$(1 - \alpha)\ln(1 - \alpha) + \alpha = kt$	0.57	Two-dimensional diffusion
$[1 - (1 - \alpha)^{1/3}]^2 = kt$	0.54	Three-dimensional diffusion (Jander equation)
$1 - 2\alpha/3 - (1 - \alpha)^{2/3} = kt$	0.57	Three-dimensional diffusion (Ginstling–Brounshtein equation)
$-\ln(1 - \alpha) = kt$	1.00	Random nucleation, one nucleus on each particle
$-\ln(1 - \alpha)^{1/2} = kt$	1.11	Phase boundary reaction, cylindrical symmetry
$-\ln(1 - \alpha)^{1/3} = kt$	1.07	Phase boundary reaction, spherical symmetry
$\alpha = kt$	1.24	Zero-order mechanism (Polany–Winger equation)
$[-\ln(1 - \alpha)]^{1/2} = kt$	2.00	Random nucleation, two-dimensional growth (Avrami–Erofeev equation)
$[-\ln(1 - \alpha)]^{1/3} = kt$	3.00	Random nucleation, three-dimensional growth (Avrami–Erofeev equation)

has been shown eloquently that the rate constant increases as particle size decreases.

$$[-\ln(1 - \alpha)]^{1/2} = kt \tag{8}$$

As this study was conducted as a function of temperature, it was also possible to calculate an activation enthalpy, thus permitting the prediction of a shelf life (based upon 90% of Form II remaining, t_{90}) for a variety of conditions and storage temperatures. For example, they predicted that t_{90} for the transition of Form II to Form III was approximately 3900 days at 63% RH and 298 K, and 17.5 days at 63% RH and 313 K.

Determination of Solubility

The quantification of solubility is not easily achieved by isothermal calorimetric means. Indeed, only one technique is described in the literature for an isothermal calorimetric determination of solubility (62). The basis of their measurement is the determination of the heat generated by the addition of a pure solvent (in this instance water) to a saturated solution containing excess solid until a homogeneous solution is formed. In their treatment, they define the dissolution process for a pure phase going to an infinitely dilute solution as consisting of two steps. The first step is transfer of a molecule from the pure phase to a saturated solution. The second step is then the transfer of a molecule from the saturated solution to an infinitely dilute solution. Knowledge of the heat of addition and the quantities of the solute and solvent permit the calculation of not only the saturated solubility but also the dissolution enthalpy and the enthalpic interaction coefficient of the excess enthalpy. For the purposes of this text, only the calculation of the solubility and enthalpy of solution will be discussed.

The experimental setup is essentially that of a titration experiment. The calorimetric chamber is charged with 0.5 to 1.0 mL of the saturated solvent along with some excess solid and allowed to reach equilibrium. To this biphasic system injections of pure solvent (15–50 μL depending on the heat generated) are added until a homogenous solution is formed (i.e., all the solid is dissolved). Further additions of pure solvent are then performed in order to measure the heat of dilution of that homogeneous solution.

For the constant addition of pure solvent to the biphasic system, a constant heat response is generated, corresponding to the heat produced by the dissolution of a fraction of the excess solid up until the ith injection. At the $(i + 1)$th injection, only a portion of the pure solvent goes to dissolving the excess solid, the remaining solvent then dilutes the resulting saturated solution. The next injection $(i + 2)$ onwards results only in heats of dilution and can be described by Equation (9):

$$Q_{dil(i+1)} = [a + b(i + n)^x + c(i + n)^y + d(1 + n)^z + \cdots] \tag{9}$$

where $n \geq 2$, x, y, z are whole or fractional numbers and a, b, c, d, and so on are merely fitting parameters.

The solubility of the compound can then be found from the average value of the heat associated with dissolution, Q_{av}, when k kg of solvent per addition is added to the biphasic solution and the heat associated with forming the monophasic system, Q_s.

In the experiment described above, up to the ith addition, all the added solvent is required for the dissolution of the excess solid. At the $(i + 1)$th addition, only a fraction, f, of the solvent is required. The $(1 - f)$th fraction of the solvent dilutes the monophasic, saturated solution and therefore the value of Q_s can be given by:

$$Q_s = fQ_{av} \tag{10}$$

where f is a value between 0 and 1.

The total heat in the $(i + 1)$th step, Q_{i+1}, is equal to the sum of Q_s and $Q_{dil(i+1)}$ and can be described by Equation (11).

$$Q_{i+1} = fQ_{av} + [a + b(i + n)^x + c(i + n)^y + d(1 + n)^z + \cdots] \tag{11}$$

The fraction, f, can be determined by least squares minimization. If f is known, the saturated solubility (mol kg^{-1}) can be found from Equation (12):

$$m_{sat} = \frac{n_T}{[k_c + (i + f)k]} \tag{12}$$

where k_c is the number of kilograms of solvent present initially.

Using this method, Castronuovo et al. (62) calculated values for the saturated solubility of potassium chloride (4.839 ± 0.02 mol kg^{-1}) as well as a number of cyclic dipeptides: (glycyl–glycyl)-diketopiperazine (0.157 ± 0.003 mol kg^{-1}) (alanyl–alanyl)-diketopiperazine (0.191 ± 0.005 mol kg^{-1}), and (leucyl–glycyl)-diketopiperazine (0.056 ± 0.005 mol kg^{-1}). It is reported that these values compare well with published values [see references within (62)]. This method seems therefore to be robust and reliable for the determination of solubility from isothermal calorimetry. The method described by Castronuovo et al. (62) for the determination of solubility was also extended to a calculation for the enthalpy of solution. The standard enthalpy of solution can be determined from Equation (13):

$$\Delta H = \frac{(1+f)Q_{av}}{n_T - m_{sat}k_c} \tag{13}$$

The limiting enthalpy of solution ΔH^0 can be then calculated from Equation (14):

$$\Delta H^0 = \left[\frac{(i+f)Q_{av}}{n_T - m_{sat}k_c}\right]\left[A_1\, m_{sat} + A_2\, m_{sat}^2 + \cdots\right] \tag{14}$$

where A_1, A_2, and so on are polynomial coefficients.

Determination of Enthalpy of Solution by Isothermal Calorimetry

An added advantage of IC, over other techniques, is the capacity to derive the enthalpy of solution of a given compound. The enthalpy of solution is important for the preformulation scientist as it permits, among other factors, characterization of the compound (63,64) determination of the extent of crystallinity (65), and interactions between drug and carrier (66,67).

In general, the enthalpy of solution obtained from solubility data is gained from a van't Hoff plot of log solubility versus reciprocal temperature (68). This methodology suffers from the limitation that the van't Hoff plot is not linear over wide ranges of temperature. It is possible to sidestep this problem by determining directly the enthalpy of solution by calorimetric measurement.

Determination of Enthalpy of Solution by Solution Calorimetry

The methodology for measuring the enthalpy of solution using a solution calorimeter is essentially the same regardless of the measurement principle: a known quantity of material is dissolved in a known volume of solvent and any associated change in the system (temperature change or power) is measured. It is this direct measure of change in the heat content of the system that allows the quantification of the enthalpy of solution (see Chap. 1 for more details). There exist a number of different types of solution calorimeter, all of which have been used to determine the enthalpy of solution of a wide variety of materials (69–76). Depending upon the instrument employed, the sample quantity can vary from a few milligrams up to 200 mg, and the solvent volume can be as little as a 10 mL to as much as 120 mL.

A recent example of the application of solution calorimetry includes the investigation of materials for calibration/validation, which concluded that the dissolution of KCl into water was the most robust test routine (70). Ramos et al. (69) re-examined these materials, using a newly designed solution insert (Thermometric 2265) for the thermometric TAM (Thermal Activity Monitor, Thermometric AB). Their conclusion was that the best calibration material depended upon the type of instrument used. For example, they argue that for the new insert (based upon a heat conduction principle), the best reference material was sucrose and not KCl. The topic of reference materials has attracted considerable interest in recent years and is dealt with in detail in Chapter 2.

Solution calorimeters have also been used to determine the enthalpy of solution of a number of pharmaceuticals and to investigate whether interactions exist between materials in binary mixtures. Jain et al. (74) used solution calorimetry to probe the possible interactions between ampicillin and amoxicillin, a combination known to act synergistically. Traditional spectroscopic analysis is not possible because of the proximity of their λ_{max} values. Using the Setaram C80 solution calorimeter, they were able to determine the enthalpies of solution at a variety of pH as well as the excess molar enthalpies of solution of the binary mixture. From the low values of the excess molar enthalpies, they concluded that the two antibiotics do not interact with each other to any great extent.

A similar study (72) reports the measurement of the enthalpy of solution of diclofenac sodium and paracetamol individually and the excess molar enthalpy of binary mixtures in order to determine whether there was any interaction when in solution. It is reported that the enthalpies of solution of diclofenac sodium and paracetamol, at pH 7.0, are 50.24 ± 0.04 kJ mol^{-1} and 24.76 ± 0.04 kJ mol^{-1}, respectively. Again the values of the excess molar enthalpy are small and indicate that little or no interaction exists. These examples highlight some of the systems that have been studied using solution calorimeters. The scope of solution calorimetry is such that a comprehensive review is not possible within the scope of this text. However, the principles of solution calorimetry and its applications can be found in the review by Royall and Gaisford (77).

Determination of the Enthalpy of Solution Using Step-Isothermal Methods

With the advent of fast-scan DSC instruments, it is possible to effect reasonably large changes in temperature in relatively short periods of time (approximately 10°C/sec) (37). Such instruments permit, in principle, the calculation of enthalpies of solution by quasi-isothermal methods. Here, a known volume of a saturated solution of a compound is placed in a DSC cell along with a known excess of solid. The temperature is raised until the solid completely dissolves (evidenced by a peak in the DSC curve). The solution is then crash-cooled to the temperature of interest, 25°C for example, inducing the solute to crystallize and the area under the resulting peak is measured, Q_1. The experiment is then repeated but this time with a slightly higher mass of excess solid. Because there is now a larger mass of excess solid, the temperature increase required to dissolve all the excess material will be slightly higher than for the first experiment. It will also be the case that the area under the DSC curve associated with the crystallization will be slightly higher, Q_2. The difference, Q_{diff}, Equation (15), between the heats of crystallization can be used to calculate the enthalpy change associated with the crystallization of the surplus solid in the second experiment, Equation (16).

$$Q_{diff} = Q_2 - Q_1 \tag{15}$$

$$\Delta_{sol}H = \frac{-Q_{diff}\text{RMM}}{m_2 - m_1} \tag{16}$$

where RMM is the relative molecular mass and m_1 and m_2 are the masses of the excess solid in each experiment.

The limitation with this approach is the fact that the assumption must be made that the solute does not form a supersaturated solution, which persists beyond the lifetime of the measurement. It is also necessary that, because of volume limitations, the material under study has a suitably large enthalpy of solution to be within the sensitivity limits of the calorimeter.

Enthalpy of Solution by Flow Calorimetry

In some instances, the sample may be so sparingly soluble that the quantities required to ensure its complete dissolution (both solvent volume and sample mass), in a conventional solution calorimeter, may be impractical. One solution to this problem is to employ a modified flow calorimeter, which, because of its design, allows the sample to be dissolved in an essentially infinite volume of solvent. The use of flow calorimetry in this manner was first described by Gill and Seibold in 1976 (78). The instrument arrangement was such that the solid sample was contained within a glass capillary tube, which was then inserted into a specially designed holder within the Tian–Calvet calorimetric flow cell. The sample is separated from the solvent by an air gap until the instrument equilibrates. Once equilibrated, the solvent is then allowed to flow through the capillary, over the sample and then out to a holding vessel separate from the calorimeter. The corresponding power signal is then measured as a function of time. This signal is then integrated to yield the number of joules of heat produced in the dissolution process and, if an accurate value for the mass (and its RMM) of the sample used is known, then the enthalpy of solution can be determined.

Using this calorimeter, Gill and Seibold measured the heats of solution of a number of materials (78). They investigated sucrose and succinic acid as materials for chemical calibrants (see also Chap. 2). Sucrose returned a value with a large standard deviation ($6.2 \pm 0.3\ \text{kJ mol}^{-1}$); the large error was attributed to its low enthalpy of solution and hence was considered to be unsuitable. The alternative, succinic acid, was found to return values that agreed to $\pm 1\%$ of the known literature values at that time (28.6 kJ mol^{-1}). Within this study, the enthalpies of solution for various other compounds of biological interest were determined, including diketopiperazine (26.7 kJ mol^{-1}).

It should be noted that here the reported values for the enthalpy of solution correspond to that enthalpy change in forming the saturated solution and not the infinitely dilute solution. This is a limitation borne out by the experimental arrangement. For instruments of this type, the saturated solution is formed while the sample is dissolved and is then carried away by the influx of fresh solvent. Hence the enthalpy of dilution of that saturated solution is not measured. Note also that no consideration of the effect of flow rate was made. (Other than that a suitable flow rate was used in order to minimize the baseline noise level.) In the article by Gill and Seibold (78), the source of the diketopiperazine or its exact nature (i.e., gly–gly, ala–ala, and so on) are not stated. If it is assumed that the compound is in fact the gly–gly diketopiperazine, then the value for the enthalpy derived here is in good agreement with that of Castronuovo et al. (62) discussed earlier.

This instrument was the precursor to a more sophisticated instrument developed by Nilsson and Wadsö but was quoted as having some shortcomings and was never reported in detail (79).

This transitional instrument was replaced by one which overcame some of the shortcomings alluded to by Nilsson and Wadsö and was based upon the perfusion and titration insert for the Thermometric TAM (79). The experimental arrangement was essentially the same as that described for Gill and Seibold's calorimeter in that the solid sample (1–10 mg) was contained within a receptacle through which a solvent was passed, causing the solid to dissolve and allowing the concomitant heat flow to be recorded.

This new design had one key difference in that the sample was covered by a saturated solution in order to separate it from the pure solvent. This arrangement introduces a level of uncertainty into the measured enthalpy and requires that two correction terms be introduced when calculating the enthalpy of solution. The first is related to any heat change associated with the dissolution of sample in, or recrystallization of solute from, the saturated solution used to separate the sample from the pure solvent. This is particularly important if there is a difference between the temperature at which the saturated solvent was prepared and the measurement temperature. If the contact area between solid and saturated solution is kept constant for all experiments, this value can be considered as constant for each measurement temperature.

The second correction that may need to be applied is to account for any dilution of the saturated solution at the start of the experiment. The time-dependent enthalpy change of the process can therefore be described by Equation (17):

$$\Delta_t H = n\Delta_{sol}H - n'\Delta_{sol}H + n''\Delta_{dil}H \tag{17}$$

where $\Delta_t H$ is the time-dependent enthalpy change measured by the calorimeter (J); n, the number of moles of material loaded into the calorimeter (mol); n', the number of moles of material dissolved/crystallized in/from the saturated solution (mol); n, the number of moles of material in the saturated solution (mol); $\Delta_{sol}H$, the enthalpy of solution (J mol^{-1}), and $\Delta_{dil}H$, the enthalpy of dilution of the saturated solution (J mol^{-1}).

This then permits the calculation of the enthalpy of solution from a plot of $\Delta_t H$ versus n. The slope of the resulting straight line will be equal to $\Delta_{sol}H$ with the intercept being equal to the sum of the two correction parameters. The molality of the solution as it leaves the calorimeter cannot be readily defined. The molality is likely to be very much lower than the saturated solubility of the material and consequently for most purposes the value of $\Delta_{sol}H^{\infty}$ can be approximated to the experimentally determined $\Delta_{sol}H$. Nilsson and Wadsö (79) measured the enthalpies of solution in water, for a variety of temperatures (and 310.15 K) of adenine (32.37 $\pm$ 0.64 kJ mol^{-1} at 288.01 K, 32.68 $\pm$ 0.66 kJ mol^{-1} at 298.15 K, 33.00 $\pm$ 0.38 kJ mol^{-1} at 308.15 K, and 33.88 $\pm$ 0.54 kJ mol^{-1} at 318.14 K) and acetanilide (15.4 $\pm$ 0.31 kJ mol^{-1} at 288.01 K, 18.13 $\pm$ 0.20 kJ mol^{-1} at 298.15 K, and 21.45 $\pm$ 0.30 kJ mol^{-1} at 310.15 K). Where available, these values were reported to be in good agreement with one reported literature value; however, they report that the values returned from this study for acetanilide did not concur with the only other reported literature value. It is not clear why this is the case, but it does highlight the need for caution when using enthalpies of solution derived from only one source.

Stability/Compatibility

Earlier in this chapter, a DSC method was described for the prediction of potential incompatibility between pharmaceutical ingredients. Although widely used, this method is fraught with difficulties, particularly, the fact that the measurement is based upon a stressed system (i.e., a temperature-induced change). This stressing of the sample can give rise to false positives and may result in the rejection of an excipient for no good reason.

IC offers the benefit that the sample can be studied under conditions closer to the real storage/processing conditions with no external stress being applied (80). Isothermal calorimetric data are recorded as a function of time and, with the high sensitivity that modern calorimeters offer, it is possible to obtain high quality quantitative data. It has been shown that the sensitivity of modern isothermal calorimeters is such that degradation rates of 2% per year can be detected for some single component systems (81).

The techniques for deriving quantitative information from calorimetric data were discussed in Chapter 3 and will be expanded on with specific reference to stability/compatibility issues in Chapter 7. It is not always necessary to obtain quantitative information; sometimes a qualitative outcome will suffice, for example, during a preliminary screen to identify excipients that may potentially interact. It is this qualitative analysis, which is the focus of the following section.

Prediction of Compatibility

It has already been described how a change in the observed calorimetric signal between a mixture and an individual output is indicative of some interaction (chemical or physical) between compounds. It is this change in signal that provides the basis for the qualitative analysis.

The basic methodology is very simple. Samples of the API and excipient(s) in question are run individually to assess their response under the conditions of interest (temperature, RH, and so on) against an inert reference (82). These materials are then run in binary mixtures (or tertiary or quaternary as necessary), usually in equi-mass ratios and under 100% RH, as this will ensure the maximum possible interaction and increase the probability of observing any interaction in the calorimeter. If no interaction is observed under these conditions, then it is usually the case that the materials will not interact under less stressed conditions.

The calorimetric outputs observed for the individual samples are then summed to give a theoretical response (remember that the calorimetric output is a summative view of all heat change within the system), which represents the calorimetric output if the two materials do not interact. This theoretical response is then compared with the calorimetric output for the real mixture. If the materials are interacting (i.e., there is some incompatibility), then the observed calorimetric response will differ from the theoretical response. The observed response may reveal an incompatibility whereby any degradation reaction is accelerated and hence the overall shape of the calorimetric curve will be the same (i.e., the mechanism is unchanged), but the rate at which the signal decays will be greater. Alternatively, the calorimetric response may reveal that a new mechanistic pathway is now possible in the presence of the excipient and the calorimetric curve is observably different (more complex), or the curve no longer conforms to a single kinetic model (see Chap. 3).

Observation of any deviation from the theoretical output is then an indicator of interaction and possible incompatibility. At this stage, the excipient can be rejected out of hand (if the goal is a preliminary screen) or investigated further if the excipient is necessary for the formulation. Schmitt et al. (41) have evaluated the capacity of calorimetry to effect a rapid and practical screen for predicting compatibility between binary mixtures of drug and excipient . They showed that it is possible to predict relative stability within a functional class (binders, diluents, and so on). Their study, although unable to provide quantitative information, predicted the worst case incompatibility for a formulation with known stability problems. However, it should be noted that the results reported in this study were obtained from systems under stressed conditions of temperature (50°C) and by adding water (20% w/w), which although not unprecedented, still represents a significant deviation from normal storage conditions. The authors of this article also make the important point:

> The ability to predict reactions in dosage forms depends on the similarity of the binary mixture to the formulation.

This is of course an inherent assumption behind any compatibility study and limits the validity of the calorimetric data if used as a sole indicator of compatibility. Such studies must therefore be treated with caution and any inferences made from the calorimetric data be backed up with information derived from ancillary sources.

Step-Isothermal Methods for the Prediction of Compatibility

Some instruments offer the capacity to run in a quasi-isothermal mode; that is, the temperature can be held isothermally for a period of time and then a rapid temperature change can be effected, and the system held isothermally at some new temperature. This capacity to change rapidly the temperature allows an alternative method of screening for compatibility. If a rapid qualitative screen of materials for potential interaction is required, then it is sometimes appropriate to stress the sample by increasing the temperature in a step-isothermal experiment (83). Here, the system is held at a number of set temperatures, usually three or more (such that an Arrhenius plot may be constructed if appropriate), in order to assess whether the system can be induced to change or that the rate of any change be accelerated to facilitate its observation calorimetrically. Any change may be indicative of interaction providing that degradation processes and physical changes can be ruled out. The methodology is essentially the same as that described earlier. The materials are first run individually to assess whether they degrade, melt, and so on over the temperature program of interest. These experiments then serve as controls. A binary mixture of the compounds of interest is made and some defined quantity is placed in the calorimeter. The observed signal is then assessed to determine whether it is identical to the summation of the control experiments. If there is a significant difference between the control experiment and the sample experiment, then it can be inferred that there is a potential interaction. This methodology has been shown to be effective in predicting the known incompatibility between aspirin and magnesium stearate (49). Under normal conditions (298 K, ambient RH), the rate of this reaction is such that it requires approximately seven days to return an observable calorimetric signal. A step isothermal technique requires less than five hours to observe and predict the incompatibility. This system is discussed in more detail in Chapter 7.

Limitations of the Step-Isothermal Technique

The limitations associated with this technique are the same as those for any experiment in which the system must be stressed in order to elicit a response, particularly that of the system being studied under conditions that are not representative of the storage or usage conditions of the product.

Packaging Material

At this point, it is worthwhile noting that stability/compatibility studies are often strongly focussed on the compounds within the formulation, but the packaging material is also part of the formulation and as such should also be considered when conducting compatibility tests. The primary function of the packaging material is to protect the formulation from its environment and hence it is implicated in the shelf life of the product (i.e., if it does not provide adequate protection from light, moisture, $O_{2(g)}$, and so on, then the shelf life of the product may be severely compromised). Moreover, the packaging itself is a chemical entity (or a combination of chemical entities) and hence may potentially interact with the formulation. Because of its invariance to physical form and its capacity to study relatively large samples, IC is ideally suited to investigate potential interactions between a formulation and its packaging.

The methodology for this aspect of formulation testing is largely the same as that described earlier for incompatibility testing. Here, the formulation is now studied (rather than the individual components) in the presence of various packaging materials. Figure 12 shows the calorimetric output for a variety of packaging for a tablet which is susceptible to oxidation. The tablet is studied individually (control experiment) and then sealed within a number of different packaging materials. It is clear that the tablet on its own is relatively unstable. For this particular formulation, Triplex packaging offers essentially no protection at all; that is, the calorimetric response is identical in both cases. Next is the polyvinyl chloride packaging. This provides little or no protection. The aluminium packaging material appears to offer a significantly better protection than all other materials

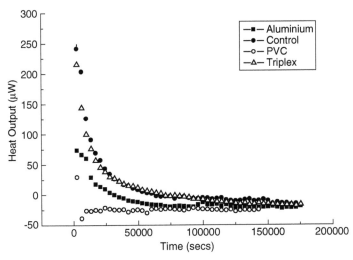

Figure 12 Comparison of packaging materials for a tablet formulation.

in that the rate of change in the calorimetric output is reduced and hence it can be inferred that the degradative process is inhibited.

Dissolution Characteristics of a Formulation

The majority of pharmaceuticals are delivered in solid formulations for a variety of reasons. Take the stability of an API, for example; the stability of a drug is usually greater in its solid-state than it is in the solution phase. Developing a drug as a solid formulation has the important implication that the formulation must break up at the appropriate time and in the appropriate location within the body. It is essential then that the release profile of a formulation can be assessed in order to best choose the most appropriate formulation with the required release characteristics. Moreover, dissolution testing plays an important part in the postmanufacturing stage of a product as it is used as part of the quality control process (e.g., to assess batch to batch uniformity). These requirements then necessitate the need for a rational design process for the solid matrix in which the drug is dispersed and some means of assessing the performance of that formulation.

Conventionally, dissolution testing is performed by dissolving the solid formulation in a large volume of solvent (usually 1 L at 37°C) in order to maintain sink conditions (i.e., the concentration of solute, following complete dissolution, should not exceed 10% of its limiting saturated solubility) (84–86). The solvent is then analyzed, usually by spectroscopic techniques, for the drug over a period of time and the release profile as a function of time can be gauged.

Although this is the only accepted method of testing for dissolution, it has some difficulties and is limited by several assumptions. Conventional dissolution tests generally show inherent variability and sensitivity to the source material and the test methodology. This variability is borne out by the range of in vitro–in vivo correlation data. In some instances, the release profiles have high correlation but also there are instances in which there is zero correlation. There are a number of reasons why this might be so. For example, conventional dissolution testing is designed to be, in principle, a simple reproducible test and hence is limited to relatively simple aqueous solvents, which often lack some of the components found in the fluids within the gastro-intestinal (GI) tract and hence are not necessarily a realistic model. Although dissolution media have been developed which are designed to mimic fluids found in the GI tract in order to improve the accuracy of dissolution tests (87–89), it is often the case that with these complex media that the solution requires manipulation in order to facilitate analysis for the drug (typically, the solution requires acidification and centrifugation in order to precipitate and remove suspended matter). This manipulation can be time consuming and also introduces a level of uncertainty into any derived data.

Dissolution experiments are often performed at a single pH or without control of pH. This is also an unrealistic view of the environment in which the formulation will be acting. In addition to these limitations, there is also the problem of volume. It is usual to use a 1 L solvent volume to perform dissolution tests and consequently it is difficult to study sparingly soluble drugs, which exceed the requirement for sink conditions. The effects of these limitations are perhaps best seen through a consideration of the dissolution process in its simplest form, which is described by the Noyes–Whitney relationship (90), Equation (18). The Noyes–Whitney equation describes the rate of change of concentration with respect to time, dC/dt, as a function of diffusion coefficient (D); surface area, A; saturated solubility, C_s; and diffusion layer thickness, h:

$$\frac{dC}{dt} = \frac{DS}{h}(C_s - C) \tag{18}$$

It should be noted that the value of C is that of the drug in solution and does not account for drug that may have been absorbed into the blood or absorbed by stomach contents. This therefore may mean that the presence of other compounds such as food products, could remove drug from solution and increase the dissolution rate. It is also the case that, if fatty components absorb to the tablet face, then the dissolution rate will also be affected.

Even with these limitations, dissolution testing is still required as a standard test to which all references should be made.

Assessment of Dissolution Characteristics by Isothermal Calorimetry

Flow-through calorimeters have been shown (91,92) to be useful for dissolution testing and allow dissolution tests to be carried out without some of the limitations encountered in conventional testing. Because of their ability to respond in the presence of complex media, without affecting the performance of the instrument, it is possible to study the dissolution process under more realistic conditions (e.g., in the presence or absence of food as well as individual components of food such as metal ions, fats, fibre, and so on). The full dissolution process is observable as well as the possibility of studying drug/drug interactions under these simulated conditions in a manner analogous to that described earlier in section "Stability/Compatibility." Flow-through techniques also have the advantage in that they can, in effect, provide a dissolution volume which is essentially infinite. It is then possible to study the dissolution characteristics of sparingly soluble compounds without the risk of exceeding sink conditions.

Ashby et al. (91) have demonstrated the use of flow-through calorimetry to assess the release profile of a controlled release tablet formulation (Phyllocontin) in the presence of a variety of dissolution media to represent different fed and fasted states. In this study, a tablet was placed in the calorimetric chamber of a flow calorimeter. Over it were passed a variety of solutions: buffer, buffer/ Ensure (a nasogastric foodstuff containing fat, protein, carbohydrate, vitamins, and minerals), and buffer/intralipid (an emulsion containing soya oil and glycerol). They were able to show that the dissolution profile of the tablet followed a process consisting of two first-order processes separated by a mixed-order region, a trend that was observed using the conventional United States Pharmacopoeia (USP) paddle method. These two first-order events were attributed to the release of drug and its subsequent diffusion from the matrix. Moreover, they demonstrated that the dissolution profiles of the drug were affected by the composition of the dissolution media such that they were able to determine, for example, that Ensure increased the initial dissolution rate but did not alter diffusion of the drug in the second phase of the process, whereas intralipid slowed the initial dissolution rate but facilitated in some way the diffusion. This experimental methodology was then adapted to probe the known interaction between tetracycline hydrochloride and calcium (92). The results of this particular study showed that the rate of dissolution of the formulation was not significantly reduced by the presence of calcium in the simulated foodstuffs, nor was there a measurable interaction with the solid matrix. It did show, however, that the calorimetric signal for the release of drug in the presence of calcium ions was significantly larger than for the control experiments. Recall from section, "Stability/Compatibility," that this would indicate some interaction between drug and some other species present.

It should be noted that any numeric values obtained from the calorimetric experiments cannot be directly compared with those obtained from the USP method because of the different hydrodynamics of the two systems. The calorimetric method, in conjunction with the USP method, does however offer the capacity to obtain valuable information on release profiles and potential interactions that might affect bioavailability for potential formulations in the early stages of development that might otherwise have not been available.

Hygroscopicity

The potential for, and the extent to which, a material will take on moisture from its surroundings is described by the equilibrium term hygroscopicity. The two parameters, potential and extent, are intrinsically linked to the RH of the system and the temperature of the atmosphere surrounding the sample. Because hygroscopicity is an equilibrium term, a sample will either lose moisture or gain moisture when placed into an atmosphere at constant RH. This transfer will continue until a new equilibrium is reached, at which point the quantity of water in the sample is called the equilibrium moisture content (EMC). The value of the EMC and the rate at which a sample attains equilibrium are governed by several factors including surface area, temperature, and RH.

Nearly every material will exchange moisture with its surrounding atmosphere and hence the understanding and regulation of hygroscopicity are vital for the preformulation phase of a pharmaceutical product.

Pharmaceutical formulations contain a multitude of components, all of which will be hygroscopic to some extent. Uptake of water by the presence of hygroscopic materials can lead to the physical and chemical properties of the formulation changing significantly. For example, sorbed (adsorbed or absorbed) water may induce or accelerate hydrolytic degradation, isomerization, and may plasticize amorphous materials thus allowing them to recrystallize. All of these effects can drastically alter the product's shelf life as well as its efficacy. The hygroscopicity of an active and any excipients is thus an important factor when considering the chemical make-up of any potential formulation and must be determined during the preformulation step.

Conventional techniques for determining the hygroscopicity of a substance evolved around the use of gravimetric methods (93). For gravimetric techniques, the extent of water uptake is measured by a mass change of the sample and the rate of uptake is monitored simply by plotting the mass change as a function of time. There exist several methods of introducing the moisture to the sample, such as passive diffusion where the sample sits in a static atmosphere. Alternatively, the sample can sit in a gas stream that can also confer the advantage of reducing the experiment run time. Gravimetric methods have been extensively used over a number of years and provide high-quality data (e.g., modern instruments have a resolution of 0.1 μg and now exist as hyphenated techniques). Lane and Buckton (94) have reported the development of an instrument which utilizes near infrared spectroscopy in combination with a gravimetric measurement, such that spectroscopic changes in the material can be followed in tandem with the more conventional measurement of mass change.

Calorimetric methods are also well established for measuring the interaction of solids with vapors, have been around for some 20 years (95,96), and still continue to generate interest. Calorimeters are ideally suited to the study of interactions of surfaces and absorbates, as they allow a measurement to be performed within a controlled environment without the need for invasive sampling.

Markova et al. (97) recently described a novel isothermal sorption microcalorimeter that permits the measurement of a variety of thermodynamic and kinetic parameters associated with hygroscopicity. It is reported that the instrument can reliably measure the differential enthalpy of sorption and condensation of solvents in capillaries, study the recrystallization of amorphous materials (lactose in this instance), investigate lyotropic phase transitions, and monitor solvate formation.

SUMMARY

A robust preformulation process is vital for any pharmaceutical product but, as a consequence, it can be complex, time consuming, and expensive. Coupled with an ever-increasing number of drugs and modes of delivery, it is also a highly dynamic phase of product development. This dynamism is such that there is a large emphasis placed on the optimization of existing techniques and development of new techniques for the determination of the physicochemical parameters important to the preformulation scientist. This chapter has highlighted a few of the important preformulation parameters, including enthalpy of solution, dissolution parameters, hygroscopicity, and purity and some of the methods used to measure them. For the most part, spectroscopic methods of analysis are the norm for these measurements, mainly because they provide direct molecular information and are accepted in regulatory documents. DSC data are accepted in regulatory submissions, whereas IC data, in isolation, are generally not, although calorimetric data can be used as supporting evidence. In spite of the reluctance of regulatory authorities to accept calorimetric data, calorimetry has historically been used for a variety of measurements. DSC, in particular, has been popular for the measurement of a number of physicochemical parameters, such as purity and melting point and, with the advent of fast-scan instruments, it is finding popularity for the determination of polymorphs and polymorphic composition of pharmaceutical systems. Isothermal techniques too have been used historically. The measurement of the enthalpy of solution of pharmaceutically-relevant compounds by IC was first reported in the 1960s. Again with the development of more sophisticated instruments, the potential application of IC in preformulation studies is always increasing.

Rapidity of measurement with minimal operator involvement is a gold standard for industry and hence high throughput is an essential feature of any analytical instrument. The fact that many isothermal calorimetric instruments are not capable of high-throughput measurements is an obvious problem. However, this need for high throughput has been recognized by the calorimetric community and there are now appearing instruments capable of performing many multiples of experiments at one time. With this increase in turnover comes, in some instances, another significant benefit: a significant reduction in the quantity of material required for each measurement (these instruments are discussed in Chap. 9).

REFERENCES

1. Wells JI. Pharmaceutical preformulation: the physicochemical properties of drug substances. In: Aulton MA, ed. Pharmaceutics. The Science of Dosage Form Design. 2nd ed. Edinburgh, Scotland: Churchill and Livingston, 2002:113–138.
2. Akers MJ. Preformulation testing of solid oral dosage forms. Methodology, management and evaluation. Can J Pharm Sci 1976; 11:1–10.
3. Buckton G, Russell SJ, Beezer AE. Pharmaceutical calorimetry: a selective review. Thermochimica Acta 1990; 193:195–214.
4. Giron D. Contribution of thermal methods and related techniques to the rational development of pharmaceuticals—Part 1. Pharm Sci Tech Today 1998; 5:191–199.
5. Giron D. Contribution of thermal methods and related techniques to the rational development of pharmaceuticals—Part 2. Pharm Sci Tech Today 1998; 6:262–268.
6. Giron D. Applications of thermal analysis and coupled techniques in pharmaceutical industry. J Therm Anal Cal 2002; 68:335–357.
7. Giron D. Thermal analysis in pharmaceutical routine analysis. Acta Pharm Jugosl 1990; 40:95–157.
8. Craig DQM. A review of thermal methods used for the analysis of the crystal form, solution thermodynamics and glass transition behaviour of polyethylene glycols. Thermochimica Acta 1995; 248:189–203.

9. Beezer AE, Gaisford S, Hills AK, Willson RJ, Mitchell JC. Pharmaceutical microcalorimetry: applications to long-term stability studies. Int J Pharm 1991; 179:159–165.
10. Gaisford S, Hills AK, Beezer AE, Mitchell JC. Thermodynamic and kinetic analysis of isothermal microcalorimetric data: applications to consecutive reaction schemes. Thermochimica Acta 1999; 328:39–45.
11. Phipps MA, Winnike RA, Long ST, Viscomi F. Excipient cmpatibility as assessed by isothermal microcalorimetry. Pharm Pharmacol 1998; 50(suppl):6–9.
12. Angberg M, Nyström C, Cartensson S. Evaluation of heat-conduction microcalorimetry in pharmaceutical stability studies VII. Oxidation of ascorbic acid in aqueous solution. Int J Pharm 1993; 90:19–33.
13. Ohta M, Tozuka Y, Oguchi T, Yamamoto K. Water vapor adsorption properties of amorphous cefditren pivoxil evaluated by adsorption isotherms and microcalorimetry. Drug Dev Ind Pharm 2000; 26:643–649.
14. Buckton G, Darcy P. Water mobility in amorphous lactose below and close to the glass transition temperature. Int J Pharm 1999; 179:141–158.
15. Hogan, SE, Buckton G. The quantification of small degrees of disorder in lactose using solution calorimetry. Int J Pharm 2000; 207:57–64.
16. de Angelis NJ, Papariello GJ. Advantages and limitations for absolute purity determinations. J Pharm Sci 1968; 57:1868–1873.
17. Gray AP, Fyans RL. Methods of analysing stepwise data for purity determinations by differential scanning calorimetry. Therm Anal Appl Study 1973; 10. Perkin Elmer, Norwalk, Connecticut, U.S.A.
18. Brennan WP, DiVito MP, Fyans, RL, Gray AP. An overview of the calorimetric purity measurement. In: Blaine RL, Schoff CK, eds. Purity Determinations by Thermal Methods. American Society for Testing and Materials ASTM STP838 (Philadelphia), 1984:5–15.
19. van Dooren AA. Purity determinations of drugs with differential scanning calorimetry (DSC)—a critical review. Int J Pharm 1984; 20:217–233.
20. Yoshii K. Application of differential scanning calorimetry to the estimation of drug purity: Various problems and their solutions in purity analysis. Chem Pharm Bull 1997; 45:338–343.
21. Giron D, Goldbronn C. Place of DSC purity analysis in pharmaceutical development. J Therm Anal 1993; 44:217–251.
22. Atkins PW, de Paula J. Atkins' Physical Chemistry. 7th ed. Oxford, England: Oxford University Press, 2001:166.
23. van Dooren AA. Effects of experimental variables on purity determinations with differential scanning calorimetry. Thermochimica Acta 1983; 66:161–186.
24. Gray AP. Thermal Analysis Newsletter (no. 6). The Perkin Elmer Corporation Norwalk CT.
25. Plato C, Glasgow AR. Differential scanning calorimetry as a general method for determining purity and heat of fusion of high purity organic chemicals. Application to 95 compounds. Anal Chem 1969; 41:330–336.
26. Giron D. Thermal analysis and calorimetric methods in the characterisation of polymorphs and solvates. Thermochimica Acta 1995; 248:1–59.
27. Haleblian J, McCrone W. Pharmaceutical applications of polymorphism. J Pharm Sci 1969; 58:911–929.

28. Leonidov N. Polymorphism and qualitative differences in drug effects. Eur J Pharm Sci 1996; 4:S95.
29. Burger A, Lettenbichler A. Polymorphism and preformulation studies in lifibrol. Eur J Pharm Biol 2000; 49:65–72.
30. Garcia E, Veesler S, Boistelle B, Hoff C. Crystallization and dissolution of pharmaceutical compounds: an experimental approach. J Cryst Growth 1999; 198–199:1360–1364.
31. Rustichelli C, Gamberini G, Ferioli V, Gamberini MC, Ficarra R, Tommasini S. Solid-state study of polymorphic drugs: Carbamazepine. J Pharm Biomed Anal 2000; 23:41–54.
32. Hino T, Ford JL, Powell MW. Assessment of nicotinamide polymorphs by differential scanning calorimetry. Thermochimica Acta 2001; 374:85–92.
33. Kawakami K, Ida Y. Application of modulated DSC to the analysis of enantiotropically related polymorphic transitions. Thermochimica Acta 2005; 427:93–99.
34. Boldyreva EV, Drebushchak VA, Paukov IE, Kovalevskaya YA, Drebushchak TN. DSC and adiabatic calorimetry study of the polymorphs of paracetamol: an old problem revisited. J Therm Anal Cal 2004; 77:607–623.
35. Koester LS, Mayorga P, Pereira VP, Petzhold CL, Bassani VL. Carbamazepine/beta CD/HPMC solid dispersions. II. Physical characterisation. Drug Dev Ind Pharm 2003; 29:145–15.
36. Chidavaenzi OC, Buckton G, Koosha F, Pathak R. The use of thermal techniques to assess the impact of feed concentration on the amorphous content and polymorphic forms present in spray dried lactose. Int J Pharm 1997; 159:67–74.
37. McGregor C, Saunders MH, Buckton G, Saklatvala RD. The use of high-speed differential scanning calorimetry (Hyper-DSC^{TM}) to study the thermal properties of carbamazepine polymorphs. Thermochimica Acta 2004; 417:231–237.
38. Serajuddin AT, Thakur AB, Ghoshal RN, Fakes MG, Ranadive SA, Morris KR, et al. Selection of solid dosage form composition through drug excipient compatibility testing. J Pharm Sci 1999; 88:696–704.
39. Wells JI. in Pharmaceutical Preformulation—The physicochemical properties of drug substances. Chichester: Ellis Horwood.
40. Monkhouse DC, Maderich A. Whither compatibility testing? Drug Dev Ind Pharm 1989; 15:2115–2130.
41. Schmitt EA, Peck K, Sun Y, Deoffroy JM. Rapid, practical and predictive excipient compatibility screeing using isothermal microcalorimetry. Thermochimica Acta 2001; 380:175–183.
42. M'Bareck CO, Nguyen QT, Metayer M, Saiter JM, Garda MR. Poly (acrylic acid) and poly (sodium styrenesulfonate) compatibility by Fourier transform infrared and differential scanning calorimetry. Polymer 2004; 45:4181–4187.
43. Sapoundjiev D, Lorenz H, Seidel-Morgenstern A. Determination of solubility data by means of calorimetry. Thermochimica Acta 2005; 436:1–9.
44. Mura P, Bettinetti GP, Faucci MT, Manderioli A, Parrini PL. Differential scanning calorimetry in compatibility testing of picotamide with pharmaceutical excipients. Thermochimica Acta 1998; 321:59–65.
45. Mura P, Faucci MT, Manderioli A, Bramanti G, Ceccarelli L. Compatibility study between ibuproxam and pharmaceutical excipients using differential scanning calorimetry, hot-stage microscopy and scanning electron microscopy. J Pharm Biomed Anal 1998; 18:151–163.

46. Zaharescu T. Assessment of compatibility of ethylene-propylene-diene terpolymer and polypropylene. Polym Degrad Stab 2001; 73:113–118.
47. Balestrieri F, Magrì AD, Magrì AL, Marini D, Sacchini A. Application of differential scanning calorimetry to the study of drug-excipient compatibility. Thermochimica Acta 1996; 285:337–345.
48. McDaid FM, Barker SA, Fitzpatrick S, Petts CR, Craig DQM. Further investigations into the use of high sensitivity differential scanning calorimetry as a means predicting drug–excipient interactions. Int J Pharm 2003; 253:235–240.
49. Mroso PN, Li Wan Po A, Irwin WJ. Solid state stability of aspirin in the presence of excipients: kinetic interpretation, modelling and prediction. J Pharm Sci 1982; 71:1096–1101.
50. Lloyd GR, Craig DQM, Smith A. An investigation into the melting behaviour of binary mixes and solid dispersions of paracetamol and PEG 4000. J Pharm Sci 1997; 86:991–996.
51. Kaplan SA. Biopharmaceutical considerations in drug formulation design and evaluation. Drug Metab Revs 1972; 1:15–34.
52. Frank TC, Downey JR, Gupta SK. Quickly screen solvents for organic solids. Chem Eng Prog 1999; 95:41–61.
53. Matsuoka M, Ozawa R. Determination of solid-liquid phase equilibria of binary organic systems by differential scanning calorimetry. J Cryst Growth 1989; 96:596–604.
54. Shibuya H, Suzuki Y, Yamaguchi K, Arai K, Saito S. Measurement and prediction of solid: liquid phase equilibria of organic compound mixtures. Fluid Phase Equilib 1993; 82:397–405.
55. Young PH, Schall CA. Cycloalkane solubility determination through differential scanning calorimetry. Thermochimica Acta 2001; 368:387–392.
56. Lorenz H, Seidel-Morgenstern A. Binary and ternary phase diagrams of two enantiomers in solvent systems. Thermochimica Acta 2002; 382:129–142.
57. Gaisford S, Buckton G. Potential applications of microcalorimetry for the study of physical processes in pharmaceuticals. Thermochimica Acta 2001; 380:185–198.
58. O'Neill MAA. Calorimetric investigation of complex systems: Theoretical developments and experimental studies. Ph.D. dissertation, University of Greenwich, 2002.
59. Christian G. Analytical Chemistry. 5th ed. New York: Wiley and Sons, 1994.
60. Urakami K, Beezer AE. A kinetic and thermodynamic study of seratrodast polymorphic transition by isothermal microcalorimetry. Int J Pharm 2003; 257: 265–271.
61. Hancock JD, Sharp JH. Method of comparing solid-state kinetic data and its application to the decomposition of kaolinite, brucite, and $BaCO_3$. J Am Ceram Soc 1972; 55:74–77.
62. Castronuovo G, Elia V, Niccoli M, Velleca F. Simultaneous determination of solubility, dissolution and dilution enthalpies of a substance from a single calorimetric experiment. Thermochimica Acta 1998; 320:13–22.
63. Craig DQM, Newton JM. Characterisation of polyethylene glycols using solution calorimetry. Int J Pharm 1991; 74:43–48.
64. Grant DJW, York P. A disruption index for quantifying the solid state disorder induced by additives or impurities. II. Evaluation from heat of solution. Int J Pharm 1986; 28:103–112.
65. Gao D, Rytting JH. Use of solution calorimetry to determine the extent of crystallinity of drugs and excipients. Int J Pharm 1997; 151:183–192.

66. Fini A, Fazio G, Feroci G. Solubility and solubilization properties of non-steroidal anti-inflammatory drugs. Int J Pharm 1995; 126:95–102.
67. Lloyd GR, Craig DQM, Smith A. A calorimetric investigation into the interaction between paracetamol and polyethylene glycol 4000 in physical mixes and solid dispersions. Eur J Pharm Biopharm 1999; 48:59–65.
68. Chadha R, Kashid N, Jain DVS. Microcalorimetric studies to determine the enthalpy of solution of diclofenac sodium, paracetamol and their binary mixtures at 310.15 K. J Pharm Biomed Anal 2003; 30:1515–1522.
69. Ramos R, Gaisford S, Buckton G, Royall PG, Yff BTS, O'Neill MAA. A comparison of chemical reference materials for solution calorimeters. Int J Pharm 2005; 299: 73–83.
70. Yff BTS, Royall PG, Brown MB, Martin GP. An investigation of calibration methods for solution calorimetry. Int J Pharm 2004; 269:361–372.
71. Bastos M, Milheiras S, Bai G. Enthalpy of solution of α-cyclodextrin in water and in formamide at 298.15 K. Thermochimica Acta 2004; 420:111–117.
72. Alves N, Bai G, Bastos M. Enthalpies of solution of paracetamol and sodium diclofenac in phosphate buffer and in DMSO at 298.15 K. Thermochimica Acta 2005; 441:16–19.
73. Xu N, Ma L, Lin RS. Enthalpies of solution of L-threonine in aqueous alcohol solutions. Thermochimica Acta 2005; 439:119–120.
74. Jain DVS, Kashid N, Kapoor S, Chadha R. Enthalpies of solution of ampicillin, amoxycillin and their binary mixtures at 310.15 K. Int J Pharm 2000; 201:1–6.
75. Hill JO, Öjelund G, Wadsö I. Thermochemical results for "tris" as a test substance in solution calorimetry. J Chem Therm 1969; 1:111–116.
76. Piekarski H. Calorimetry—an important tool in solution chemistry. Thermochimica Acta 2004; 420:13–18.
77. Royall PG, Gaisford S. Application of solution calorimetry in pharmaceutical and biopharmaceutical research. Curr Pharmaceut Biotechnol 2005; 6:215–222.
78. Gill SJ, Seibold ML. Flow calorimeter cell for measuring heats of solutions of solids. Rev Sci Instrum 1976; 47:1399–1401.
79. Nilsson SO, Wadsö I. A flow-microcalorimetric vessel for solution of slightly soluble solids. J Chem Thermodyn 1986; 18:1125–1133.
80. Gaisford S. Stability assessment of pharmaceuticals and biopharmaceuticals by isothermal calorimetry. Curr Pharmaceut Biotechnol 2005; 6:181–191.
81. Pikal MJ. Application note AN335, Thermometric AB, Järfälla, Sweden.
82. Skaria CV, Gaisford S, O'Neill MAA, Buckton G, Beezer AE. Stability assessment of pharmaceuticals by isothermal calorimetry: two component systems. Int J Pharm 2005; 292:127–135.
83. Wissing S, Craig DQM, Barker SA, Moore WD. An investigation into the use of stepwise isothermal high sensitivity DSC as a means of detecting drug–excipient incompatibility. Int J Pharm, 2000; 199:141–150.
84. Siewert M, Weinandy L, Whiteman D, Judkins C. Typical variability and evaluation of sources of variability in drug dissolution testing. Eur J Pharmaceut Biopharmaceut 2002; 53:9–14.
85. Qureshi SA, Shabnam J. Cause of high variability in drug dissolution testing and its impact on setting tolerances. Eur J Pharmaceut Biopharmaceut 2001; 12:271–276.

86. Perng CY, Kearney AS, Palepu NR, Smith BR, Azzarano LM. Assessment of oral bioavailability enhancing approaches for SB-247083 using flow-through cell dissolution testing as one of the screens Int J Pharm 2003; 250:147–156.
87. Galia E, Nicolaides E, Hörter D, Löbenberg R, Reppas C, Dressman RB. Evaluation of various dissolution media for predicting in vivo performance of Class I and Class II drugs. Pharm Res 1998; 15:698–705.
88. Pedersen BL, Brøndsted H, Lennernas H, Christensen FN, Müllertz A, Kristensen HG. Dissolution of hydrocortisone in human and simulated intestinal fluids. Pharm Res 2000; 17:183–189.
89. Luner PE, Vander Kamp D. Wetting behavior of bile salt-lipid dispersions and dissolution media patterned after intestinal fluids. J Pharm Sci 2001; 90:348–359.
90. Willson RJ, Sokoloski TD. Ranking of polymorph stability for a pharmaceutical drug using the Noyes–Whitney titration template method Thermochimica Acta 2004; 417:239–243.
91. Ashby LJ, Beezer AE, Buckton G. In vitro dissolution testing of oral controlled release preparations in the presence of artificial foodstuffs. I. Exploration of an alternative methodology: microcalorimetry. Int J Pharm 1989; 51:245–251.
92. Buckton G, Beezer AE, Chatham SE, Patel KK. In vitro dissolution testing of oral controlled release preparations in the persence of artificial foodstuffs. II. Probing drug/food interactions using microcalorimetry. Int J Pharm 1989; 56:151–157.
93. Stubberud L, Arwidsson HG, Graffner C. Water–solid interactions: I: a technique for studying moisture sorption/desorption. Int J Pharm 1995; 114:55–64.
94. Lane RA, Buckton G. The novel combination of dynamic vapour sorption gravimetric analysis and near infra-red spectroscopy as a hyphenated technique. Int J Pharm 2000; 207:49–56.
95. Duisterwinkel AE, Vanbokhoven JJ. Water sorption measured by sorption calorimetry. Thermochimica Acta 1995; 256:17–31.
96. Jakobsen DF, Frokjaer S, Larsen C, Niemann H, Buur A. Application of isothermal microcalorimetry in preformulation. I. Hygroscopicity of drug substances. Int J Pharm 1997; 156:67–77.
97. Markova N, Sparr E, Wadsö L. On application of an isothermal sorption microcalorimeter. Thermochimica Acta 2001; 374:93–104.

5

Lead Compound Optimization

INTRODUCTION

The selection of a lead compound can be based on rational design, the result of screening thousands of compounds randomly drawn from a compound library or a serendipitous event (recall the discovery of penicillin for instance). Irrespective of its method of discovery, identification of a lead compound is merely the start of the drug development process. The structure of the lead compound must be optimized to produce the final active moiety. There are many properties that can be altered: solubility, stability, efficacy, toxicity, and taste, forming a non-exhaustive list. Good solubility is an essential requirement for nearly all drugs and is discussed in Chapter 4. Stability (fully discussed in Chap. 6) and taste masking are also important and can be controlled to a large extent by careful formulation. Efficacy and toxicity, conversely, are governed almost entirely by the structure of the molecule and its interaction with biological entities (although formulation and dosing regimens can be used to improve/mitigate such effects). Hence, the measurement (and if possible quantification) of such interactions plays an essential role in lead compound optimization and is an area where calorimetric techniques can play a vital role.

Two principal areas where calorimetric measurements can drive lead compound development present themselves. Both areas exploit the fact that calorimeters can measure processes as they occur in situ. As well as the fact that this means the data obtained relate directly to the process under study, the need for specific assay development is obviated. These benefits, along with the recent development of sensitive and user-friendly instruments, make calorimetric techniques central to drug design programs. The first application is the direct measurement of binding interactions and thermodynamics, an area that is fundamental to structure-based drug design (SBDD) and optimization of efficacy. The second application is the direct quantification of efficacy against microorganisms.

In the former, isothermal titration calorimetry (ITC) is usually employed and in the latter flow calorimetry is used. This chapter thus considers both areas, starting with the applications of ITC.

APPLICATION OF ISOTHERMAL TITRATION CALORIMETRY TO RATIONAL DRUG DESIGN

At the core of the drug discovery process is the detection of an interaction between a potential drug molecule and a target-binding site on a macromolecule. Screening, commonly in high-throughput mode, has re-emerged as a key approach to the detection of this crucial interaction, the job of the primary screen being, simply, to alert the scientist to this sought-after interaction. Currently, the search for this signal of interaction generally involves perturbing the system with markers or labels and thus suffers from the drawback of specificity to a particular assay. Primary recognition of a "hit" sets off a cascade of events resulting in the evaluation of properties both involving the target and separate from it. The initial observation is followed by dose-range studies and, importantly, the determination of binding constants thus moving the objective to a quantitative assessment of binding potency capable of comparative evaluation.

Combinatorial chemistry techniques make the generation of vast numbers of new chemical entities (NCEs) extremely simple. However, the assessment of the likely therapeutic effect of each NCE is a considerably time consuming and expensive process. It is simply not a cost-effective approach to quantify the efficacy of each NCE against a range of diseases. As noted earlier, it is increasingly the case that the development strategies for pharmaceuticals are thus structure-based, the design of a particular lead compound being optimized through a series of rational structural changes for a specific biological target. When quantitative changes in a parameter are determined as a function of structural changes, the result is a quantitative structure–activity relationship (QSAR). Of the myriad of parameters upon which a QSAR could be based, it is usually the case that the parameter being optimized is the binding affinity (K_a). The reason for this is simply that K_a provides an overall view of the binding interaction, its value encompassing all of the events that have occurred upon binding (note here that the binding of a ligand to a substrate is an event that can include a multitude of separate processes, including changes in structural configuration, the loss or formation of hydrogen bonds and electrostatic interactions, and changes in the properties of the solvent). Whether the binding affinity is the most appropriate parameter upon which to rationalize structure is a topic that will be discussed subsequently.

When a ligand (L) binds to a substrate (S), an equilibrium state is reached between the free species and the complex. Thus:

$$nL_{(aq)} + S_{(aq)} \leftrightarrow L_nS_{(aq)}$$

The binding affinity thus represents the position of the equilibrium reached and is expressed by an equilibrium constant. Assuming a 1:1 stoichiometry between the two components (i.e., $n = 1$), K_a would be represented by:

$$K_a = \frac{[LS]}{[L][S]} \tag{1}$$

where [ ] indicates that concentration terms are defined (strictly activities should be defined, but it shall be assumed that the system behaves ideally and that concentration terms are valid). Alternative equilibrium constants can be constructed for other binding stoichiometries. Note that this expression simply defines the equilibrium state; it does not consider the rate at which it is reached. Where the binding affinity is high, the equilibrium lies toward the complex and the concentration of free ligand is small. When the binding affinity is small, the equilibrium lies toward the free species and the concentration of the complex is low. Thus, the structure of a lead compound is optimized to maximize the value of K_a.

The reasons for the widespread use of ITC for the determination of binding affinities are numerous and were alluded to earlier. Briefly, (*i*) it does not require optical clarity of the solutions, (*ii*) it is rapid, (*iii*) it does not require the development of a specific assay for each interaction, and (*iv*) it directly measures the enthalpy of binding ($\Delta_b H$). Indeed, it is the only analytical tool capable of measuring a thermodynamic quantity directly and obviates the need for indirect determinations via van't Hoff analyses. In addition, careful experimental design allows further parameters to be determined from the same data set; the entropy of binding ($\Delta_b S$), the Gibbs energy of binding ($\Delta_b G$), the change in heat capacity upon binding (ΔC_p), and the stoichiometry of binding (n).

Calorimetry is, therefore, ubiquitous and usable, in principle, from system to system. Furthermore, it can be a single, common thread running through compound evaluation in systems of increasing bio-complexity, for example, in progressing from macromolecules to cells [the incorporation of cells into high-throughput screening (HTS) is of increasing importance to ensure the quality of lead compounds] (1). Other off-target properties (such as solubility and degradation kinetics, for instance) are also potentially accessible via calorimetric evaluation. It is surprising, therefore, that the direct, universal, and versatile property of enthalpy change has not featured more prominently in the drug discovery literature as a primary signal of interaction in HTS, its application in this area having been only recently discussed (2).

The reasons for the absence of calorimetry from the screeners' arsenal often reduce simply to factors of scale and time. Hitherto, commercially-available instruments have demanded significant quantities of protein and drug molecules and were relatively time-demanding (requiring several hours for equilibration and operation), although ITC is often used to characterize binding in the later stages of the drug discovery process (3).

ITC, thus, has a significant role in lead compound optimization, as one can opt to rationalize the design of the active against one or more of these parameters.

The following sections discuss various aspects of the use of ITC, considering both experimental factors and the interpretation of the binding parameters obtained.

Experimental Design

As noted in Chapter 2, a proper understanding of the origin of the signal and its correction to remove experimental artifacts is essential when undertaking ITC experiments. Critically, it must be ensured that the correct blank experiments are performed and subtracted from the binding data. Usually, a minimum of three blank experiments are required.

1. Dilution of the ligand by the solvent.
2. Dilution of the substrate by the solvent.
3. Solvent mixing.

The first experiment is performed by injecting ligand solution into solvent, the second by injecting solvent into substrate solution, and the third by injecting solvent into solvent. Of the three experiments, the correction for the dilution of the ligand is often most significant, because its initial concentration is high as small aliquots are titrated. If a series of experiments are being performed with the same substrate then this blank needs only be recorded once, as it is then common to every subsequent experiment. The third blank is unnecessary if the same solvent is used for both the ligand and the substrate solutions. If not, large enthalpies of mixing (different solvents) or ionization (different buffers) may be recorded that are likely to be of a magnitude significantly larger than the binding interaction.

   The rate of mixing is also a factor (a faster stirring rate will mix more efficiently but cause a larger disruption to the baseline). Other factors vary depending upon the type of instrument used. For instance, if an instrument uses a fixed volume of sample (such as the VP-ITC, Microcal Inc), then a correction must be made for the dilution of the substrate, both as the experiment progresses and after loading (the cell will never be totally dry before loading the sample). If an instrument uses a cannula to introduce the ligand solution [such as the Thermal Activity Monitor (TAM), Thermometric AB], then it must be decided whether the cannula is placed above or below the surface of the substrate solution. If the cannula is below the surface then diffusion of the ligand into the solution can occur during equilibration, an effect that often manifests itself in a smaller than expected first injection peak.

   In order to derive meaningful K_a values, it is also essential that the concentrations of the two species are selected such that the binding passes through its saturation point. An easy way to confirm this is that the titration peaks at the end of the experiment should have settled to a constant and a minimum value (note here that as these peaks essentially represent the dilution of the ligand and substrate, it is possible to use the value to correct the earlier peaks, thus obviating the need for blank experiments, although this is an approximation).

The shape of the binding isotherm generated (discussed subsequently) will vary depending on the magnitude of the binding affinity. It is important that the binding isotherm contains sufficient resolution to allow successful analysis. One way of characterizing binding isotherms is through the use of a c-value, where c is a unitless parameter derived from the product of the binding affinity, the concentration of binding sites $[S_{tot}]$, and the binding stoichiometry:

$$c = K_a[S_{tot}]n \tag{2}$$

The importance of c-values is discussed subsequently.

Measurement of the Binding Enthalpy

A value for the binding enthalpy ($\Delta_b H$) is produced automatically when the binding isotherm is analyzed by least squares minimization to determine K_a (see subsequent discussion) but it is also possible to derive its value by direct experimental measurement. To do so, the experiment must be performed under conditions where the substrate is present at a much higher concentration than the ligand; this then allows the assumption to be made that total binding occurs with each injection of ligand. Thus, the area under each peak for sequential injections should be equal. Knowledge of the number of moles of ligand titrated thus allows the simple determination of $\Delta_b H$. In order to have confidence in this type of analysis, the c-value should be between 1 and 1000.

It is also possible to determine a value for the binding enthalpy noncalorimetrically through a van't Hoff analysis. A van't Hoff isochore plots the temperature dependence of the binding affinity and allows the derivation of a van't Hoff enthalpy ($\Delta_b H_{vH}$). Such an analysis is sometimes useful because the ratio of the calorimetric and van't Hoff enthalpies correlates to the co-operativity of the system. If there is no co-operativity, the two enthalpies are equal. However, it should be noted that van't Hoff analyses invoke many assumptions and are often open to significant experimental error or other artifacts; this being so, the use of the ratios must be undertaken with caution.

Measurement of the Binding Stoichiometry

The value of n is easily determined from knowledge of the molar ratio of ligand to substrate at the equivalence point of the titration, although there is the requirement that the concentrations of the species are known accurately. This simple method for determining n is a powerful tool in determining the nature of the binding interaction. Ward and Holdgate (4) note that such measurements may reveal, for instance, the presence of nonfunctional protein, which is not easily determinable from any other assay.

Measurement of the Binding Affinity

The output from a conventional ITC experiment is a plot of power versus time, showing a series of peaks corresponding to sequential injections of ligand solution into substrate solution (Fig. 1, top). Integration of each peak results in a binding isotherm of enthalpy (kJ mol^{-1}) versus molar ratio of ligand to substrate (Fig. 1, bottom). The binding isotherm is then fitted to a binding model, by least squares minimization, in order to derive the value of K_a (and, in the process, $\Delta_b H$). With modern instruments (such as the VP-ITC, Microcal Inc.), this

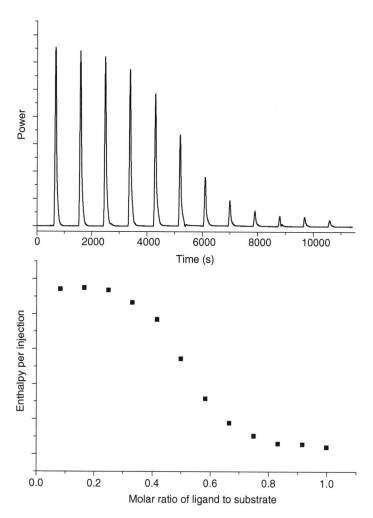

Figure 1 Typical power–time data for the sequential titration of aliquots of a ligand into a substrate solution (*top*) and the subsequent binding isotherm (*bottom*).

model fitting is done automatically; a "black-box" approach like this undoubtedly increases the appeal and the usage of the technique, but does not necessarily mean that the user is informed of the assumptions that have been made in deriving, and precise meaning of, the data so produced. The following text provides a brief summary of the principles of the approach. The method, first derived by Wiseman et al. (5) for Microcal instruments, has been eloquently examined by Indyk and Fisher (6) and it is to this reference that readers seeking further details and clarification are directed.

It was noted earlier that the equilibrium constant for a simple single-site binding is given by:

$$K_a = \frac{[LS]}{[L][S]} \tag{1}$$

Assuming that $[S_T]$ is the total concentration of substrate (both bound and unbound) and $[L_T]$ is the total concentration of ligand (both bound and unbound), then it can be written that:

$$[S_T] = [S] + [LS] \tag{3}$$
$$[L_T] = [L] + [LS] \tag{4}$$

Equation (1) may be rearranged to give:

$$[LS] = [L][S]K_a \tag{5}$$

Substitution of Equation (5) into Equations (3) and (4) yields:

$$[S_T] = [S] + [L][S]K_a \tag{6}$$
$$[L_T] - [L] = [L][S]K_a \tag{7}$$

Both Equations (6) and (7) can be rearranged in terms of [S]:

$$\frac{[S_T]}{1 + [L]K_a} = [S] = \frac{[L_T][L]}{[L]K_a} \tag{8}$$

Thus:

$$[S_T][L]K_a = [L_T][L](1 + [L]K_a) \tag{9}$$

Equation (9) can be expanded to a polynomial form:

$$[L]^2 + \left([S_T] - [L_T] + \frac{1}{[L]K_a}\right)[L] - [L_T] = 0 \tag{10}$$

Solving Equation (10) with the quadratic formula results in two roots, only one of which provides a sensible value for [L]:

$$[L] = \frac{[L_T] - [S_T] - (1/K_a) + \sqrt{([L_T] - [S_T] - (1/K_a))^2 + 4[L_T]}}{2} \tag{11}$$

Dividing through by $[S_T]$ and introducing the substitutions $X_R = [L_T]/[S_T]$ and $r = 1/(K_a[S_T])$ yields:

$$\frac{[L]}{[S_T]} = \frac{X_R - 1 - r + \sqrt{(X_R + 1 + r)^2 - 4X_R}}{2} \tag{12}$$

Note here that r is equivalent to $1/c$ (for single site binding, n is 1 and $[S_{tot}] = [S_T]$), where c is the unitless parameter that defines the shape of the binding isotherm defined earlier.

In ITC, the derivative of heat with respect to $[L_T]$ is the measured parameter. Hence, knowledge of the derivative of the reaction species (L and S) with respect to $[L_T]$ is required in order to effect a meaningful analysis. Starting with the ligand, the derivative of Equation (12) is:

$$\frac{d[L]}{dX_R} = [S_T]\left(\frac{1}{2} + \frac{-1 + X_R + r}{2\sqrt{(X_R + 1 + r)^2 - 4X_R}}\right) \tag{13}$$

Recognizing that:

$$\frac{d[L]}{d[L_T]} = \frac{d[L]}{dX_R}\frac{dX_R}{d[L_T]} \tag{14}$$

and:

$$\frac{dX_R}{d[L_T]} = \frac{1}{[S_T]} \tag{15}$$

Then:

$$\frac{d[L]}{d[L_T]} = \frac{1}{2} + \frac{-1 + X_R + r}{2\sqrt{(X_R + 1 + r)^2 - 4X_R}} \tag{16}$$

For a single site binding titration, the heat measured per injection in the ITC is equivalent to:

$$q = [LS]\Delta_b HV \tag{17}$$

where V is the volume of the sample. The experimental signal is recorded as a differential:

$$\frac{dq}{d[L_T]} = \Delta_b HV \frac{d[LS]}{d[L_T]} \tag{18}$$

Recalling from Equation (4) that $[LS] = [L_T] - [L]$, then:

$$\frac{d[LS]}{d[L_T]} = \frac{d[L_T]}{d[L_T]} - \frac{d[L]}{d[L_T]} = \frac{1}{2} + \frac{-1 + X_R + r}{2\sqrt{(X_R + 1 + r)^2 - 4X_R}} \tag{19}$$

Substitution of Equation (19) into Equation (18) then results in:

$$\frac{dq}{d[L_T]} = \Delta_b H \left(\frac{1}{2} + \frac{-1 + X_R + r}{2\sqrt{(X_R + 1 + r)^2 - 4X_R}} \right) \tag{20}$$

Equation (20) is fitted to the binding isotherm data by least squares minimization to recover values for K_a and $\Delta_b H$. Similar derivations can be made for other binding stoichiometries.

Earlier it was noted that the value of c defines the shape of the binding curve and that its value should range between 1 and 1000. As c is equivalent to $1/r$, Equation (20) can be used to illustrate the reasoning behind this statement (values of c from 1 to 1000 correspond to values of r from 1 to 0.001). Figure 2 shows the shapes of the binding isotherms that would be obtained for various r values, from 1 to 0.00001. At values of r smaller than 0.001, the isotherms show very high binding affinities; in effect, the ligand binds tightly and instantaneously until all the binding sites are occupied. It thus becomes impossible to differentiate between binding affinities. At r values greater than 1, the isotherms lose resolution and approach straight lines; binding is thus very weak and, again, it becomes impossible to resolve between binding affinities. Hence, experimental design is critical to the successful interpretation of ITC binding isotherms.

A number of strategies exist for dealing with poor binding isotherms (discussed in detail subsequently). If binding is very strong, then competitive

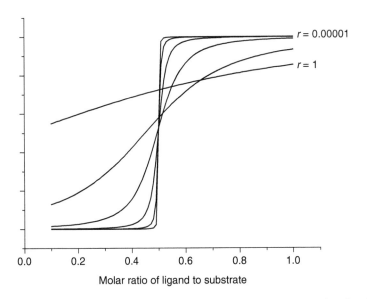

Figure 2 Binding isotherms generated using Equation (20) showing the effect of varying r values (1, 0.1, 0.01, 0.001, 0.0001, and 0.00001) on the shape of the curve.

displacement titrations may be employed. If binding is weak, then simply swapping the positions of the ligand and substrate (i.e., placing the substrate in the syringe and the ligand in the ampoule) may be effective, although this approach is often limited by the solubility of the substrate (which needs to be high in the syringe).

Binding Thermodynamics

Ladbury (7) notes that the promise of designing the perfect drug for a biological target, using computer software, has failed for three primary reasons: the limitation of computer technology, the requirement for structural detail in atomic resolution, and the determination of how structural features can be related to the thermodynamics of interactions. An ITC experiment returns the value of $\Delta_b H$ directly (indeed, it is the only analytical tool capable of doing so) as well as the binding affinity. Although either term could be used as an indicator to drive the rational development of an active, usually the binding affinity is optimized, as this is taken to reflect the overall magnitude of the interaction. However, it is simple to obtain the other thermodynamic parameters governing binding ($\Delta_b S$ and $\Delta_b G$). The value of $\Delta_b G$ is obtained from K_a:

$$K_a = e^{(-\Delta_b G)/RT} \tag{21}$$

where T is temperature and R is the universal gas constant. The value of $\Delta_b S$ is then derived from:

$$\Delta_b G = \Delta_b H - T\Delta_b S \tag{22}$$

If the experiment is conducted at more than one temperature, then it is also possible to derive the change in heat capacity upon binding, ΔC_p:

$$\Delta C_p = \frac{d(\Delta_b H)}{dT} = T\frac{d(\Delta_b S)}{dT} = \frac{\Delta_b H_{T_2} - \Delta_b H_{T_1}}{T_2 - T_1} \tag{23}$$

The heat capacity is an important parameter, as it allows calculation of enthalpy and entropy as a function of temperature (although there is the assumption that ΔC_p does not itself change with temperature).

It is important to recognize that the magnitude of the binding affinity is thus governed by the Gibbs energy. As the Gibbs function comprises both enthalpic ($\Delta_b H$) and entropic ($\Delta_b S$) contributions, there are many combinations of $\Delta_b H$ and $\Delta_b S$ values that will result in the same $\Delta_b G$ value and, hence, binding affinity (2). The binding of a ligand to a receptor is accompanied by changes in inter- and intramolecular interactions of all the components in the system, including water molecules and any other intermediary compounds that may be present. Such changes manifest themselves in the reaction enthalpy and entropy but, necessarily, are not resolved in the binding affinity. Consequently, while the binding affinity is a good measure of the strength of the interaction between a specific ligand and target and can successfully be used to discriminate between many potential drug candidates for which binding data are available, it cannot be used to drive structure–activity relationships because it is insensitive to the molecular basis of binding.

In addition, the phenomenon of enthalpy–entropy compensation must also be considered when assessing binding thermodynamics, as this often lessens the change in the Gibbs energy as the mechanism of binding changes, further reducing the sensitivity of $\Delta_b G$ (and hence K_a) as a basis for QSAR relationships. The principle of enthalpy–entropy compensation is easily explained. The value of $\Delta_b H$ reflects changes in the strength and number of all the interactions between atoms during binding: an increase in the strength or number of interactions is associated with a greater release of heat and, hence, a more negative $\Delta_b H$ value. It is apparent from Equation (22) that a negative enthalpy makes binding more favorable (i.e., to have a negative Gibbs energy). However, the processes of binding usually involve the formation and breakage of a multitude of bonds. In some instances, the binding of the ligand to the substrate will occur with a concomitant release of several waters of hydration; this will lead to an overall increase in the disorder of the system and a positive $\Delta_b S$ value. This increases the negative value of $\Delta_b G$ and makes binding more favorable. Alternatively, it may be the case that binding actually increases the order of the system (when water molecules are not displaced). In this case, the entropy value would be negative. Thus, the increase in enthalpy is balanced by a decrease in entropy (enthalpy–entropy compensation), an effect that acts to reduce the change in $\Delta_b G$.

The interpretation of thermodynamic binding data is therefore not straightforward and must be undertaken with caution. In addition to enthalpy–entropy compensation, there are numerous events that can contribute both to the enthalpic and entropic values. Such events include hydrogen bonding, hydrophobic interactions, electrostatic interactions, changes in conformation, and molecular motion. To drive the rational development of an active, it is thus necessary to understand the binding process on a molecular level. This usually requires ancillary structural data in addition to the thermodynamic parameters. It is also important to consider at least three species when interpreting binding thermodynamic data; the ligand, the substrate, and the solvent. A comprehensive treatment of binding thermodynamics and their relation to biological systems can be found in Cooper et al. (8).

Practical Considerations

The prediction of ligand-binding thermodynamics from first principles is difficult and QSARs based on Gibbs functions are (as noted earlier) insensitive to binding mechanisms, meaning that it is usually the case that the only reliable method for determining binding affinities, even for series of ligands or targets that differ structurally in a logical manner, is by direct experimental measurement. A good example of this is provided by the interactions of carbonic anhydrase II with

4-carboxybenzenesulfonamide (CBS) and 5-dimethyl-amino-1-napthalene-sulfonamide (DNSA), two compounds that mimic variations on a common drug pharmacophore (9). The compounds bind with similar free energies ($\Delta_b G_{CBS} = -8.4$ kcal mol^{-1}; $\Delta_b G_{DNSA} = -8.8$ kcal mol^{-1}), but the binding enthalpies differ by nearly 2.5 fold ($\Delta_b H_{CBS} = -11.9$ kcal mol^{-1}; $\Delta_b H_{DNSA} = -4.8$ kcal mol^{-1}). In this case, the binding of one compound is entro-pically favored, whereas the binding of the other is entropically disfavored. This complexity and lack of predictability mean that SBDD is time consuming and uses considerable resources, as it is preferable to determine binding enthalpies exper-imentally wherever possible.

The binding response of a ligand can be affected by changes in its environ-ment or the protein target and these changes will be different for enthalpically or entropically driven binding, even if the ligands have similar binding affinities. The binding enthalpy reflects the strength and nature of the interaction between the ligand and the receptor (including contributions from van der Waal's forces and hydrogen bonding, for instance), relative to those with the solvent, whereas changes in the binding entropy derive from two sources; changes in solvation entropy and changes in conformational entropy. Upon binding, solvent molecules are released from the receptor, resulting in an increase in solvent entropy, while the ligand and some functional groups of the receptor lose conformational freedom. As a result, conformationally constrained ligands exhibit improved affinity compared with otherwise similar but conformationally relaxed ligands because of a smaller loss in conformational freedom upon binding. Knowledge of $\Delta_b H$ and $\Delta_b S$ there-fore gives a fundamental insight into the nature of binding that assays based on $\Delta_b G$ or binding affinities do not.

It was seen earlier that it is difficult to use ITC to characterize binding, if the binding affinity is very high. Once the binding affinity increases beyond approxi-mately $10^8 - 10^9$ M^{-1}, isotherms lose their characteristic curves and become indistinguishable (5). Recently, an elegant solution to this problem, which had previously been a significant limitation of the technique, has been presented (10). In this approach, a high-affinity ligand is titrated into substrate that has been pre-bound to a weaker inhibitor. Over time, competitive displacement results in the stronger ligand replacing the weaker. The approach requires three titrations: a titration with the weak inhibitor, in order to characterize its binding thermodynamics, a titration with the high-affinity ligand, and the displacement titration. Once the binding characteristics of the weak inhibitor are known, under a given set of conditions, this titration does not need to be repeated. The titration with the high-affinity ligand confirms the binding enthalpy and allows a more rigorous analysis of the displacement titration. This method-ology has allowed the analysis of systems previously outside the scope of ITC, such as the inhibition of HIV-1 protease (11). In addition, selection of a weak inhibitor that has a binding enthalpy of opposite sign to the high-affinity ligand amplifies the calorimetric signal.

If binding is accompanied by a change in protonation, then the interaction will be pH-dependent and the measured binding enthalpy will be dependent upon the ionization enthalpy of the buffer system in which the reaction is occurring (12). The ITC experiment must then be carefully constructed in order that this effect is correctly compensated for. Initially, this is done by performing titrations in buffers with different (and known) ionization enthalpies ($\Delta_i H$). The measured enthalpy ($\Delta_e H$) is then given by (12):

$$\Delta_e H = \Delta_b H + n_H \Delta_i H \tag{24}$$

where n_H is the number of protons absorbed or released. A methodology designed to dissect protonation contributions from binding has been published (13) and applied to several protein systems including aspartic protease (14), stromelysin-1 (15), and HIV-1 protease (11).

Applications in Drug Development

Holdgate and Ward (16) note a number of areas in which ITC measurements can influence drug discovery programs.

1. *Characterization of proteins.* The binding parameters recovered from an ITC experiment are sensitive to the specific protein being studied and to its purity, without the need for the development of additional assays. This allows a quantitative comparison of various purification and storage protocols for instance.

2. *Assay validation.* The success of any assay designed for identification of hits and secondary screening is dependent upon its sensitivity to a range of ligands. When combined with the relentless drive for HTS, this means that assay protocols are sometimes less rigorous than desired. The precision and general applicability mean that at best ITC methods should be the assay of choice, and at worst mean that ITC data can be used to verify data from other assays.

3. *Construct characterization.* Short protein constructs (often containing point mutations) are often used as substitutes for large, wild-type proteins. The binding characteristics of such substitutes must be verified in order to ensure that misleading conclusions are not drawn. ITC, not requiring specific assay development for each protein, is ideal for this assessment.

4. *Characterization of functional binding groups.* QSARs are developed through sequential alteration of specific functional groups on a common moiety and, often, computational packages are used to identify and optimize the functional group to be changed. However, it is always desirable to be able to correlate these "in silico" data with actual in vitro binding data. ITC can be used to assist these studies,

because good experimental design can demonstrate the ability of a secondary ligand to influence the binding of the primary ligand, thus identifying the important functional groups involved in an interaction.

5. *Protonation events.* Performing experiments in buffers with different ionization enthalpies allows an assessment to be made of the number and direction of protons involved in the binding mechanism.

In addition to these (routine) uses of ITC, there are a number of areas that offer future potential. The first, as discussed earlier, is that it is possible to optimize structures, using enthalpic or entropic data, rather than using binding affinities. The potential benefits of such an approach should be clear by this point: the enthalpic and entropic parameters are much more directly affected by changes in the binding mechanism than the Gibbs energy (and hence binding affinity).

Velazquez-Campoy et al. (17) discuss this point in relation to drug-design rationales, taking the enthalpy of binding as a focus. They note that most design programs follow a classic lock and key approach: the structure of the ligand is constrained to fit tightly within a binding site on the target substrate. Thus, the more conformationally constrained the drug molecule is, the less the Gibbs energy penalty becomes upon binding (where, for instance, a bond that is freely able to rotate becomes fixed upon binding), and the more favorable the binding affinity. They further note that an alternative strategy to increase the favorability of binding to a solvent is to decrease the affinity of the ligand for the solvent (usually, by increasing the hydrophobicity of the drug). Selecting a lead compound on the basis of a large binding enthalpy gives confidence that a good interaction occurs between it and the substrate. Subsequent optimization is then possible by introducing conformationally constrained groups, additional enthalpic interactions, or hydrophobic groups. It is much more difficult to add enthalpically-favorable interactions to a lead compound whose binding interaction is primarily entropic in character.

One major drawback of ITC is that it is not really amenable to high-throughput analysis, each experiment being individually set up and run (and lasting an hour or more). This fact, to a large degree, means that alternative assays are used for screening compound libraries and ITC is used to verify the results of the top hits. The realization of the potential of calorimetry in HTS has awaited the availability of instrumentation capable of detecting small heat changes in large numbers of samples on a time scale that is both sympathetic to protein stability and conducive to high-throughput operation. Recently developed solid-state (chip) calorimeter technologies, such as the liquid nanocalorimeter [Xensor Integration BV, (18,19)], the MiDiCal® (Vivactis NV), and the enthalpy array (20), now offer grounds for optimism among investigators: (These new technologies have the sensitivity and capacity to determine extents of reaction for quantities as low as 2.5×10^{-10} mole.) The Xensor calorimeters are used commercially in the Setsys™ range of calorimeters (Setaram, France), although they are limited to an array of four calorimeters and can only be purchased as chips (the user being required to incorporate them into a calorimeter design). In the MiDiCal system, an array of calorimeters is arranged in an automated handling system such that multiple experiments can be conducted simultaneously. The MiDiCal is based on an 8×12 arrangement, as this is a standard layout used in other high-throughput techniques. A demonstration of the use of calorimeter arrays, and the type of information that can be obtained from them, have recently been published by Torres et al. (20). A further discussion of the principles and applications of solid-state calorimeter arrays can be found in Chapter 8.

LEAD OPTIMIZATION USING ISOTHERMAL FLOW CALORIMETRY

The prevalence, adaptability, and sheer tenacity of microorganisms present a number of unique challenges. They are responsible for blockages in oil-carrying pipelines. They have been found growing happily in nuclear reactors, and one particular robust *Streptococcus* bacterium was found to be still viable after an extended stint, sealed in a camera case, left on the moon several years after the Apollo landing (21). They are, of course, also responsible for everyday problems such as spoilage of food (22) and pharmaceuticals as well as being the primary cause of death and illness in the world.

Consequently, detection of contamination and control and prevention of infection are of utmost importance. The pharmaceutical industry is therefore seeking constantly to improve methods for the detection of bacteremia, and the effective and rapid control of infection is the focus of many pharmaceutical treatment regimens. The regular emergence of resistant microorganisms (23–25) requires a constant search for new therapies. These new therapies are made increasingly possible by rapid developments in the fields of genomics and proteomics (26), both of which offer the possibility of tailoring molecules to target specific microorganisms (27).

Combinatorial chemistry has been the vogue in recent years with many thousands of NCEs being synthesized. It is equally possible then that new antimicrobials may have already been synthesized but not yet recognized. There is thus a drive for HTS of these compounds against target organisms to determine their efficacy. To this end, various techniques have been developed by researchers in the field.

Early methodologies for antimicrobial testing revolved around incorporation of the antimicrobial agent to be tested into agar (26). This technique, agar diffusion, although a breakthrough in susceptibility testing, was quickly found to be time consuming and cumbersome in application. Moreover, the inherent variability of the technique often gave rise to spurious data if not used properly. One of the ways in which this variability was reduced was to introduce replication devices and to replace serial dilutions (28) with a critical concentration of antimicrobial, which separated resistant strains from susceptible strains; the so-called breakpoint test (29).

These classical methods for assessing susceptibility of microorganisms still suffer from the major drawback of being relatively time consuming. A typical assay can take 18 to 24 hours to perform. It has been reported (30), however, that some methods can return data after as little as four to six hours in some instances. A further and perhaps more serious limitation of these techniques is that they all rely on phenotypic testing (i.e., the bacteria are isolated from a pure culture on an unselective medium). Here, a single colony is selected and inoculated onto a secondary plate, which is then used to provide the inoculum for the analysis. Although a relatively simple procedure, this can take as long as two days to achieve. This time delay in a clinical setting is intolerable, and it is usual for the treatment regimen to begin before the identity of the bacteremic organism is known.

Further criticisms of this testing technique were voiced by Bergeron and Ouellette (31). In particular, they noted that the outcome of such tests can be highly dependent upon the experimental conditions and that in many cases more than one test needs to be performed to obtain accurate susceptibility profiles. They also report that different bacterial species can exhibit different susceptibilities to the same antibiotic and that no international standard exists on breakpoints for the interpretation of susceptibility tests. Furthermore, these tests are also subject to the vagaries of classical microbiological testing, for example the use of nonstandardized inocula, which can lead to conflicting results within or between laboratories. They also often require large numbers of observations for satisfactory results, are labour-intensive, and cannot always be automated.

A more wide-ranging concern of these testing methodologies is that the microorganism/drug interactions are often studied under conditions that are far from being representative of in vivo conditions, where interactions between any number of species (proteins, metal ions, other cells, and so on) may become important. These classical microbiological tests therefore all exhibit specific disadvantages. Those tests that employ photometric methods of analysis require that the antibiotic does not interfere with the turbidity measurement. Respirometric methods rely upon the fact that any indicators used must not interfere with the assay and that the antibiotic preparation is homogenous. Agar diffusion methods are not effective at assessing synergistic or antagonistic effects of drugs.

These techniques are now being succeeded by genotype testing with DNA-based assays being developed specifically to probe bacterial resistance genes (23). These analyses too have come under criticism from Bergeron and Ouellette (31), who encourage the notions that the presence of a resistant gene may not always be indicative of a resistant bacterium, or that if a gene coding for resistance is not detected, then it does not necessarily follow that the bacterium will be susceptible to that particular agent. Genotypic testing is hailed as a major breakthrough, although it does still require a good understanding of the mechanism of resistance and the genes involved. Genotypic testing is not cost-effective for routine use at present, and the process needs to be developed further in order to identify the resistant organism if it is to become the norm in a clinical setting.

A consequence of the disadvantages encountered in classical techniques is that alternative methods of analysis for the interaction between organisms and drugs are constantly sought. Isothermal calorimetry (IC) has shown great promise in this area of analysis since the early 1950s (32). These early studies tried initially to correlate observed heat effects with thermodynamic data and were largely descriptive in nature. The potential capacity of calorimetry to identify and discriminate between organisms has also been recognized (33). This identification may be possible from comparison of heat profiles produced by growing organisms in some defined medium.

Calorimetry for Following Microbial Growth

As has been discussed in the preceding chapters, nearly all processes result in exchange of heat energy to (or from) the environment; growth and metabolism of microorganisms are no exceptions.

The use of calorimetry for the study of micro-biological systems was widely reported in the early days of calorimetry and a number of pioneering studies were conducted [for a comprehensive review, see (34)]. This initial enthusiasm for studying cellular systems waned in the early 1970s to be replaced by measurements of more fundamental thermodynamic and kinetic parameters from less complex systems. Recent years have seen a resurgence in the popularity of IC for use in microbiological studies (35–37). Several factors may be responsible for this: (*i*) the development of more sensitive instruments, (*ii*) the availability of instruments with high-throughput capabilities, or (*iii*) the pharmaceutical industry seeking improved methods for the investigation of such complex systems.

IC techniques have been historically (and recently) used successfully to study a number of systems including (*i*) assessment of the microbiological load of soils (38), (*ii*) spoilage of foods (39), (*iii*) efficacy of pharmaceuticals (40), and (*iv*) action of drugs on animal (including human) cells (41) as well as studies on living animals [see Ref. 42 and references therein]. It is interesting to note that the majority of recent papers, describing applications of IC to microbial systems, center on the assessment of soil quality and food spoilage rather than new developments in the field of antibiotic susceptibility testing, although this imbalance is slowly being rectified. The description of all of these applications is beyond the scope of this book. Here, only the capacity of IC for use in susceptibility testing of microorganisms will be considered.

Calorimetry is then a powerful tool for the study of microorganisms and their interactions with drugs. There are many commercial instruments available capable of studying such systems (see Chap. 1 for more details). Modern instruments are very sensitive and, if used correctly, offer good reproducibility and precision between measurements. They have the capability to operate at different temperatures and specifically designed inserts allow a large degree of control over the conditions within the calorimetric cell (e.g., stirring rate, pO_2, addition of substrates). The baseline stability of modern instruments permits long-term experiments (days or even weeks) to be performed. It is possible, with some instrument designs, to perform parallel analytical measurements. Perhaps the most important features of all calorimeters are that they are noninvasive in their measurement and are invariant to the physical form of the sample. This latter feature permits the interactions of interest to be studied under conditions which are as close to in vivo as possible; that is, complex/heterogeneous media can be used, and interactions can be monitored in the presence of multiple species (e.g., bacterial load in whole blood can be assessed).

There are, of course, many different calorimeter designs, all of which have some capacity to investigate microbial systems. For the purposes of the discussion, here they will be classified as either ampoule techniques (i.e., a static system in a closed ampoule) or flow techniques. Each approach has its advantages and disadvantages and, as such, appropriate selection of the correct instrument is vital. In general, ampoule instruments exhibit more disadvantages than observed for flow instruments.

Ampoule Calorimeters

Ampoule calorimeters generally employ removable glass or stainless steel ampoules, which are hermetically sealed. Normally, the contents of these ampoules are unstirred. This poses the problem of how to deal with issues such as sedimentation of cells, which can cause concentration gradients of pH, nutrients, cells, and so on. These effects can affect significantly the observed power-time trace obtained and are a major limitation associated with all ampoule-type calorimeters. It is, however, possible to purchase specially adapted inserts that permit stirring of the reaction medium for some instruments.

In all ampoule instruments, it is the case that the ampoules are hermetically sealed. This means that it is very difficult to maintain control over the atmosphere inside the calorimetric ampoule (43). This poses particular problems if the organism under study is an aerobe. It is possible in some instances to control the atmosphere within the calorimetric ampoule by the use of a gas perfusion module. A further limitation associated with having sealed ampoules is that it is difficult to sample the system under study during the course of the experiment, and it is not convenient to add other material to the medium (e.g., drug or substrate) without upsetting the thermal equilibrium of the calorimeter. These limitations can make the analysis of the power-time curves difficult.

The fact that the ampoules are hermetically sealed does, of course, provide some benefits. The samples are completely removed from the environment, making it less problematic to study high-risk organisms. Because the ampoules are removable and can be disposable, the risk of cross contamination between experimental runs is minimized.

Flow Calorimeters

Flow calorimeters differ from ampoule instruments in that the reaction medium is held externally from the calorimeter in some sort of thermostatic bath and is pumped through the calorimetric cell either in a return loop or to waste. This arrangement then offers some significant advantages. For example, a flowing system (if flowing quickly enough) will minimize sedimentation in the calorimetric cell and thus can avoid the issues with pH gradients seen for ampoule techniques. Moreover, because the reaction medium is held separately from the calorimeter, it is possible to control more precisely the gaseous environment to which the system is exposed. It is also the case that this separation provides easy access to the medium and therefore allows the facile addition of cell modifiers to the incubation vessel without upsetting the balance of the thermal equilibrium of the calorimeter.

This flexibility does have its drawbacks. It is often the case, in order to ensure good thermal equilibration between the reaction medium and the calorimeter, that a long lead time is required. This can be circumvented, in part, by employing higher flow rates, but this in itself can cause problems (see Chap. 3 for a detailed discussion of flow calorimetry). Although minimized, there is still the possibility of sedimentation (especially of larger cells) and hence blocking (or partially blocking) the flow lines with cellular matter. These issues have been recognized by Guan et al. (44) in their studies on Chinese hamster ovary cells, to which end they have reported the design and application of a modified flow insert with wide bore tubing and high flow-rate capacity. It is also important to ensure that the external incubation vessel does not become contaminated with other organisms (from the atmosphere, for instance) while the experiment is conducted. There is the added limitation that the reaction medium must be able to be pumped through calorimetric tubing. This can be an issue if the medium is highly viscous. In addition, flow instruments are, on the whole, less sensitive than their ampoule equivalents. In general, the heat generated by a concentration of cells of $<10^4$ colony-forming units (CFUs) mL^{-1} is insufficient to be detected by the calorimeter (although this value depends on the organism under study and the conditions under which it is studied).

It should also be noted that the physical effects associated with absorption of cellular matter or medium constituents to the exposed inner surfaces of the calorimeter may also pose problems for both modes (ampoule or flow) of operation. This issue can be mitigated somewhat by either coating the inner surfaces with an inert material (e.g., Repelcote, a silicone-based material) or by conditioning the calorimetric ampoule in some way prior to starting the experiment (for instance, introducing medium into the ampoule before the start of the experiment).

Even though calorimetric techniques suffer limitations of their own, the fact remains that they offer many desirable advantages over and above conventional assays (Table 1). The capacity to record thermodynamic and kinetic

Table 1 Comparison of the Classical Agar Plate Diffusion Technique with Isothermal Calorimetry

	Microcalorimetry	Agar plate diffusion
Reproducibility (%)	3	5–10
Lowest determinable concentration (unit mL^{-1})	0.1	20
Range (unit mL^{-1})	0.1–100	20–100
Time per assay (hr)	1–6	16–24

Source: From Ref. 35.

information, in real time, of complex heterogeneous systems under conditions that are closer to those in vivo than can be achieved with other techniques, is of vast importance if quantitative information is to be derived.

Form of Calorimetric Output

The complex sequence of reactions that comprise metabolism, cell replication, and utilization of nutrient sources as well as cell death, all contribute to the power–time signal. In general, the shape of a typical power–time curve, for a growing organism, exhibits an initial exponential increase in heat followed by a series of peaks and troughs, which can be attributed, in part, to the sequential utilization of energy sources (45); Figure 3 shows a typical example. If some modifier is added now to these growing systems, an antimicrobial drug for example, then the recorded power–time curve will change according to the modifier added, its concentration, and the time at which it was added. In a simpler

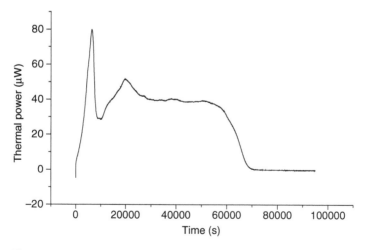

Figure 3 Typical calorimetric response for the growth of a microorganism.

environment, for example, where a drug is metabolizing a single nutrient source but is not growing, then often a less complex power–time trace is observed, but the premise remains that this curve will change if the system is modified in some way. It is the way in which the power–time trace changes in the presence of these modifiers that provides clues as to the efficacy and sometimes the mode of action of the drug.

Characterization of Microorganisms

Earlier in this chapter it was noted that the potential capacity of IC to identify microorganisms from their thermal profile had been identified. Although this is an attractive notion, the feasibility of such an identification procedure has been criticized by Newell (33). The premise of the identification is that individual species of organism will, under strictly defined conditions, yield distinct power-time curves. In reality, it is nearly impossible to reproduce experimental conditions between experiments exactly. Even very small variations can significantly affect the form of the data. In particular, the largest hurdles to overcome are the standardization of viable cell inoculum and inoculum history. The former has been dealt with to some extent by Beezer et al. (46) by developing standardized freezing and storage protocols for large volumes of inocula (essentially allowing the inoculum to be treated as a standard reagent). The latter still poses significant problems in that it immediately restricts the analysis to isolate from similar types of background, for example, blood, sputum, and so on.

There also exists the problem of variation between calorimetric instruments and calorimetric procedures, giving rise to inconsistencies between nominally-identical reaction systems (47). Newell (33) concludes that to claim IC is capable of achieving anything other than an academic identification is inappropriate, but that it is its use for characterization that provides the most interesting application.

Calorimetric Analysis of Drugs

Beezer (34) notes that the analysis of drugs can be subdivided into three broad objectives. The first is the quantitative and qualitative assay of the drug in a variety of matrices. The second is the investigation of the mode of action of the drug (this is important for three reasons: if the action of the drug is to be selective in nature, to aid in the design of new drug entities, and so that the drugs can be used as "tools" in scientific experiments). The third objective is the design and clinical evaluation of new drugs. At all stages of the process, the qualitative and quantitative analyses of the drug(s) in question is important.

Qualitative analysis: Some of the earliest experiments, studying the effects of antimicrobials on organisms, were conducted by Prat (32) and Zablocki et al. (48). For example, Prat (using a Calvet batch instrument) demonstrated that the effect of adding 30 mg of streptomycin to a culture of *Escherichia coli* in broth was to reduce the power output from the organism to one-tenth of its original value after approximately six hours. It was also noted that the observed response of the organism to streptomycin was much more rapid if the culture was stirred, indicating that the effectiveness of the streptomycin was limited by the rate of diffusion through the culture medium.

These early studies were followed by Zablocki et al. (48), who, in 1962, reported the results of a study in which the mode of action of penicillin on *E. coli* was probed. They reported that the response of the organism varied according to the time at which the penicillin was added. In particular, they noted that there was a reduction in the latent period for cells in the exponential phase compared with those in the stationary phase. These studies were repeated for a series of drugs, streptomycin, chlortetracycline, and oxytetracycline, all of which caused changes in the latent period of *E. coli*. Vine and Bishop (35) expanded on these conclusions and postulated that it may be that the cells in the exponential phase are more susceptible to the drug(s), because they are "younger" compared with those in the stationary phase and perhaps are less hardy. This early indication that the mode of action of these drugs varies, according to the position in the life cycle at which they are employed, is an important observation and is the driver for further microbiological studies (perhaps in conjunction with complementary IC techniques) to elucidate the exact mechanism of action.

There then followed a number of studies using flow calorimetric techniques. The effect of ampicillin on growing cultures of *E. coli* has been reported (48). It was shown that, at 310 K, the observed power–time responses for *E. coli* were significantly different in the presence and absence of ampicillin, thus indicating some sensitivity to the drug.

The relative efficacy of various tetracycline drugs has also been measured by flow calorimetry (49). The authors report that the sensitivity of *E. coli* under aerobic conditions, in the logarithmic phase of growth at 310 K, above and below the minimum inhibitory concentration (MIC), could be observed clearly. The sensitivity of *E. coli* to the tetracyclines used under these conditions was minocycline > doxycycline > oxytetracycline > tetracycline. The authors also report that this information could be used in conjunction with pharmacokinetic data to establish dosing regimen.

Beezer et al. (50) have also utilized flow calorimetry to study the effects of a series of antibiotics on different strains of *E. coli*. They were able to show that at concentrations below the MIC the effect on the organism was marginal or nonexistent, while for concentrations greater than the MIC the heat output from the organisms was reduced.

Kjeldsen et al. (51–54) reported the results of a series of investigations to develop assays of the effects of an antimicrobial spray (Trisep®)[a] on three organisms: *Bordetella bronchiseptica*, *Bacillus pumilus*, and *Micrococcus luteus*. The

---

[a]Trisep® is a registered trademark of ICI PLC, Macclesfield, UK, Aquacel AG Hydrofiber® is a registered trademark of ConvaTec Ltd, Uxbridge, UK, Acticoat 7 with SILCRYST® is a registered trademark of Smith and Nephew Healthcare Ltd, Hull, UK.

organisms were so chosen because they are the test organisms for the components of the Trisep spray (polymixin B sulfate, neomycin sulfate, and zinc bacitracin). They observed that it was likely that antagonistic behavior between the antibiotics, when in equimolar ratios, resulted in increased effectiveness against the organisms compared with the formulated mixture. Parallel studies were also conducted on *E. coli* in order to determine the feasibility of developing assays for these drugs using a single test organism. They reported significant improvements in terms of time, sensitivity, and reproducibility over the conventional microbiological assays for these drugs but were unsuccessful in developing assays using *E. coli*. The authors concluded, however, that a screening program could, in principle, identify a set of more sensitive organisms better suited to a rapid and reproducible calorimetric assay.

A more recent study (40) has highlighted the use of IC to distinguish qualitatively the efficacy of the silver containing wound dressings between Aquacel Ag Hydrofiber® and Acticoat™ 7 with SILCRYST®. The wound dressings studied were all (with the exception of Acticoat which swelled slightly but retained its physical form) reported to form gels in the presence of aqueous media, making any study of their effect by conventional means highly challenging. This challenge is further increased, as it is thought that the gel forming dressings (without the presence of silver) may also provide some inhibitory effect by trapping the microorganisms within the forming gel, thus limiting ingress of nutrients and egress of toxic metabolites. The dressings were all designed to act by releasing silver to the wound environment in order to provide antibiotic protection against common wound pathogens. The organisms tested were *Pseudomonas aeruginosa* and *Staphylococcus aureus*. Figure 4 shows the growth curves for *P. aeruginosa* and *S. aureus* at 310 K. Figure 5 shows the response of *P. aeruginosa* to the Aquacel dressing (nonsilver).

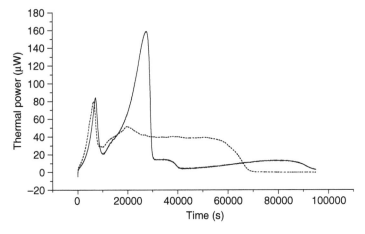

Figure 4 Calorimetric response of *Pseudomonas aeruginosa* (*solid line*) and *Staphylococcus aureus* (*dotted line*) at 310 K in a batch experiment. *Source*: From Ref. 40.

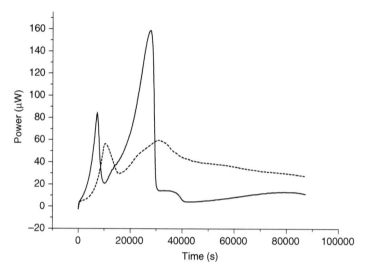

Figure 5   Calorimetric response of *Pseudomonas aeruginosa* control experiment (*solid line*) and in the presence of Aquacel (nonAg) dressing (*dotted line*). *Source*: From Ref. 40.

The marked reduction in the calorimetric output is a clear indication that the hypothesis that the dressing itself is inhibitory to the growth of the organism is valid. Figure 6 shows the response of *S. aureus* in the presence of the two wound dressings. The power–time curves from both dressings imply that

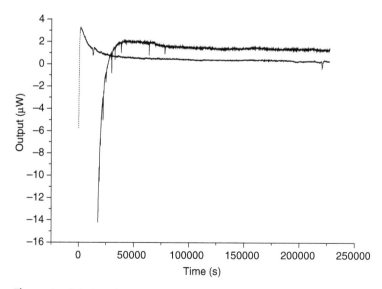

Figure 6   Calorimetric response of *Pseudomonas aeruginosa* (*solid line*) and *Staphylococcus aureus* (*dotted line*) in the presence of Aquacel Ag dressing. *Source*: From Ref. 40.

complete kill is achieved (an observation confirmed by agar plate tests on samples of residual growth medium from the calorimetric ampoule). The low-level responses are postulated by the authors to be interactions between metal ions and proteins and associated cell detritus, and not microbial growth, again confirmed by the agar plate tests. The long term efficacy of the Aquacel Ag Hydrofiber dressing was examined over a seven day period with a reinoculation to the original cell density (approximately, 10^6 CFU mL^{-1}) after five days: again, complete inhibition of growth was reported.

The application of IC to the qualitative analysis of drugs is not limited to antimicrobials. Several studies have investigated the effects of antifungal and antiviral species. Beezer et al. (55) and Cosgrove et al. (56) used combinations of clotrimazole and 5-fluorocytosine with amphotericin B in the growth and res-piration phases of *Saccharomyces cerevisiae* SQ1600. They were able to dis-tinguish between antagonistic and synergistic effects. They reported that the combination of 5-fluorocytosine and amphotericin B did not exhibit antagonistic or synergistic behavior compared with amphotericin B alone. In contrast, they showed that the combination of clotrimazole and amphotericin B showed anta-gonistic effects in both the respiration and growth phase of the organism. These observations were further confirmed by measurements made on the absorp-tion of clotrimazole and amphotericin B to the yeast cells. The results indicated that there exists a competition between the two drugs for sites on the cell membrane, hence giving rise to the antagonistic behavior.

A relatively recent article reported the results of a study on the metabolism of Vero cells infected by Herpes simplex virus type I (HSV-1) (57). It is reported that HSV-1 interferes with the metabolism of the infected Vero cells, resulting in an increase in heat production. This observation is attributed to the virus growing and replicating within the living host cell. When acyclovir, an antiviral, was added to the system, the power–time curves became consistent with the normal metabolism of the Vero cells. It was reported that acyclovir had no detri-mental effect on the normal metabolism of Vero cells. Although an interesting application of IC, it should be noted that the conditions within the calorimetric ampoule were not controlled: a cell concentration of 2×10^5 CFU mL^{-1} was given, but there is no comment as to the accuracy or reproducibility of this inocu-lum. Therefore, the results of this study should be treated with caution, as should the claim that this particular method could yield quantitative information.

The examples given earlier (and many more not reviewed here) are all qualitative in nature; this is perhaps unsurprising, given the complexity of living cellular systems. It is possible however, if the experimental conditions are carefully controlled, to gain quantitative information from complex systems.

Quantitative analysis: Because of the extremely complex nature of the power–time curves obtained from growing cultures, it is exceptionally difficult to obtain quantitative data on the organism itself and, consequently, very few examples exist in the literature. Quantitative analysis is proving elusive because of the lack of precise analytical data and the thermodynamic/kinetic equations that describe those data. Another barrier to quantitative analysis of microbial systems is the variation exhibited in the metabolic performance of the inocula. These variations can arise from inoculum size and history as well as the phase of growth which they occupy at the time of testing (33). It is in some instances possible to constrain the system in some way so as to simplify the response, such as forcing the organisms to respire only and not increase in biomass. This can be achieved by supplying only an energy source. Under such conditions, the kinetic profile of the organism respiring is governed by Michaelis–Menten kinetics and thus quantitative kinetic information can be gleaned. The analysis of IC data using Michaelis–Menten kinetics is described elsewhere (58). Some researchers have reported that it is also possible to obtain quantitative data on the drug being used to probe the microbial system.

As a consequence of the limiting factors described earlier, Beezer et al. (46) described a method that increased significantly the reproducibility between inocula, allowing greater faith to be had in quantitative results obtained from IC studies. They report that it is essential to have a defined growth protocol, combined with a rigorous freezing and storage procedure in liquid nitrogen, to have a highly reproducible inoculum.

Using this standardized procedure for organism preparation, they undertook a study to make a quantitative analysis of the interaction of a series of polyene antibiotics (nystatin, filipin, pimaricin, amphotericin B, candicin, and lucensomcin) and the antifungal species clotrimazole and 5-fluorocytosine with the yeast *S. cerevisiae* NCYC 239 (55). The experiments were conducted at 303 K and under anaerobic conditions, using an LKB 10700-1 flow calorimeter. The yeast was inoculated into a pH 4.5 buffer, containing 10 mM glucose, in order to ensure only respiration (hence zero-order kinetics) took place. The addition of the drugs resulted in a marked decline in the respiration of the yeast with the exception of 5-fluorocytosine. This drug acts against DNA replication and hence its effects would be observed only in growing culture. They reported that the dose–response relationship for the polyenes and clotrimazole, between concentrations of 10^{-5} M and 10^{-6} M, was linear and that this relationship was dependent upon the nature of the drug and/or its concentration. As a consequence, they claim that a quantitative analysis of the efficacy of these drugs is possible.

The realization that microbial systems forced into zero-order kinetic conditions could be monitored by IC allowed the quantitative analysis of drug concentration. This bioactivity determination is effected through a simple determination of the response of the calorimeter from a drug-treated system at a fixed time after inoculation with the drug (Fig. 7). From these simple measurements, it is possible to construct log dose–response curves (Fig. 8). Such measurements are intrinsically linked to the kinetic response of the organism to the drug.

The kinetic information that can be derived from calorimetric data has been shown to be useful in generating QSARs. QSARs have been described using equations of varying complexity and the discussion of which could form the basis of an entirely separate text. For an introduction to QSARs, see Ref. 59.

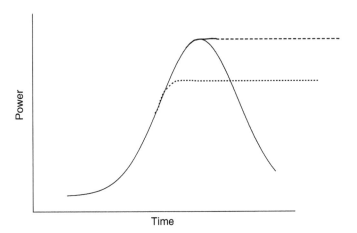

Figure 7 Schematic power–time curves for a control system-buffered glucose inoculated with a microbial suspension (*dashed line*), microbial system treated with drug A (*dotted line*) and drug B (*solid line*). The solid and dotted lines reveal different modes of action of drugs on interaction with the organism. *Source*: From Ref. 34.

Equation (25) gives one example of the form a QSAR equation could take.

$$\log\left(\frac{1}{C}\right) = k_1 \log(P^2) + k_2 \log(P) + k_3\sigma + k_4 \tag{25}$$

Here, C represents the concentration required to elicit a response in a specific time (and hence $\log(1/C)$ can be considered to be a biological response); the k

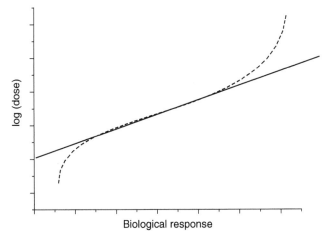

Figure 8 A schematic representation of a log(dose) versus biological response curve obtained from calorimetric data. The dotted line represents the actual log(dose) response curve of the microbial system to the added antimicrobial. The intercept of the solid line with the y-axis represents $\log(\text{dose})_{\text{max}}$ (i.e., the minimum dose below which no effect is observed). *Source*: From Ref. 34.

values are correlation constants; P is the partition coefficient and σ the Hammett sigma function. As the form of this equation defines the biological response in terms of time, it is essentially a kinetic model. Quantitative information such as $\log(\text{dose})_{\max}$ (the maximum concentration of a drug that can be added without eliciting a response) can be derived from such measures of biological response and can be used as a sensitive measure of drug potency.

Montanari et al. (60,61) have published the results of a study determining QSARs from IC techniques. They claim that biological activity can be measured by calorimetry for structurally related compounds whose activity is not readily measured by more conventional methods. The benefit that calorimetry offers in the elucidation of QSARs is that it can, in part, help to overcome the problems associated with the measurement of biological properties (IC 50, K_i etc.). The capacity to study complex systems and the possibility of fast and reproducible studies all contribute to its attractiveness for the elucidation of QSARs. Montanari et al. (61) assayed a series of antimicrobial hydrazides against *E. coli* and *S. cerevisiae*. They report that the potency of the compounds could be ranked according to their structural and chemical characteristics and that these potencies could be quantitatively determined by IC and thus could be used in QSAR studies.

SUMMARY

Calorimetry can play a central role in lead compound optimization. Here, two principal areas of application have been discussed. In the first, the role of ITC in lead compound development was explored; the technique is unique in its ability to measure a thermodynamic quantity directly and, hence, allows the derivation of QSARs based on a number of parameters. Although the binding affinity is usually the parameter optimized, there are many arguments in favor of the use of the binding enthalpy as an alternative. In any case, the fact that the ITC measurements can be made directly on complex systems means that the technique will remain as a fundamental tool in pharmaceutical development strategies.

Calorimetry is perhaps one of the most attractive techniques for the measurement of complex biological systems. Its noninvasive and nondestructive nature coupled with a universal measurement means that it is ideal for the measurement of microorganisms whose measurement by conventional techniques usually requires some form of extraction and analysis in media remote from in vivo conditions. Its versatility is also, in part, a burden to quantitative analysis. Despite its shortcomings, IC offers significant advantages over conventional techniques for the study of microorganisms and their interaction with antimicrobial agents. In particular, the potential timesaving is highly desirable in the clinical setting. Many researchers have contributed to the field of microbiological calorimetry but, despite this vast array of knowledge, there are still significant steps to be taken. For example, truly quantitative assays of organism and organism response are still elusive, although this may be ameliorated by new developments in the ways in which calorimetric data are analyzed (62). It is also the case that calorimetry still lacks the high-throughput capacity that would make it universally appealing. To this end, new instruments are appearing on the market allowing multiple parallel experiments to be conducted. As researchers begin to realize the potential capabilities for such instruments, it is likely that a renewed vigor in the field of microbiological calorimetry will be observed.

REFERENCES

1. Fox S. Accommodating cells in HTS. Drug Discovery World 2003; 5:21–30.
2. Freire E. Isothermal titration calorimetry: controlling binding forces in lead optimization. Drug Discov Today Technol 2004; 1:295–299.
3. Holdgate GA. Making cool drugs hot: isothermal titration calorimetry as a tool to study binding energetics. Biotechniques 2001; 30:164–184.
4. Ward WHJ, Holdgate GA. Isothermal titration calorimetry in drug discovery. Prog Med Chem 2001; 38:309–376.
5. Wiseman T, Williston S, Brandts JF, Lin LN. Rapid measurement of binding constants and heats of binding using a new titration calorimeter. Anal Biochem 1989; 179: 131–137.
6. Indyk L, Fisher HF. Theoretical aspects of isothermal titration calorimetry. Methods Enzymol 1998; 295:350–364.
7. Ladbury JE. Isothermal titration calorimetry: application to structure-based drug design. Thermochimica Acta 2001; 380:209–215.
8. Cooper A, Johnson CM, Lakey JH, Nollmann M. Heat does not come in different colours: entropy-enthalpy compensation, free energy windows, quantum confinement, pressure perturbation calorimetry, solvation and the multiple causes of heat capacity effects in biomolecular interactions. Biophys Chem 2001; 93:215–230.
9. Day YSN, Baird CL, Rich RL, Myszka DG. Direct comparison of binding equilibrium, thermodynamic and rate constants determined by surface- and solution-based biophysical methods. Protein Sci 2002; 11:1017–1025.
10. Sigurskjold BW. Exact analysis of competition ligand binding by displacement isothermal titration calorimetry. Anal Biochem 2000; 277:260–266.
11. Velazquez-Campoy A, Kiso Y, Freire E. The binding energetics of first- and second-generation HIV-1 protease inhibitors: implications for drug design. Arch Biochem Biophys 2001; 390:169–175.
12. Gomez J, Freire E. Thermodynamic mapping of the inhibitor site of the aspartic protease endothiapepsin. J Mol Biol 1995; 252:337–350.
13. Baker BM, Murphy KP. Evaluation of linked protonation effects in protein binding reactions using isothermal titration calorimetry. Biophys J 1996; 71:2049–2055.
14. Xie D, Gulnik S, Collins L, Gustchina E, Suvorov L, Erickson JW. Dissection of the pH dependence of inhibitor binding energetics for an aspartic protease: direct measurement of the protonation states of the catalytic aspartic acid residues. Biochemistry 1997; 36:16166–16172.
15. Parker MH, Lunney EA, Ortwine DF, Pavlovsky AG, Humblet C, Brouillette CG. Analysis of the binding of hydroxamic acid and carboxylic acid inhibitors to the stromelysin-1 (matrix metalloproteinase-3) catalytic domain by isothermal titration calorimetry. Biochemistry 1999; 38:13592–13601.

16. Holdgate GA, Ward WHJ. Measurements of binding thermodynamics in drug discovery. Drug Discov Today 2005; 10:1543–1549.
17. Velazquez-Campoy A, Luque I, Freire E. The application of thermodynamic methods in drug design. Thermochimica Acta 2001; 380:217–227.
18. Herwaarden AW, Sarro PM, Gardner JW, Bataillard P. Liquid and gas microcalorimeters for (bio)chemical measurements. Sensor Actuator 1986; 10:321–346.
19. Lerchner J, Wolf A, Wolf G. Recent developments in integrated circuit calorimetry. J Therm Anal Cal 1999; 57:241–251.
20. Torres FE, Kuhn P, De Bruyker D, et al. Enthalpy arrays. Proc Nat Acad Sci USA 2004; 101:9517–9522.
21. Bryson B. A Short History of Nearly Everything. USA: Broadway Books, 2003.
22. Gardea AA, Caravajal-Millán, Higuera-Ciapara, et al. Calorimetric assessment of microbial growth in milk as affected by different conditions. Thermochimica Acta 2002; 394:179–184.
23. Poupard JA. Is the pharmaceutical industry responding to the challenges of increasing bacterial resistance? Clinical Microbiology Newsletter 2005; 28:13–15.
24. Levy SB. The challenge of antibiotic resistance. Sci Am 1998; 278:32–39.
25. Walsh C. Molecular mechanisms that confer antibacterial drug resistance. Nature 2000; 406:775–781.
26. Wheat PF. History and development of antimicrobial susceptibility testing methodology. J Antimicrob Chemother 2001; 48(S1):1–4.
27. McDevitt D, Rosenburg M. Exploiting genomics to discover new antibiotics. Trends Microbiol 2001; 9:611–617.
28. Steers E, Foltz EL, Graveds BS. An inocula replicating apparatus for routine testing of bacterial susceptibility to antibiotics. Antibiot Chemother 1959; 9:307–311.
29. Ericsson H, Sherris JC. Antibiotic sensitivity testing. Report of an international collaborative study. Acta Pathol Microbiol Scand Suppl 1971; 217:1–90.
30. Aldrige C, Jones PW, Gibbson S, et al. Automated microbiological detection/identification system. J Clin Microbiol 1977; 6:406–413.
31. Bergeron MG, Ouellette M. Preventing antibiotic resistance through rapid genotypic identification of bacteria and of their antibiotic resistance genes in the clinical microbiology laboratory. J Clin Microbiol 1998; 36:2169–2172.
32. Prat H. Observations on bacterial thermogenesis. Rev Can Biol 1953; 12:19–34.
33. Newell RD. The identification and characterization of micro-organisms by microcalorimetry. In: Beezer AE, ed. Biological Microcalorimetry. London: Academic Press, 1980.
34. Beezer AE, ed. Biological Microcalorimetry. London: Academic Press, 1980.
35. Vine GJ, Bishop AH. The analysis of microorganisms by microcalorimetry in the pharmaceutical industry. Curr Pharm Biotech 2005; 6:223–238.
36. Torres A, Garedewa A, Schmolz E, Lamprecht I. Calorimetric investigation of the antimicrobial action and insight into the chemical properties of "angelita" honey—a product of the stingless bee *Tetragonisca angustula* from Colombia. Thermochimica Acta 2004; 415:107–113.
37. Critter SAM, Freitas SS, Airoldi C. Microbial biomass and microcalorimetric methods in tropical soils. Thermochimica Acta 2002; 394:145–154.
38. Prado GS, Airoldi C. The influence of moisture on microbial activity of soils. Thermochimica Acta 1999; 332:71–74.
39. Alklint C, Wadsö L, Sjöholm I. Effects of modified atmosphere on shelf-life of carrot juice. Food Control 2004; 15:131–137.

40. O'Neill MAA, Vine GJ, Beezer AE, et al. Antimicrobial properties of silver-containing wound dressings: a microcalorimetric study. Int J Pharm 2003; 263:61–68.
41. Wadsö I. Isothermal microcalorimetry for the characterization of interactions between drugs and biological materials. Thermochimica Acta 1995; 267:45–59.
42. Lamprecht I. Calorimetry and thermodynamics of living systems. Thermochimica Acta 2003; 405:1–13.
43. Winkelmann M, Hüttl R, Wolf G. Application of batch-calorimetry for the investigation of microbial activity. Thermochimica Acta 2004; 415:75–82.
44. Guan YH, Lloyd PC, Kemp RB. A calorimetric flow vessel optimised for measuring the metabolic activity of animal cells. Thermochimica Acta 1999; 332:211–220.
45. Russell WJ, Farling SR, Blanchard GC, Boling EA. Interim review of microbial identification by microcalorimetry. In: Schlessinger D, ed. Microbiology-1975. Washington D.C.: American Society for Microbiology, 1975.
46. Beezer AE, Newell RD, Tyrrell HJV. Application of flow microcalorimetry to analytical problems: the preparation, storage and assay of frozen inocula of *Saccharomyces cerevisiae*. J Appl Bact 1976; 41:197–220.
47. Monti M, Wadsö I. Microcalorimetric measurements of heat production in human erythrocytes. 4. Comparison between different calorimetric techniques, suspension media and preparation methods. S Scand J Clin Lab Invest 1976; 36:573–580.
48. Sedlaczek L, Czerniawski E, Radziejewska J, Zablocki B. Thermogenesis and growth yield of *Escherichia coli* in media containing various nitrogen sources. Bull Acad Pol Sci Biol 1996; 14:613–619.
49. Mardh PA, Ripa T, Andersson KE, Wadsö I. Kinetics of the actions of tetracyclines on Escherichia coli as studied by microcalorimetry. Antimicrobial Ag Chemother 1976; 10:604–609.
50. Beezer AE, Chowdhry BZ, Newell RD, Tyrrell HJV. Bioassay of antifungal antibiotics by flow microcalorimetry. Anal Chem 1977; 49:1781–1784.
51. Kjeldsen NJ, Beezer AE, Miles RJ. Flow microcalorimetric assay of antibiotics—I. Polymyxin B sulphate and its combinations with neomycin sulphate and zinc bacitracin on interaction with *Bordetella bronchiseptica* (NCTC 8344). J Pharm Pharmac 1989; 7:851–857.
52. Kjeldsen NJ, Beezer AE, Miles RJ. Flow microcalorimetric assay of antibiotics—II. Neomycin sulphate and its combinations with polymyxin B sulphate and zinc bacitracin on interaction with *Bacilluspumilus* (NCTC 8241). J Pharm Pharmac 1989; 7:859–864.
53. Kjeldsen NJ, Beezer AE, Miles RJ. Flow microcalorimetric assay of antibiotics—III. Zinc bacitracin and its combinations with polymyxin B sulphate and neomycin sulphate on interaction with *Micrococcus luteus*. J Pharm Pharmac 1989; 7:865–869.
54. Kjeldsen NJ, Beezer AE, Miles RJ, Sodha H. Flow microcalorimetric assay of antibiotics—IV. Polymyxin B sulphate, neomycin sulphate, zinc bacitracin and their combinations with *Escherichia coli* suspended in buffer plus glucose medium. J Pharm Pharmac 1989; 7:871–875.
55. Beezer AE, Newell RD, Tyrrell HJV. Bioassay of nystatin bulk material by flow microcalorimetry. Anal Chem 1977; 49:34–37.
56. Cosgrove RF, Beezer AE, Miles RJ. In vitro studies of amphotericin B in combination with the imidazole antifungal compounds clotrimazole and miconazole. J Infect Dis 1978; 138:681–685.

57. Tan AM, Lu JH. Microcalorimetric study of antiviral effect of drug. J Biochem Biophys Methods 1999; 38:225–228.
58. O'Neill MAA, Beezer AE, Mitchell JC, Orchard JA, Connor JA. Determination of Michaelis-Menten parameters obtained from isothermal flow calorimetric data. Thermochimica Acta 2004; 417:187–192.
59. Florence AT, Attwood D. Physicochemical Principles of Pharmacy. 3rd ed. Basingstoke: Macmillan, 1998.
60. Montanari MLC, Beezer AE, Montanari CA. QSAR based on biological microcalorimetry: the interaction of *Saccharomyces cerivisiae* with hydrazides. Thermochimica Acta 1999; 328:91–97.
61. Montanari MLC, Andricopulo AD, Montanari CA. Calorimetry and structure–activity relationships for a series of antimicrobial hydrazides. Thermochimica Acta 2004; 417:283–294.
62. O'Neill MAA. Recent developments for the analysis of data obtained from isothermal calorimetry. Curr Pharm Biotechnol 2005; 6:205–14.

6

Investigation of the Amorphous State

INTRODUCTION

Many active pharmaceutical ingredients (APIs) exist in the solid-state or are formulated into solid dosage forms. This confers several advantages: stability in the solid-state is usually greater than in the liquid state; solids are easier to process, transport, package, and store; and the majority of medicines, at least in the United Kingdom, are formulated as tablets as the oral route is both quick and has a high degree of patient compliance.

Formulation in the solid-state does have some drawbacks, however. Principally, solid-state formulations are usually slower acting than liquid medicines containing the same active ingredients, because the first event that must occur following administration is that the API must dissolve in a suitable biological fluid (such as saliva or gastric juices). On a molecular level, this requires the intramolecular forces holding the solid lattice together to be overcome (endothermic) and the formation of new interactions with the solvent (generally exothermic). This event is (usually) rate limiting, and thus controls the observed dissolution rate of the API from the formulation.

Furthermore, while chemical stability is not generally an issue for solid-state formulations (reactions are much more likely to occur in solution), physical (i.e., phase) changes can be a major issue. For pharmaceuticals, the biggest concern is the detection and quantification of any polymorphic forms present. The intramolecular forces in different polymorphs of a material (the lattice energy) will vary. It is therefore imperative that the crystal form of any solid-state API is known, because the dissolution rates of each polymorph will be different. Indeed, patents may be broken by registering a different polymorph of an existing drug. Moreover, over time, all the metastable polymorphs will convert to the stable polymorph, which may have a disastrous effect on the efficacy of the formulated product. A discussion of the detection and quantification of polymorphic forms can be found in Chapter 4.

In addition to polymorphism, solid APIs may also exhibit amorphicity (most simply defined as a lack of long-range molecular order). As amorphous materials have no lattice energy and are essentially unstable (over time, they will convert to a crystalline form), they usually have appreciably faster dissolution rates than their crystalline equivalents, which makes them especially suited for formulation into fast-acting medicines. Thus, it may often be the case that the conditions used to prepare a pharmaceutical are selected to ensure a high proportion of the amorphous form. These may include spray- or freeze-drying or quench-cooling. However, other processing steps such as milling or compression, may also result in the formation of amorphous material (1,2). The accidental inclusion of amorphous content, in what is otherwise presupposed to be a crystalline material, is thus a hazard that should be borne in mind when processing any solid-state material. Although the percentage of amorphous content introduced in this way is usually low (of the order of 1% w/w), its location primarily on the surface of what are usually small particles often gives it a disproportionate control on the properties of the material (3). It is clear, then, that the detection and quantification of (often small) amorphous contents is of the utmost importance during the characterization of a solid pharmaceutical.

If the amorphous form is the desired state of the pharmaceutical, then demonstration (and control) of its stability upon storage is critically important. If a crystalline material has been processed, then understanding whether any amorphous material has been formed, and its proportion of the bulk material, is essential for the implementation of automated process control. Isothermal calorimetry (IC) can be used for either type of investigation, and these topics form the basis of this chapter; preceding these discussions, however, is a brief overview of the formation and properties of the amorphous state.

AMORPHOUS MATERIAL

Theory of Formation

Before considering the effects that an amorphous material can exert on a pharmaceutical, and the ways in which IC can be used to quantitate its stability or proportion in a sample, it is appropriate to discuss the mechanisms by which amorphous materials can be formed, because this underpins the following text.

Before starting any discussion, it must be noted that the amorphous state is (by definition) complex and, despite much research effort, remains poorly understood. In particular, because the amorphous state is thermodynamically unstable there is much focus on its stability (which is usually correlated with molecular mobility; discussed further in this chapter). Indeed, a discussion of the mechanisms involved in molecular mobility could form the basis of a separate book. However, the theory of formation of an amorphous material is easier to understand and more widely accepted.

Formation starts from the molten state, in which the molecules of a material are free to move and there is no underlying structure. As the liquid is cooled, it will reach a temperature at which it will crystallize; the melting temperature (T_m), equal and opposite to the crystallization temperature (T_c). Assuming the cooling occurs very slowly, complete crystallization will occur at T_m and the material will form its most stable (assuming there is more than one crystal form) polymorph. By definition, this must be the form in which the molecules are packed as efficiently as possible; the total energy of the system is minimized, the lattice energy is at a maximum, and the volume is at a minimum. This process is represented in Figure 1.

At fast cooling rates, the material does not have time to crystallize, and a supercooled liquid (i.e., a liquid that is below its melting temperature) will exist. Relative to the crystalline state, the supercooled liquid is in a nonequilibrium state; however, the relatively short time scales for molecular mobility during cooling means that it can be said to be at equilibrium because any change in temperature will result in an essentially immediate change in structural and thermodynamic properties. The supercooled liquid is thus said to be in structural equilibrium (4).

If the properties of the supercooled liquid were to keep changing linearly with a reduction in temperature, at some point its energy and volume would

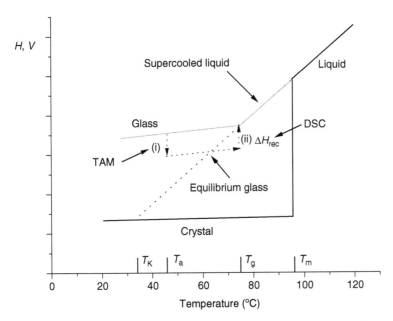

Figure 1 A schematic representation of the processes involved in the formation of an amorphous material. *Abbreviations*: DSC, differential scanning calorimetry; TAM, thermal activity monitor.

become smaller than that of the perfect crystal (represented by the dotted extrapolation in Fig. 1); the temperature at which the intersection would occur is denoted the Kauzmann temperature (T_K). In practice, the cooling of the supercooled liquid occurs with a concomitant reduction in volume; correspondingly, there is an increase in viscosity and decrease in molecular mobility. Thus, there is a point at which the properties of the supercooled liquid cannot change in response to a change in temperature because the time scale for molecular mobility becomes too large. At this point, the material is said to become a glass, and the temperature is defined as the glass transition temperature (T_g). Below T_g, the material is no longer in structural equilibrium; essentially, the glass has the structure of a liquid, but the viscosity (and hence external appearance) of a solid. This state is known as a real glass and is represented in Figure 1.

The extrapolated line in Figure 1, which represents the state the system would have attained had it not formed a glass, is thus a "virtual" state and can be thought of as essentially that of the glassy state in structural equilibrium. It is a virtual state because although it can be conceived, it does not actually exist. It is referred to as the equilibrium glassy state and is the state to which a real glass will tend if it is held at any particular temperature less than T_g. The temperature at which the glass is held is known as the annealing temperature, T_a, and the approach from the glassy state to the equilibrium glass state is known as relaxation [represented by process (i) in Fig. 1]. The relaxation of amorphous pharmaceuticals is an extremely important topic because it is related to molecular mobility and, hence, stability.

It is notable here that two types of relaxation are usually defined: structural relaxation and Johari–Goldstein relaxation. Structural relaxation (termed α-relaxation) describes whole molecule movements (such as diffusional motion and viscous flow) while Johari–Goldstein relaxation (termed β-relaxation) is usually taken to represent intramolecular motion (such as rotation of a side-chain). In the context of pharmaceuticals, α-relaxation events are usually considered to be more important than β-relaxation events, and are thus the subject of this chapter.

Being in a nonequilibrium state, any real glass is unstable and will eventually change form, the rate of conversion being generally dependent upon the degree of molecular mobility. Other than relaxation to the equilibrium glassy state, an amorphous material can recrystallize. This requires sufficient molecular mobility for the molecules to rearrange to a crystalline form and is generally attained with either an increase in temperature (to greater than T_g) or by the addition of a plasticizer (plasticizers are small molecules, often water or organic vapors, that can easily penetrate the spaces between the molecules in an amorphous material, altering its physical and thermodynamic properties). Effectively, the plasticizer lowers the T_g of the material; when sufficient plasticizer has been added to reduce T_g to less than storage temperature, the material will recrystallize. Water is very often a plasticizer for amorphous pharmaceuticals, which is why control of relative humidity (RH) is critical in determining pharmaceutical stability.

Formation of Amorphous Materials During Processing

There are several ways in which amorphous material can be formed during processing. These can be broadly split into those which are intended to form amorphous material (i.e., which create amorphous contents approaching 100%) and those that result in the formation of amorphous material unintentionally (i.e., which create amorphous contents of approximately 1–10%).

Processes which can be used intentionally to form amorphous material include:

1. *Spray-drying*. A solution of material is forced through an atomizing nozzle at high pressure, producing aerosolized droplets, which are dried in a vortex of air before being separated and collected.
2. *Freeze-drying*. An aqueous solution of material is frozen before being heated under vacuum. The water sublimes to leave the freeze-dried material.
3. *Quench-cooling*. A molten material is rapidly reduced in temperature, often using liquid nitrogen.

The detection and quantification of amorphous content in materials processed in these ways is not such a great problem, because the percent of amorphous content is usually high and its presence is expected.

The unintentional formation of amorphous material can be much more important pharmaceutically, because it is often not analyzed for but may have a dramatic effect on the property of the material (discussed more fully subsequently). Furthermore, even small (approximately 1%) amorphous contents can be important but the number of techniques that can quantify to this level is small. This is one area where calorimetric measurements offer significant advantages over other techniques. Typical processes that can result in the formation of small amounts of amorphous material include:

1. milling or micronization—processes that are used to reduce particle size by mechanical abrasion;
2. compaction, such as during tablet manufacture; and
3. any other process that exerts a force upon a material.

Effects of Amorphous Materials

It is well known that the properties of an amorphous material will differ from its crystalline counterpart. Some of these are pharmaceutically advantageous and others are detrimental; a careful balance of these factors can result in a product with excellent characteristics. The property change most commonly associated with amorphous material, an increase in solubility is, in fact, a misnomer. Solubility is an equilibrium state and is, therefore, constant for a given solute in a given solvent under a specified set of conditions. Expressed another way, there are no solute–solute interactions in an ideal solution and, hence, one could not determine the solid-state of the solute prior to dissolution by measuring the properties of the solution. The misnomer arises because many amorphous materials have appreciably faster dissolution rates than their crystalline counterparts (primarily because amorphous materials do not have a lattice energy and therefore one of the principal barriers to dissolution is overcome). Indeed, amorphous materials may dissolve so fast that a supersaturated solution may form (i.e., the solute is dissolved above its solubility). Over time, the supersaturated solute must precipitate, resulting in a saturated solution (these events are represented schematically in Fig. 2). However, within the time frame of an experimental measurement, an amorphous material may be said to exhibit a higher apparent solubility than its crystalline counterpart. Of course, the absorption of any drug given orally is dependent upon both its dissolution rate and its apparent solubility, and it is usually the case that an amorphous form will maximize these parameters.

Other properties include the lack of a melting point (again, there is no crystal lattice; upon heating, an amorphous material will crystallize and it is the crystal form that will subsequently melt), an inherent instability (being a nonequilibrium state), and a propensity to absorb humidity or other solvents. The latter two properties are usually disadvantageous, causing instability. Structural relaxation and crystallization of the amorphous form will often have a dramatic and disastrous effect on the bioavailability of the drug, because the dissolution rate and apparent solubility benefits already noted will no longer exist. Any product application must therefore demonstrate the stability of the

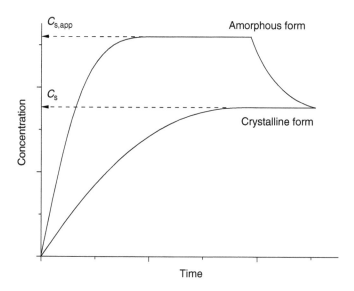

Figure 2   A schematic representation of the dissolution profiles of amorphous and crystalline forms of a drug. Also shown are the solubility (C_s) and apparent solubility ($C_{s,app}$) values.

amorphous form during storage. The rate of relaxation and likelihood of crystallization are often increased with the absorption of humidity or solvents because these substances can act as plasticizers. Furthermore, the absorption of humidity, in particular, can have consequences for the chemical stability of the drug, which may be susceptible to hydrolysis.

It is clear, then, that an understanding of the amorphous state of an API and, more importantly, control over it are essential in ensuring the safety and efficacy of a medicine. Calorimetric methods are becoming ever more central to the detection and quantification of amorphous materials [e.g., the United States Pharmacopoeia (USP) monograph for α-lactose monohydrate for dry-powder inhalation includes a calorimetric method for assessment of amorphous content], and their range of application in this area can only increase. The remainder of this chapter focuses on its two main applications: monitoring the aging of amorphous materials undergoing relaxation and quantification of small amorphous contents in processed pharmaceuticals.

AGING OF AMORPHOUS MATERIAL

Amorphous pharmaceuticals undergo two types of degradative processes: physical and chemical. Chemical degradation is defined as a change in the molecular form of the material via some form of chemical reaction while physical degradation is taken to mean any event that changes the properties of a material but not its chemical form (such as relaxation, crystallization, wetting, and so on). Usually, analytical techniques are easily able to separate these processes, because they are designed to follow a change in one specific event; UV spectroscopy or high performance liquid chromatography will follow chemical degradation while X-ray powder diffraction (XRPD) will follow changes in physical properties, for instance. Calorimetry, uniquely, measures both chemical and physical changes and does not differentiate between them. As should be apparent from earlier chapters, this is often an advantage but in the specific case of relaxation of amorphous materials, care must be taken to ensure that the contribution of any chemical reactions to the measured heat flow is accounted for. This is not straightforward, as both chemical and physical stabilities can be correlated with molecular mobility (5).

Relaxation can be considered as the sum of all the configurational changes that must occur in a sample for it to change from its "frozen" glassy state to its equilibrium glassy state at a given annealing temperature (T_a). The decay function that characterizes relaxation can therefore be expressed as:

$$\Phi(t, T_a) = \sum_n a_i \exp\left[-\frac{t}{\tau_i(T_a)}\right] \tag{1}$$

where $\Phi(t, T_a)$ is the decay function at time t and temperature T_a, n is the total number of states that are changing independently, and a_i is a weighting factor for each given state i with an exponential decay constant τ_i (5).

In general, a good approximation of the multi-exponential decay function of Equation (1) can be made using a "stretched exponential" function [known as the Kohlrausch–Williams–Watts (KWW) equation]:

$$\Phi(t, T_a) = \exp\left[-\left(\frac{t}{\tau(T_a)}\right)^{\beta}\right] \tag{2}$$

where τ represents the relaxation time and β is known as the stretch power. It is important to note that the KWW equation is simply an empirical relationship, although it is possible to derive theoretical frameworks that attempt to rationalize its terms (6). The value of β (which usually has a value between 0 and 1) can be taken to reflect the distribution of independently relaxing states; when β is small, the distribution of microstates is wide (i.e., there would be a significant number of microstates with widely differing relaxation times) and when β approaches unity the distribution of microstates is very narrow (7). Other interpretations have been suggested, and although the KWW relationship is the most widely used approach for the analysis of the relaxation of amorphous materials, other theories exist. In order to describe the dynamics of relaxation in amorphous materials, both constants (β and τ) are required.

Before considering the application of IC to this area, it is worth discussing the use of differential scanning calorimetry (DSC), because this technique is widely used to follow the relaxation of amorphous materials and it is important to be able to compare data from the two techniques. Using DSC, the enthalpy recovery is measured as the sample is heated above its T_g; this enthalpy is assumed to be equal and opposite to the enthalpy of relaxation ($\Delta_{rec}H$), and its origination is represented by process (ii) in Figure 1. The value of $\Delta_{rec}H$ is dependent upon the temperature at which the sample has been annealed and the period of annealing. The decay function can then be determined from $\Delta_{rec}H$ by:

$$\Phi(t) = 1 - \frac{\Delta_{rec}H(t)}{\Delta_{rec}H(\infty)} \tag{3}$$

where $\Delta_{rec}H(t)$ and $\Delta_{rec}H(\infty)$ are the enthalpies of recovery at time t and time ∞, respectively. The value of $\Delta_{rec}H(\infty)$ can be determined by measuring the glass transition:

$$\Delta_{rec}H(\infty) = (T_g - T_a)\Delta C_p \tag{4}$$

where ΔC_p is the heat capacity change through the glass transition. Thus, a plot of $\Phi(t)$ as a function of time can be constructed; fitting these data to the KWW equation [Equation (2)] by least-squares minimization allows the derivation of τ and β.

Other than the relative simplicity of this approach, one advantage of making DSC measurements is that the instrument records all of the heat lost over the annealing time (in other words, the sample may relax at a relatively slow rate, but the DSC measures only the cumulative heat loss at the end of the time period of storage). This maximizes the calorimetric signal, compensates for the slight lack of sensitivity inherent to most DSCs, and means that the technique offers a precise method with which to follow relaxation. The major drawback, of course, of the DSC approach is that it only monitors the relaxation enthalpy at specific time points; the number of data points available to fit to the KWW equation is thus limited to the number of times a DSC measurement has been made on the sample.

The alternative approach would be to monitor the enthalpy of relaxation in real time as a continuous measurement, which is the option offered by IC. From the previous discussion, this can sometimes be hampered by the fact that the technique measures relaxation directly (and, hence, the power signal is minimized). However, the increase in sensitivity of modern isothermal instruments means IC is becoming a real alternative to DSC measurement.

Isothermal power–time data must be fitted to the time derivative of the KWW equation:

$$\frac{dq}{dt} = \Delta H_r(\infty)\left(\frac{\beta}{\tau}\right)\left(\frac{t}{\tau}\right)^{\beta-1}\exp\left[-\left(\frac{t}{\tau}\right)^{\beta}\right] \tag{5}$$

However, consideration of the KWW equation reveals that the decay function [or power in Equation (5)] will approach infinity as time approaches zero. Equation (1), conversely, approaches a finite value as time approaches zero. This dichotomy arises because the KWW equation merely approximates Equation (1); it does not mean that there is an infinitesimally fast decay function at time zero and, thus, the KWW equation is unsuitable for the study of relaxation occurring over short time scales.

A convenient way of circumventing this problem was the development of the modified stretch exponential (MSE) function, which was originally derived to aid interpretation of nuclear magnetic resonance (NMR) data (8):

$$\Phi(t) = \exp\left[-\left(\frac{t}{\tau_0}\right)\left(1+\frac{t}{\tau_1}\right)^{\beta-1}\right] \tag{6}$$

When $t/\tau_1 \gg 1$:

$$\Phi \rightarrow \exp\left[-\left(\frac{t}{\tau_D}\right)^{\beta}\right] \tag{7}$$

where τ_D is a parameter equal to τ in the KWW equation. The value of τ_D can be calculated from:

$$\tau_D = \tau_0^{1/\beta}\tau_1^{(\beta-1)/\beta} \tag{8}$$

As time approaches zero:

$$\Phi \rightarrow \exp\left[-\left(\frac{t}{\tau_0}\right)\right] \rightarrow 1 - \frac{t}{\tau_0} \tag{9}$$

and:

$$\frac{d\Phi}{dt} \rightarrow -\frac{1}{\tau_0} \tag{10}$$

Thus, the time derivative of the decay function approaches a finite value as time approaches zero, but at long times the MSE equation reduces to the KWW equation. For analysis of IC data, the time derivative of the MSE equation is required:

$$\frac{dq}{dt} = \frac{\Delta H_r(\infty)}{\tau_0}\left(1 + \frac{\beta t}{\tau_1}\right)\left(1 + \frac{t}{\tau_1}\right)^{\beta-2}\exp\left[-\left(\frac{1}{\tau_0}\right)\left(1 + \frac{t}{\tau_1}\right)^{\beta-1}\right] \tag{11}$$

The development of Equation (11) and its application to the study of the relaxation of amorphous pharmaceuticals has been demonstrated on a number of systems, including a series of disaccharides (4,5) and an amorphous maltose formulation (9).

By way of an example of the use of the KWW and MSE equations, Figure 3 shows some simulated data created using Equations (5) and (11). In the case of the KWW simulation (dotted line), the time constant used was τ_D (in other words, it was derived from the values of τ_0 and τ_1 used in the MSE simulation).

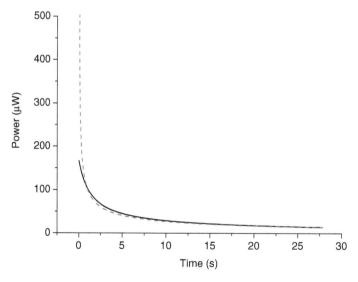

Figure 3   A comparison of the power–time data that would be obtained by simulating data using the Kohlrausch–Williams–Watts equation [Equation (5), dotted line] and the Modified Stretch Exponential equation [Equation (11), solid line]. The following values were used in the simulation: $\Delta_{rec}H(\infty)$, 10 J/mol; β, 0.5; τ_0, 6×10^4; τ_1, 6×10^3; τ, 6×10^5.

Although strictly the values of τ and τ_D are not identical, in practice, they are assumed to be roughly equivalent and this approach provides a convenient set of data for comparison. The values of the parameters used to simulate the data are given in the figure legend. As implied from the previous discussion, it can be seen that the KWW data approach ∞ as time approaches 0, while the MSE data approach a finite value. This highlights the fact that, over shorter relaxation times, the MSE equation is a more appropriate analysis tool.

QUANTIFICATION OF AMORPHOUS CONTENT

The discussion so far in this chapter has focused on quantifying stability in what is an entirely, or largely, amorphous material. Materials with such high amorphous contents are usually created because their properties fulfill a specific role and, hence, the amorphous content is intended. Of perhaps greater concern is the case where a material accidentally becomes partially amorphous during processing, because here the amorphous material is not expected and can be present at relatively small levels (levels which can be difficult to detect, let alone quantify). While this section focuses, in the main, on various isothermal calorimetric methods that can be used for amorphous content quantification, the first section considers the impact of what can be very small amorphous contents, and shows why amorphous content screening should be a routine for many processed samples. Furthermore, there are a wide variety of methods for quantifying amorphous contents, some of which are more suited to certain samples than others. These techniques are briefly considered so that the reader can draw comparison with isothermal techniques.

Potential Impact of Small Amorphous Contents

Any discussion about the potential impact of small amorphous contents must first define what constitutes a small amorphous content. Although the exact effect will vary from sample to sample, the following example (lactose[a]) is both illustrative and typical.

Lactose is a solid material at room temperature and is often micronized to produce a product with ideal characteristics for formulation. During micronization, because a mechanical force is being applied, lactose can become partially amorphous. It can be shown that the amount of material formed depends upon the processing conditions, but can be as low as 0.5% to 1% w/w (10). Although this amount appears negligible, its location is of paramount importance. Because it was produced from the mechanical action of micronization, the

---

[a]Although lactose is a fairly unimaginative choice of subject for a discussion of amorphous contents, it exhibits a number of properties that make an ideal subject. It exists in two anomeric forms (α and β) and as a monohydrate (α-lactose monohydrate); it is easily made amorphous; it is plasticized by water; and, most importantly within the context of this text, it is widely used in pharmaceuticals.

amorphous material resides primarily upon the surfaces of the particles. Consequently, this means that because solid materials can interact only through their surfaces, the micronized lactose sample will behave as if it were almost entirely amorphous while the original (unmicronized) sample will behave like a crystalline material.

The surface energy data in Table 1 [recorded using inverse-phase gas chromatography (IGC)] serve to illustrate this point perfectly (11). The surface energies for entirely crystalline and entirely amorphous lactose are 31.2 and 37.1 mJ/m², respectively. The surface energy of a micronized lactose sample (which was determined to be approximately 1% w/w amorphous) is 41.6 mJ/m². In other words, the micronized sample has a higher surface energy than even the entirely amorphous sample. In order to demonstrate that location, and not proportion, of amorphous content is the principal factor, Table 1 also shows the surface energy of a partially amorphous lactose sample (1% w/w, prepared by dry-blending appropriate mass quantities of entirely amorphous and entirely crystalline samples); it can be seen that in this case the surface energy is 31.5 mJ/m², exactly the same as the crystalline material.

Of course, the consequences for the product of having such a low amorphous content may not be dramatic. One common occurrence with amorphous material on a surface is the absorption of water which causes recrystallization; this usually results in the agglomeration of the micronized particles, exactly the opposite of the intended outcome of the process. Lactose is, however, a special case in that its role in a formulation is sometimes more than as a simple diluent. In the case of dry-powder inhalers, micronized lactose is often used as a carrier for the active ingredients. This is because a powder for inhalation must be 5 μm or smaller in order to reach the lower lung, but this particle size is difficult to aerosolize; thus, the drug particles adhere to the lactose carrier, producing a larger particle that is easily aerosolized. Upon inhalation, the turbulent airflow separates the drug from the carrier, the drug being carried to the lower lung and the carrier impacting in the throat. The force required for separation is entirely dependent upon the strength of the interaction between drug and

Table 1 Surface Energy Data for Various α-Lactose Monohydrate Samples Determined by Inverse-Phase Gas Chromatography

Sample	Dispersive surface energy (mJ/m²)
Amorphous (spray-dried) lactose	37.1 ± 2.3
Crystalline lactose	31.2 ± 1.1
Micronized lactose (~1% w/w amorphous)	41.6 ± 1.4
99:1 crystalline:amorphous lactose (dry-mixed)	31.5 ± 0.4

Source: From Ref. 11.

carrier; as already noted, this will differ significantly if the surface of the carrier is amorphous or crystalline and there will be a consequential change in bioavailability of the active ingredient. Indeed, this issue is significant enough that the USP monograph for α-lactose monohydrate for dry-powder inhalation includes a gas-perfusion methodology (discussed later) for ensuring the sample does not contain >0.5% w/w amorphous material.

These are not the only consequences that can arise from the accidental inclusion of a small amount of amorphous material. Sebhatu et al. (12) showed the effect of storage on the strength of tablets made from milled (and, hence, partially amorphous) lactose. The data are summarized in Figure 4. It can be seen that if the lactose was stored for up to four hours at 57% RH, the tablet strength increased with storage time. However, at storage times longer than four hours, the tablet strength was constant and at a minimum. They correlated these data with the water content of the lactose (which was absorbed by the amorphous material). As the water content increased, the mobility of the lactose increased, meaning that the compressibility of the lactose got higher. After a critical time (four hours), the amount of water absorbed was sufficient to cause crystallization of the amorphous lactose, and compressibility reduced to a minimum. The knowledge and quantification of amorphous content in processed pharmaceuticals is therefore of considerable importance and should not be overlooked during preformulation or formulation characterization.

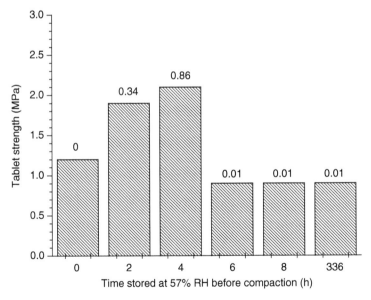

Figure 4 Tablet strength as a function of storage time at 57% RH for tablets made from spray-dried amorphous (15%) lactose. The numbers above each column represent the water content of the powder immediately prior to compression. *Abbreviation*: RH, relative humidity. *Source*: From Ref. 12.

Methods to Quantify Amorphous Contents

Noncalorimetric Methods

Many techniques are available to quantify amorphous contents, with the industry "gold-standard" being XRPD. However, in routine use, XRPD cannot quantify amorphous contents $<5\%$ w/w, so it is not suitable for the analysis of processed materials. Other techniques that have been shown to quantify amorphous contents to 1% w/w or better include Dynamic Vapor Sorption-Near Infra-red Spectroscopy (DVS-NIR) (13), Inverse Phase Gas Chromatography (IGC) (14), and Fourier-Transform (FT)-Raman Spectroscopy (15).

Calorimetric Methods

Solution calorimetry: The use of solution calorimetry to quantify the degree of crystallinity in a solid sample is predicated on the relationship shown in Equation (12), where the measured heat of solution ($\Delta_{sol}H$) is given by the sum of the enthalpies and weight fractions for the crystalline and amorphous states present (16):

$$\Delta_{sol}H = (X_a\Delta_aH) + (X_c\Delta_cH) \tag{12}$$

where X_a and X_c are the fractions the amorphous and crystalline forms, respectively, and Δ_aH and Δ_cH are the heats of solution of the amorphous and crystalline forms, respectively. The usual methodology is to prepare a calibration line of $\Delta_{sol}H$ against degree of crystallinity using a number of known standards (usually prepared by blending the appropriate mass quantities of wholly amorphous and wholly crystalline material); the calibration plot should be linear and can thus be used to determine the degree of crystallinity in an unknown sample.

There are two potential drawbacks to using this approach. First, the sample may exhibit polymorphism, in which case there may be more than one crystalline form present; if this is the case then, for the calibration plot to be linear, it must be ensured that the crystalline material used to prepare the standards and the unknown sample contain the same proportions of the polymorphs. Second, it is likely that the interactions in a particle that has a crystalline core and amorphous material on its surface (the likely situation for a processed pharmaceutical) differ from those of wholly amorphous and wholly crystalline particles, which may result in the calibration plot producing spurious results.

The first use of solution calorimetry for the quantitative measurement of the degree of crystallinity of pharmaceuticals was by Pikal et al. (17), who measured the heat of solution of various β-lactam antibiotics. Subsequent studies include the analysis of sulfamethoxazole from different sources (18), sucrose (19), and clathrate warfarin sodium (19).

In many of these studies, the aim was to quantify relatively large mass percentages of crystalline material, giving heat changes easily within the detection limit of the technique. The issue of detection limits becomes more important if the objective of the study is to assess the quantity of amorphous material present in what is a predominately crystalline sample because, as stated earlier, in a milled sample amorphous material may typically only be present up to 1% w/w. An assessment of the applicability of solution calorimetry to study small amorphous contents in solid pharmaceuticals was conducted by Hogan and Buckton (3), who prepared a calibration curve for lactose between 0% and 10% w/w amorphous content in the same way as already described. They found that the technique could quantify amorphous content to ±0.5% w/w, but noted that care needed to be taken when preparing the ampoules, because ingress of even small amounts of humidity caused partial recrystallization of the sample before measurement.

Usually, in experiments designed to measure the degree of crystallinity or amorphous content, a solvent is selected in which the solute is freely soluble. This ensures complete dissolution of the sample within the time frame of the experiment. Taking an alternative approach, Harjunen et al. (20) studied the dissolution of lactose into saturated aqueous lactose solutions, a system where clearly the solute would not completely dissolve. Interestingly, they observed a linear relationship between the amorphous content of the lactose solute and the measured heat of solution in the saturated lactose solution ($\Delta_{sat}H$). Similarly, a linear relationship was found between the amorphous content of lactose and $\Delta_{sat}H$ in methanolic saturated solutions of lactose (21).

Differential scanning calorimetry: Two events can occur during a DSC experiment that are associated with the presence of amorphous material: a step-change in the baseline as the material passes through its glass transition and an exothermic peak as the material crystallizes. Either event can be used to quantify amorphous contents, assuming a linear calibration curve can be derived.

When the glass transition is used to quantify amorphous contents, it is assumed that the step height is linearly proportional to the amorphous content. However, it is sometimes difficult to measure this parameter accurately in a DSC experiment, for two reasons. First, the small energy changes that accompany the event make it inherently difficult to measure and second, as already discussed, the enthalpy of recovery is recorded at the same moment. In practice, this means that the glass transition is often masked by the enthalpy of recovery, and the raw data signal can comprise what appears to be a single peak. It can be difficult even to recognize that a glass transition has occurred in these circumstances (one characteristic sign is that there is a step-change in the baseline before and after the peak), let alone assign a transition temperature.

One approach that allows the separation of these events is to use temperature-modulated DSC (TMDSC) (22). As discussed in Chapter 1, TMDSC allows the deconvolution of the raw power data into two parts, often termed "reversing" and "nonreversing." The glass transition is a reversible event, while the enthalpic recovery is a nonreversible event. Using TMDSC, therefore, it is possible to separate the enthalpy of recovery from the underlying glass transition. Some raw TMDSC data are shown in Figure 5. The data reflect the modulation that

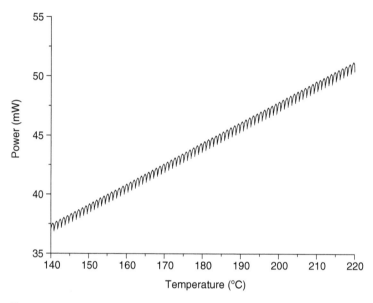

Figure 5   Raw temperature-modulated differential scanning calorimetry data (in this case for a film of HPMC E4M). *Source*: From Ref. 54.

was imposed upon the sample. The separation of these data is shown in Figure 6; the position of the glass transition is now clear. Examples of the use of TMDSC to quantify small amorphous contents in processed pharmaceuticals are provided by Guinot and Leveiller (23) and Saklatvala et al. (24).

   An alternative option is to use fast-scan DSC, where heating rates of up to $500°C/min$ are employed. Here, the principle is that because the heat change through the glass transition occurs over a much shorter time scale, the magnitude of the step-change must increase. Saunders et al. (25) have demonstrated how fast-scan techniques can be used to assess small amorphous contents in powders. Using lactose as a model example, they showed that the height of the step-change increased from approximately 1 W/g at $100°C/min$ to 3 W/g at $250°C/min$ and to 10 W/g at $500°C/min$. They also showed that the magnitude of the step-change increased linearly with amorphous content, quoting a limit of quantification of 1.89% w/w and a theoretical limit of detection of 0.57% w/w. A similar study by Lappalainen et al. (26) on sucrose showed limits of detection and quantification of 0.062% and 0.207% w/w, respectively.

   Isothermal calorimetry: The most common approach to determine amorphous contents by IC is to measure the heat changes that result when a partially crystalline material is exposed to a specific vapor (often humidity) (10,27). The vapor selected should act as a plasticizer for the sample material, with the result that as it is absorbed by the sample, the glass transition temperature is

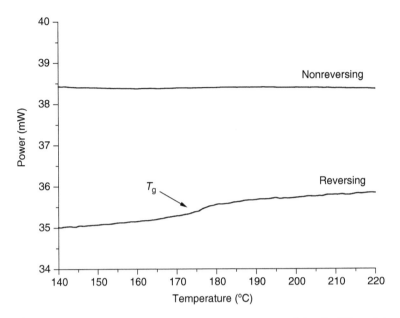

Figure 6 The separation of the raw temperature-modulated differential scanning calorimetry data shown in Figure 5 into reversing and nonreversing components. *Source*: From Ref. 54.

lowered. When T_g drops lower than the experimental temperature, the sample will crystallize. Using this approach, the detection of as little as 0.5% w/w amorphous content in a micronized pharmaceutical has been demonstrated (28).

The relative ease with which such interactions can be studied has been made easier by the commercial availability of perfusion apparatus that allows the introduction of water vapor directly into the sample ampoule. Such apparatus allows the RH of a gas flowing over a sample to be varied as a function of time.

An alternative, but cheaper, experiment is to study a sample under conditions of a specific RH, by placing a small glass tube (a Durham tube or hydrostat), containing a small quantity of a saturated salt solution, directly within the ampoule. Saturated salt solutions maintain a constant RH at a specific temperature. Table 2 in Chapter 1 lists the RHs obtainable at different temperatures with various saturated salt solutions (29). The hydrostat method gives rise to the generation and use of humidity within the measuring site of the calorimeter and hence much (but not all) of the wetting response for the sample is matched by the (almost) equal and opposite response for the generation of the humid air. The generation and use of humidity in the hydrostat method means that both processes are measured by the calorimeter, giving rise to a very different response from gas-flow experiments, where the humidity is generated remotely from the measuring site. The construction of "blank" experiments for these experiments is therefore not straightforward and has been discussed by Buckton (30). Consequently, the hydrostat method can be very useful for measurements in which a sample changes after exposure to humidity, whereas the gas-flow method is better suited to the measurement of wetting (and possibly subsequent changes if they are large enough to be visible in the presence of a large wetting response).

This approach has been widely used to determine the amorphous contents of many pharmaceuticals and is not limited to water as a plasticizer. For instance, Kawakami et al. (31) used IC to quantify the amorphous content of erythromycin using acetonitrile as a plasticizer. They showed that the time taken to crystallize the sample could be controlled by using water/acetonitrile mixtures to control the vapor pressure above the sample. Vemuri et al. (32) used gas perfusion calorimetry to study amorphous contents in pharmaceutical powders, using ethanol to induce crystallization. They note that although the baseline noise was low, quantification <5% w/w was hampered by the presence of additional events in the calorimetric data. The effects of some of these events are discussed in more detail subsequently.

Al-Hadithi et al. (33) used IC to measure the amorphous content of trehalose when mixed with a crystalline diluent (lactose). It was shown that as long as the amount of trehalose was greater than a minimum level (4 mg), the calorimetric response was not affected by the amount of the crystalline diluent. Other examples from the literature include quantitative analyses of salbutamol sulfate (34) and nifedipine (35), but lactose (10,31,36–39) will be discussed as a case study.

Case study: Lactose. As mentioned already, lactose, in particular, is an interesting subject, both as a model system and because it is a commonly used pharmaceutical excipient. A survey of the literature reveals that a wide range of values for the net enthalpy of lactose crystallization has been reported (Table 2). Two principal factors account for these differences. First, different researchers integrate different sections of the calorimetric power–time response (because, as already noted, the data are a composite response, comprising of a large number of competing exothermic and endothermic processes). Second, changes in experimental conditions can impact upon the observed response in the calorimeter, because the balance of events occurring is altered. For

Table 2 Reported Apparent Enthalpies of Crystallization for Lactose Using Isothermal Calorimetry at 25°C

Reference	Enthalpy (J/g)
(36)	32
(10)	45–48
(39)	50
(40)	48

example, experiments conducted at 60°C by Darcy and Buckton (40) resulted in a larger net crystallization enthalpy, because less plasticizing water needed to be absorbed, and subsequently expelled.

Clearly, in order to use calorimetric data to determine quantitatively small amorphous contents, there must be an accepted value for the net enthalpy of crystallization of 100% amorphous lactose (or any other material that is used). An important step in deriving this value is the definition of the starting and ending points of the integration. Such a study was recently reported by Dilworth et al. (41), who proposed several methods for integrating complex peaks and then applied the analysis to a number of partially amorphous lactose samples.

A typical thermal response generated by amorphous lactose under high RH conditions (>50%) is represented in Figure 7 (in this case, a 33 mg sample of amorphous lactose under 75% RH at 25°C). The sample has several distinct regions, denoted A–E. The initial sharp exotherm, which results from the friction of lowering of the ampoules and does not form part of the thermal response of the sample, is followed by a more prolonged exotherm lasting for almost one hour (region A). This comprises contributions from evaporation of water from the hydrating reservoir in order for the atmosphere in the ampoule to reach the desired RH (endothermic), wetting of the lactose as it absorbs water (exothermic), and, possibly, structural collapse of the sample. The likelihood that the rates of water evaporation and absorption are slightly out of balance explains both the presence of the exotherm (region A; absorption predominates) and the subsequent endothermic baseline (region B; evaporation predominates) before the commencement of crystallization. It should be noted that while in region B

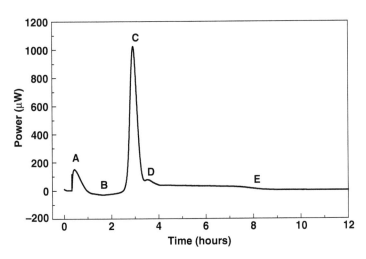

Figure 7 A typical power–time response for a sample of amorphous lactose held at 25°C under a relative humidity of 75%. The data comprise of five main regions, denoted as A to E (discussed in the text). *Source*: From Ref. 41.

it is usually assumed that the processes of water evaporation and absorption are considered to be equal in rate and, thus, generate no heat response (42,43), close inspection of the data reveals this not to be the case.

Usually, an equivalent saturated salt solution is placed in both the sample and reference ampoules of the calorimeter. As intended, this will compensate for some of the response associated with generating water vapor in the ampoules; however, the presence of amorphous material will cause a greater generation of water vapor in the sample side as the lactose effectively desiccates the air space. The choice of reference for such an experiment is therefore not straight-forward, as it is clear that a hydrostat reference is not perfect. It is also clear that the data generated using this methodology will differ from those obtained when an empty ampoule (no saturated salt solution reservoir) is used as the reference.

After approximately 2.5 hours, the sample crystallizes (region C, Fig. 7). It has been shown that this lag time correlates to the mass of amorphous material, the RH maintained by the saturated salt solution and the surface area of the hydro-stat (10,44). It has also been shown that samples removed immediately after this exothermic response are crystalline (10) and physical inspection of the sample reveals a hard, fused, solid mass instead of a free-flowing powder. Following crystallization, carbohydrates expel the previously absorbed plasticizing water (45); the same effect can be followed using dynamic vapor sorption apparatus (46). While crystallization will be an exothermic event, the expulsion of water will be endothermic and the calorimetric data will thus represent the balance of these heat changes.

The crystallization response (region C) shows a clear shoulder (region D). Sebhatu et al. (36) suggested that this shoulder should not be attributed to the expulsion of water following crystallization because that event should be can-celled out by the exothermic condensation of that water back into the hydrating reservoir, in a manner similar to that suggested by Angberg et al. (42,43) to explain the lag phase before recrystallization. However, because it has been shown that the lag phase is, in fact, an endothermic event, a similar imbalance of events could also explain this shoulder. Sebhatu et al. (36) suggested that the shoulder represented incorporation of water into the anhydrous α-lactose formed immediately after crystallization to form α-lactose monohydrate. Briggner et al. (10) attributed the event to mutarotation of β-lactose to α-lactose monohydrate, which could be possible as Angberg et al. (47) showed that such mutarotation occurs to a significant extent at RHs >94% (which would be achieved as large quantities of water desorb from the crystallizing sample).

Finally, a slow exothermic process is observed after crystallization (region E). It is possible that this event arises from water movement within the ampoule, either between the sample and the hydrating reservoir or within the sample itself, or may represent mutarotation. Angberg et al. (47) found that complete muta-rotation occurred over a period of 112 days at 94% RH. This indicates that mutarotation is a slow process, however, so it would be surprising if it went to completion over the few hours that region E extends.

The crystallization of lactose is, therefore, a complex event, comprising a number of exothermic and endothermic processes, and the calorimeter records only the balance of these events. It is thus not a trivial matter to determine the enthalpy of crystallization of amorphous lactose, because of the difficulty of assigning start- and end-points for the integration.

Dilworth et al. (41) considered the regions (A–E) in an attempt to define a reproducible integration strategy. As an endothermic region precedes the crystallization response, the start-point of region C will be negative. It must be decided whether the area should be measured from the minimum inflection of this endotherm or, perhaps, from the point at which region C first crosses the $y = 0$ axis. If the area of the endotherm is ignored, it may be the case that an event which sometimes occurs during the main recrystallization peak, and hence contributes to the measured heat change, is being omitted. This would lead to an overestimation of the enthalpy of crystallization.

The long exothermic event following recrystallization gives rise to a difficulty in assigning the end-point of the area to be integrated. Without a greater understanding of the nature of the processes occurring during this phase, it is not possible to know if this region should, or should not, be regarded as an extension of the crystallization event. As already mentioned, it seems likely that water movement, mutarotation, or a combination of both are responsible for this signal. However, inclusion of the area under region E can have a huge impact on the measured heat change, depending upon the integration method (discussed further subsequently). To omit it raises a further question regarding the assignment of the end-point to some coordinate following the main recrystallization region. This is difficult because the return to baseline after region D is masked by the onset of region E. It must be decided if the area is to be integrated to $y = 0$ or to a sloping baseline (in other words, it must be judged whether the heat capacity of the sample changes in going from an amorphous or partially amorphous form to the crystalline form).

Dilworth et al. (41) selected 12 combinations of start- and end-points and assessed the reproducibility and validity of each against 18 lactose data sets.

1. Method 1A: determination of enthalpy based upon the inclusion of region C only. Integration from $y = 0$.
2. Method 1B: as Method 1A but with the inclusion of region D.
3. Method 1C: as Method 1A but with the inclusion of regions D and E.
4. Method 2A: integration of peak C only, using a sloping baseline from the minimum inflection of region B to the end-point of region C.
5. Method 2B: as Method 2A but with the inclusion of region D.
6. Method 2C: as Method 2A but with the inclusion of regions D and E.
7. Method 3A: integration of region C only, using a flat baseline from the minimum inflection of region B.

8. Method 3B: as Method 3A but with the inclusion of region D.
9. Method 3C: as Method 3A but with the inclusion of regions D and E.
10. Method 4A: integration of regions B and C from $y = 0$.
11. Method 4B: as Method 4A but with the inclusion of region D.
12. Method 4C: as Method 4A but with the inclusion of regions D and E.

The results of applying each method are presented in Table 3. It is evident from these data that the use of the different integration methodologies results, as might be expected, in a wide range of crystallization heats ($36.2–69.2 \, \text{J/g}$). It is also evident that the errors associated with some of the integration methods are considerable and those methods were thus disregarded. Another important factor is that the response of wholly amorphous material may not be the same as that of partially amorphous samples (which, as a consequence, may not exhibit all five regions) and, hence, ease of use of the method is of importance. For these reasons, Dilworth et al. (41) selected Methods 1B and 4C to analyze partially amorphous samples. Note that Method 1B represents an attempt to integrate solely the crystallization response (and results in an average value of $45.4 \, \text{J/g}$, very close to other reported values), while Method 4C represents the determination of the net heat change for all the processes occurring during the course of the experiment (and, hence, results in a larger value of $53.4 \, \text{J/g}$).

Lactose samples containing between 1% and 75% w/w amorphous content were crystallized at 75% RH and the heat changes recorded for each were determined using integration Methods 1B and 4C. The data were used to calculate percent amorphous content by using the heat change obtained for the wholly amorphous sample, using each respective method as a reference enthalpy.

The known amorphous contents versus calculated amorphous contents determined using Method 1B are shown for amorphous contents between 1% and 10% w/w in Figure 8 and for amorphous contents between 5% and 75% w/w in Figure 9. It can be seen that, in all cases, the repeatability of measurement is poor and at most amorphous contents the average calculated values are lower than the known values, indicating the likely presence of a systematic error with this integration method, possibly associated with the difficulty in assigning an end-point to region D. In particular, the 1% to 10% w/w data exhibit a very poor linear regression fit ($R^2 = 0.769$), although this improves in the 5% and 75% w/w data ($R^2 = 0.986$).

Similar data sets are presented in Figures 10 and 11 for amorphous contents determined using Method 4C. In this case, good linear regression fits are obtained for both regions (1% to 10% w/w, $R^2 = 0.974$; 5% to 75% w/w, $R^2 = 0.984$) although calculated amorphous contents were consistently lower than the actual values. Possible reasons for these discrepancies are discussed subsequently.

The crystallization response of a sample containing a very low amorphous content is clearly much smaller than that of an entirely amorphous sample and

Table 3 Enthalpy of Crystallization for 100% Amorphous Lactose Using Gas Perfusion Calorimetry (at 25°C, 75% Relative Humidity) Determined Using the 12 Integration Methodologies Discussed in the Text ($n = 18$)

Integration method	1A	1B	1C	2A	2B	2C	3A	3B	3C	4A	4B	4C
						Enthalpy of crystallization (J/g)						
Average	41.3	45.4	53.8	36.2	42.8	55.7	43.5	46.8	69.2	39.4	45.1	53.4
SD	2.8	3.3	4.3	3.3	3.4	9.1	2.9	5.0	11.8	2.8	3.1	3.7
% Error	6.9	7.2	8.1	9.0	8.0	16.3	6.6	10.6	17.0	7.2	6.9	6.9

Abbreviation: SD, standard deviation.
Source: From Ref. 41.

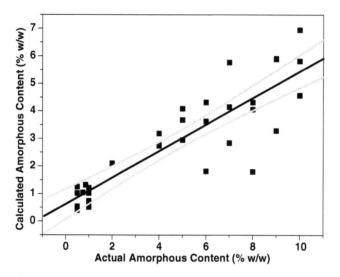

Figure 8 The calculated versus known amorphous contents for samples containing between 1% and 10% w/w amorphous content, determined by integration of power–time data (at 25°C and 75% relative humidity) using Method 1B (see p. 269), showing the linear fit to the data and the 95% confidence limits. *Source*: From Ref. 41.

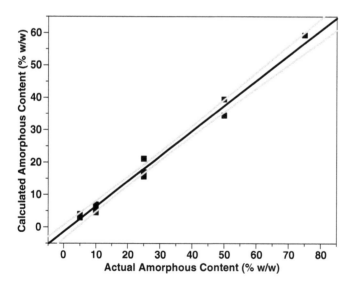

Figure 9 The calculated versus known amorphous contents for samples containing between 5% and 75% w/w amorphous content, determined by integration of power–time data (at 25°C and 75% relative humidity) using Method 1B (see p. 269), showing the linear fit to the data and the 95% confidence limits. *Source*: From Ref. 41.

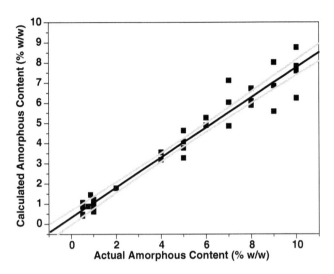

Figure 10 The calculated versus known amorphous contents for samples containing between 1% and 10% w/w amorphous content, determined by integration of power–time data (at 25°C and 75% relative humidity) using Method 4C (see p. 269), showing the linear fit to the data and the 95% confidence limits. *Source*: From Ref. 41.

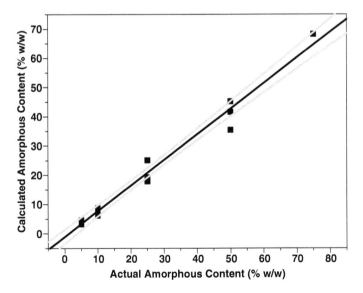

Figure 11 The calculated versus known amorphous contents for samples containing between 5% and 75% w/w amorphous content, determined by integration of power–time data (at 25°C and 75% relative humidity) using Method 4C (see p. 269), showing the linear fit to the data and the 95% confidence limits. *Source*: From Ref. 41.

will occur over a much shorter time span (because less water needs to be absorbed to initiate crystallization). Indeed, crystallization may often occur before a baseline has been achieved after the ampoules have been lowered into the instrument. One consequence of this is that it is difficult reproducibly to determine crystallization enthalpies because the starting point of the event may be lost. One way to mitigate this effect is to use a larger sample mass, although this usually impedes water absorption and expulsion, leading to multiphase crystallization peaks. An alternative approach is to lower the RH to which the sample is exposed. Dilworth et al. (41) opted to crystallize samples under an RH of 53%, because previous data have shown that this provides sufficient water to induce crystallization but noticeably increases the lag time (10).

It is notable here that recent studies of lactose crystallization using environmentally-controlled atomic force microscopy (EC-AFM) have suggested that primary nucleation cannot occur in samples maintained under an RH of 58% and that under low RH atmospheres lactose does not fully crystallize; it was suggested that complete crystallization requires an RH of $\geq 94\%$ (48). In the work of Dilworth et al. (41), all samples removed from the TAM (once the heat flow had returned to zero) were completely crystalline, irrespective of the RH under which they were maintained. Two effects may explain this apparent dichotomy. First, the presence of crystalline material in the ampoule will promote secondary nucleation and it may be the case that a small quantity of crystalline material exerts a disproportionate effect on the observed behavior of the system. Second, it is likely that the expulsion of water following crystallization, being an extremely rapid process, will result in the (temporary) formation of a saturated vapor space, which will force the complete crystallization of the sample.

The thermal response of amorphous lactose crystallized at 53% RH is shown in Figure 12. Integration using Method 4C resulted in a heat change of 57.3 J/g, which is slightly higher than that calculated for the 75% RH data. Darcy and Buckton (40) noted that changes in temperature resulted in different heat changes but did not note this effect of RH at 25°C. A plot of known versus calculated amorphous content between 1% and 100% w/w is shown in Figure 13. As in the case of experiments conducted at 75% RH, the method showed a negative deviation from ideality, returning calculated amorphous contents that are lower than the known contents.

Gravimetric data have shown that the mass increase following crystallization of amorphous lactose is of the order of 3% w/w (44). If the sample crystallized totally to α-lactose monohydrate, a mass increase of approximately 5% w/w would be expected. These data therefore suggest that amorphous lactose crystallizes to a mixture of α-lactose monohydrate and anhydrous β-lactose, and it may be the case that the ratio of the two species formed varies as a function of RH. As the enthalpy of crystallization of the α- and β-forms vary, this effect may explain both the difference in net heat recorded as a function of RH and the negative deviation in the calibration plots.

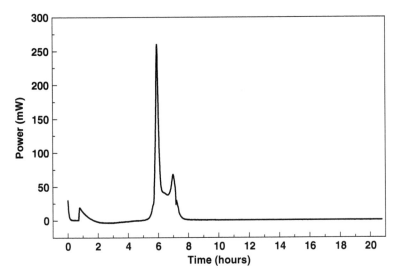

Figure 12 A typical power–time response for a sample of amorphous lactose held at 25°C under a relative humidity of 53%. *Source*: From Ref. 41.

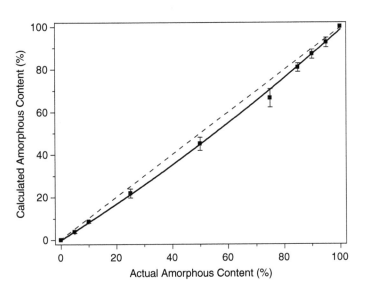

Figure 13 The calculated versus known amorphous contents for samples containing between 1% and 100% w/w amorphous content, determined by integration of power–time data (at 25°C and 53% relative humidity) using Method 4C (see p. 269). *Source*: From Ref. 41.

Preparation of Calibration Curves

From the discussion, it should be apparent that, as for nearly all analytical techniques, the accuracy of any amorphous content determination is dependent entirely upon the accuracy of the calibration curve. Unlike other analytical techniques, however, calorimetric data are uniquely susceptible to systematic errors, because of the universal nature of the measured property.

Phillips (49) suggested a method for estimating the amorphous content of pharmaceutical powders, which obviated the need for calibration standards, arguing that the conventional standards (blends of amorphous and crystalline material) were not representative of the unknown sample and hence invoked too many assumptions. The method starts with the premise that the enthalpy of crystallization of an amorphous fraction of a sample ($\Delta_c H^{\text{amor}}$) is proportional to its enthalpy of fusion ($\Delta_f H^{\text{amor}}$) at its melting temperature (T_m):

$$\Delta_c H^{\text{amor}} = \left(\Delta_f H^{\text{amor}} \frac{\Delta T}{T_m} \right) \left(\frac{T_c}{T_m} \right) \tag{13}$$

where T_c is the crystallization temperature and ΔT is the difference between T_m and T_c. The amorphous content of a material can then be calculated from the following relationship:

$$\text{Amorphous content} = \frac{\Delta_f H^{\text{amor}}}{\Delta_f H^{\text{crys}}} \tag{14}$$

where $\Delta_f H^{\text{crys}}$ is the enthalpy fusion of the crystalline material. It is then assumed that the measured heat from a calorimetric experiment is equivalent to $\Delta_c H^{\text{amor}}$. Thus, the value of $\Delta_f H^{\text{amor}}$ can be calculated if T_m is known. Knowledge of $\Delta_f H^{\text{crys}}$, which is easily determined by DSC, thus allows the determination of the amorphous content. Although only providing an estimated amorphous content, Phillips (49) showed that the model gave good agreement with literature data. Publication of the method did provoke a subsequent exchange of communication in the literature, however, which highlighted a number of the assumptions required for this approach (50,51). Specifically, the fact that preparation of the crystalline standard for the DSC measurement was likely to be difficult and any small differences in its composition would affect the outcome; that knowing the $\Delta_f H^{\text{crys}}$ for larger molecular weight or polymeric pharmaceuticals was not always possible; that the isothermal data were likely to contain other contributions from sample wetting, degradation, and so on; and that it is essentially a two-point calibration that assumes a linear relationship between the 100% and 0% crystalline responses.

Many of these issues were already noted and it is plain that if the study material has more than one isomer, anomer, or polymorph, and the enthalpies of solution or crystallization of the isomers, anomers, or polymorphs are different, then it is essential that the material used to prepare the calibration curve matches the isomeric, anomeric, or polymorphic composition of the unknown (processed) sample. This effect was recently discussed by Ramos et al. (52), who looked at the consequences of the anomeric composition of lactose on the calibration curves obtained for both isothermal and solution calorimetry, using lactose batches that were either predominantly α or predominantly β.

Isothermal Calorimetry

Ramos et al. (52) used RH perfusion to study the recrystallization of partially amorphous lactose to avoid some of the problems with water evaporation/condensation, as already noted. However, they note that the use of RH perfusion does not mean data interpretation is necessarily straightforward; this is because, once the RH increases from zero, all the internal surfaces of the ampoule, as well as the sample, are wetted. This produces a large exothermic heat signal that often occurs over a time period that is longer than the time required for the sample to crystallize. Essentially, the crystallization signal is obscured by the wetting response; an example of this is shown by the response of a 5% w/w amorphous sample of α-lactose (Fig. 14). There are four ways to interpret the data:

1. Assume that the wetting response is uniform in all samples and measure the total heat released once the RH is increased.
2. Determine the wetting response of the empty ampoule in a separate experiment and subtract this value from the experimental data.

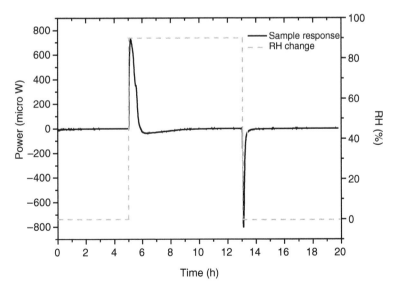

Figure 14 The wetting and drying calorimetric response of a 5% w/w amorphous sample of α-lactose using relative humidity (RH) perfusion calorimetry. *Source*: From Ref. 52.

3. Determine the wetting response of an equivalent mass of the crystalline material in a separate experiment and subtract it from the experimental data.
4. Return the system to 0% RH after crystallization and subtract the drying response from the wetting response.

Method 1 does not allow a quantitative assessment of the enthalpy of crystallization while Method 2 does not compensate for the wetting response of the sample, which may become significant with larger sample masses. Neither is, therefore, a suitable choice for analysis. Method 3 compensates for the wetting response of the crystalline component of the sample (and ignores the wetting of the amorphous material) while Method 4 compensates for the wetting response of the entire sample (although it is noted that the sample that wets, partially amorphous, differs slightly from that which dries, crystalline, and the method does not compensate for any water used to form the monohydrate); in both cases it is assumed that the errors are negligible which is reasonable for the standards used to prepare the calibration plots but may be an issue for processed samples—this point is discussed more fully subsequently.

Ramos et al. (52) thus prepared calibration curves using Methods 3 and 4. Figure 15 shows a typical wetting response of a crystalline sample of α-lactose. Subtraction of these data from those shown in Figure 14 resulted in the peak shown in Figure 16. The area under this peak was used in the construction of a calibration plot in accordance with Method 3 (Fig. 17). Alternatively,

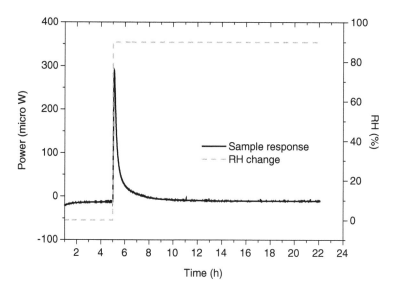

Figure 15 Power–time data for the wetting response of a crystalline batch of α-lactose using relative humidity (RH) perfusion calorimetry. *Source*: From Ref. 52.

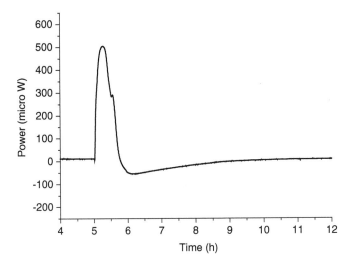

Figure 16 The power–time data obtained via Method 3 (subtraction of the crystalline wetting response from the wetting response of the partially amorphous sample) for a 5% w/w sample of amorphous α-lactose. *Source*: From Ref. 52.

subtraction of the wetting and drying peaks shown in Figure 14 resulted in the peak shown in Figure 18; the area under this peak was used in the construction of a calibration plot in accordance with Method 4 (Fig. 19). Calibration plots

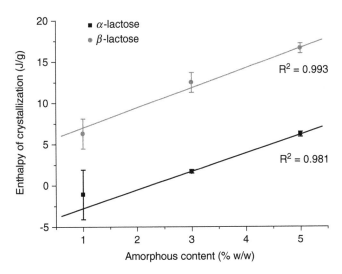

Figure 17 Calibration curves for the two lactose batches prepared using Method 3 (subtraction of the crystalline wetting response from the wetting response of the partially amorphous sample). The lines shown are linear regression fits. *Abbreviation*: R^2 = linear regression coefficient. *Source*: From Ref. 52.

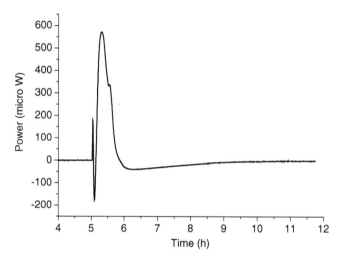

Figure 18 The power–time data obtained via Method 4 (subtraction of the drying response from the wetting response of the partially amorphous sample) for a 5% w/w amorphous lactose sample (predominately α-lactose). *Source*: From Ref. 52.

were constructed using these methods for batches of lactose that were either predominantly α or predominantly β.

It is patent that the calibration curves prepared using the two lactose batches are different. As discussed already, these dissimilarities are likely to

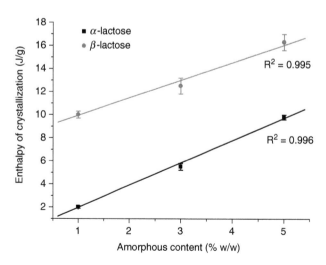

Figure 19 Calibration curves for the two lactose batches prepared using Method 4 (subtraction of the drying response from the wetting response of the partially amorphous sample). The lines shown are linear regression fits. *Abbreviation*: R^2 = linear regression coefficient. *Source*: From Ref. 52.

arise because of the different enthalpies of crystallization of the two forms (α-lactose, approximately 169 J/g and β-lactose, approximately 197 J/g; both estimated by DSC) (44).

A further interesting feature of the calibration curves is that those for the β-lactose batch do not tend to a y-intercept value of zero with no amorphous material present, the effect being more dramatic when Method 4 is used to construct the curve. Although this observation initially seems counter-intuitive, one explanation for this discrepancy may be the fact that β-lactose is known to mutarotate to α-lactose monohydrate under elevated RH conditions, an event that is exothermic (47). This being so, mutarotation of the sample during measurement would result in a net exothermic addition to the measured response during wetting. This event would not be cancelled out at all using Method 4 and, assuming mutarotation occurs faster in amorphous material than crystalline material because of its faster uptake of water, would only be partially compensated for using Method 3. This would lead to the behavior noted for the β-lactose batches.

The result of these observations is that measurement of the crystallization response of a partially amorphous lactose sample of unknown isomeric composition would give different results depending upon which of the two calibration plots was employed. The best way to resolve this problem, therefore, is to ensure that any calibration standards are prepared from the same batch of material as that to be tested during processing. It is also not sufficient to quote an enthalpy of crystallization for a material without also defining its isomeric state. This, in part, may be one of the factors that contribute to the range of stated enthalpies of crystallization for lactose although these discrepancies also arise in part through the use of different integration strategies.

Comparison of the standard deviations for the two methods reveals that Method 4 is the most precise, and the authors thus recommended it for future studies of this type.

An additional problem with isothermal calorimetric methodologies is that the nature of the amorphous standards used to prepare the calibration plot do not mimic the physical nature of processed materials [i.e., the standards comprise particles that are either wholly amorphous or wholly crystalline while a processed material is likely to comprise of smaller particles (and hence have a greater surface area) consisting of crystalline cores with an outer corona of amorphous material]. This raises a number of concerns. First, the differences in surface area mean that the rate of water absorption will be different in a processed material than from the calibration standards; necessarily, this will manifest itself as a change in the kinetic response of the samples during analysis although, as net areas are measured, should not affect the enthalpy obtained.

Second, a greater problem is likely to be that the wetting response of a processed sample will (in effect) be that of a wholly amorphous material while that of a standard approximates to the wetting response of a crystalline material. This is important because, using Method 4, the drying response (of the now crystalline material) is subtracted from the wetting and crystallization response (of the processed material). Thus, if the wetting enthalpies of the amorphous and crystalline forms are different, it is likely that the amorphous content predicted from the calibration plot will have a (small) systematic error.

Finally, the amorphous regions in a processed material are sited directly on top of a crystalline substrate, which acts as a seed, meaning that secondary nucleation predominates. In the material used for calibration, the amorphous and crystalline particles are discreet entities, which means that primary nucleation may occur. As already noted, this requires an almost saturated vapor space presumably because more water needs to be absorbed to plasticize the sample sufficiently to induce crystallization. To a large degree, the data presented suggest that the crystalline particles do act as a seed, allowing secondary nucleation, because at such low amorphous contents they considerably outnumber the amorphous particles; this effect will, however, diminish as the proportion of amorphous material increases.

As it is difficult to conceive of a method by which calibration standards that mimic processed materials could be prepared, Method 4 appears to be the best currently achievable.

Solution Calorimetry

Figure 20 shows the raw power output from the solution calorimeter for the dissolution of a 5% w/w amorphous sample of α-lactose. The solution calorimeter

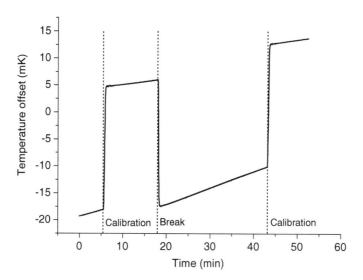

Figure 20   Raw solution calorimeter data showing the dissolution of a 5% w/w amorphous sample of α-lactose into water and the two electrical calibrations. *Source*: From Ref. 52.

used here operated on a semi-adiabatic principle, meaning the raw data recorded are of the form temperature offset versus time. Data are converted to power–time using a series of heat-balance equations [discussed previously (53)], and integrated to obtain the heat of solution ($\Delta_{sol}H$). Figure 21 shows the calibration curves constructed for both batches of lactose using the enthalpy of solution data obtained.

Two important observations can be made. First, it is immediately apparent that there are significant differences between the two batches. Using entirely crystalline samples, the enthalpy of solution of the α-lactose batch was determined to be 57.1 ± 0.2 J/g, while that of the β-lactose batch was 6.5 ± 0.2 J/g, corresponding to the y-intercepts of the two calibration plots. For both batches, the measured heats of solution decrease as the amorphous content increases, reflecting the fact that the endothermic contribution from $\Delta_{lattice}H$ is decreasing. Of course, one of the major concerns regarding the preparation of calibration curves for amorphous content determination is now clear; measurement of the enthalpy of solution of a sample of lactose that is partially amorphous would result in two different estimates of amorphous content and, without prior knowledge of its isomeric composition, it would be impossible to know the correct value.

Second, the error limits are much smaller [the greatest standard deviation (SD) in the measurements being ± 0.3 J/g] in this study than those reported by

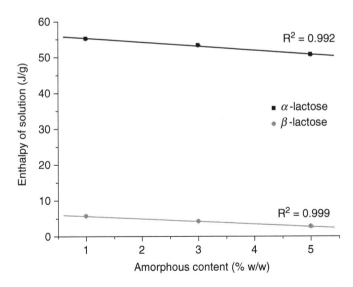

Figure 21 Calibration curves for the two lactose batches prepared from enthalpy of solution data. The lines shown are linear regression fits; error bars are plotted but fall within the symbols. *Source*: From Ref. 52.

Hogan and Buckton (3) using the same equipment. It is likely that this is a result of the different methodologies used to prepare the ampoules. Here, the appropriate masses of the crystalline and amorphous materials were weighed directly into the crushing-ampoule while in the previous study large batches of partially amorphous material were prepared by mixing (in a turbula mixer) from which samples were taken and loaded into the crushing-ampoules. In the former method, a weighing error is introduced (50 ± 0.01 mg or relative SD 0.02%), but this is likely to be much lower than the errors inherent in mixing and sampling. It therefore appears that for quantification of small (0.5% w/w) amorphous contents, this methodology is preferable.

SUMMARY

Knowledge, and more importantly control, of the amorphous form is an essential prerequisite for the successful development of a pharmaceutical. Two clear situations present themselves: the case where the inclusion of the amorphous material is intentional (and usually the mass proportion is high) and the case where amorphous material is accidentally induced through processing (with a low mass proportion). In the former case, quantification of the stability of the amorphous form is the priority. In the latter, accurate detection and quantification are more important.

The conventional calorimetric method to determine the stability of the amorphous form is DSC; here, the decay function is plotted as a function of time, and relaxation parameters are determined using the KWW equation. The number of points generated in a typical experiment is low. An alternative approach is to use IC data and fit the data to the MSE equation. In this case, relaxation is followed in real time and, hence, the number of data points is high. The use of IC in this way is relatively recent, but it offers much potential.

The detection and quantification of small amorphous contents is of growing concern; particularly in the case of materials that have been processed, because the mechanical forces imparted upon the surface of the material can result in the formation of small, and unexpected, amorphous content. Although the proportion of amorphous material is small (on the order of 1% w/w), its location on the surface of the sample means it effectively governs the way the material behaves. Of the myriad of techniques available for quantifying amorphous contents, calorimetric methods are among the most promising, primarily because they can study the entire sample noninvasively. Isothermal methods include measuring the heat of crystallization as the sample is plasticized with water and measuring the heat of solution as the sample is dissolved. Already, the discussion has shown that the techniques are capable of quantifying amorphous contents to 0.5% w/w (dependent upon the magnitude of the enthalpy change). However, care must be taken that the calibration curves are prepared with due consideration of the composition of the sample to ensure that erroneous results are not obtained.

REFERENCES

1. Grant DJW, York P. Entropy of processing—a new quantity for comparing the solid-state disorder of pharmaceutical materials. Int J Pharm 1986; 30:161–180.
2. Giron D, Remy P, Thomas S, Vilette E. Quantitation of amorphicity by microcalorimetry. J Therm Anal 1997; 48:465–472.
3. Hogan SE, Buckton G. The quantification of small degrees of disorder in lactose using solution calorimetry. Int J Pharm 2000; 207:57–64.
4. Kawakami K, Pikal MJ. Calorimetric investigation of the structural relaxation of amorphous materials: evaluating validity of the methodologies. J Pharm Sci 2005; 94:948–965.
5. Liu S, Rigsbee DR, Stotz C, Pikal MJ. Dynamics of pharmaceutical amorphous solids: the study of enthalpy of relaxation by isothermal microcalorimetry. J Pharm Sci 2002; 91:1853–1862.
6. Phillips JC. Microscopic theory of the Kohlrausch relaxation constant β_K. J Non-cryst Solids 1994; 172–174:98–103.
7. Shamblin SL, Hancock BC, Dupuis Y, Pikal MJ. Interpretation of relaxation time constants for amorphous pharmaceutical systems. J Pharm Sci 2000; 89:417–427.
8. Peyron M, Pierens GK, Lucas AJ, Hall LD, Stewart RC. The modified stretch exponential model for characterization of NMR relaxation in porous media. J Magn Reson Ser A 1996; 118:214–220.
9. Kawakami K, Ida Y. Direct observation of the enthalpy relaxation and the recovery processes of maltose-based amorphous formulation by isothermal microcalorimetry. Pharm Res 2003; 20:1430–1436.
10. Briggner L-E, Buckton G, Bystrom K, Darcy P. The use of isothermal microcalorimetry in the study of changes in crystallinity induced during the processing of powders. Int J Pharm 1994; 105:125–135.
11. Buckton G. Unpublished data, 2000.
12. Sebhatu T, Elamin AA, Ahlneck C. Effect of moisture sorption on tabletting characteristics of spray dried (15% amorphous) lactose. Pharm Res 1994; 9:1233–1238.
13. Hogan SE, Buckton G. The application of near infrared spectroscopy and dynamic vapour sorption to quantify low amorphous contents of crystalline lactose. Pharm Res 2001; 18:112–116.
14. Newell HE, Buckton G, Butler DA, Thielmann F, Williams DR. The use of inverse-phase gas chromatography to measure the surface energy of crystalline, amorphous and recently milled lactose. Pharm Res 2001; 18:662–666.
15. Taylor LS, Zografi G. The quantitative analysis of crystallinity using FT-Raman spectroscopy. Pharm Res 1998; 15:755–761.
16. Royall PG, Gaisford S. Application of solution calorimetry in pharmaceutical and biopharmaceutical research. Curr Pharm Biotech 2005; 6:215–222.
17. Pikal MJ, Lukes AL, Lang JE, Gaines K. Quantitative crystallinity determinations for β-lactam antibiotics by solution calorimetry: correlations with stability. J Pharm Sci 1978; 67:767–772.
18. Guillory JK, Erb DM. Using solution calorimetry to quantitate binary mixtures of three crystalline forms of sulfamethoxazole. Pharm Manuf 1985; Sep:28–33.
19. Gao D, Rytting JH. Use of solution calorimetry to determine the extent of crystallinity of drugs and excipients. Int J Pharm 1997; 151:183–192.

20. Harjunen P, Lehto V-P, Koivisto M, Levonen E, Paronen P, Järvinen K. Determination of amorphous content of lactose samples by solution calorimetry. Drug Dev Ind Pharm 2004; 30:809–815.
21. Katainen E, Niemelä P, Päällysaho M, Harjunen P, Suhonen J, Järvinen K. Evaluation of the amorphous content of lactose by solution calorimetry and Raman spectroscopy. Eur J Pharm Sci Abstr 2003; 19:S36.
22. Reading M, Luget A, Wilson R. Modulated differential scanning calorimetry. Thermochimica Acta 1994; 238:295–307.
23. Guinot S, Leveiller F. The use of MTDSC to assess the amorphous phase content of a micronised drug substance. Int J Pharm 1999; 192:63–75.
24. Saklatvala R, Royall PG, Craig DQM. The detection of amorphous material in a nominally crystalline drug using modulated temperature DSC—a case study. Int J Pharm 1999; 192:55–62.
25. Saunders M, Podluii K, Shergill S, Buckton G, Royall PG. The potential of high speed DSC (Hyper-DSC) for the detection and quantification of small amounts of amorphous content in predominantly crystalline samples. Int J Pharm 2004; 274:35–40.
26. Lappalainen M, Pitkänen I, Harjunen K. Quantification of low levels of amorphous content in sucrose by hyperDSC. Int J Pharm 2006; 307:150–155.
27. Byström K. Microcalorimetric testing of physical stability of drugs in the solid state. Application note 22004, Thermometric AB, Järfälla, Sweden, 1990.
28. Mackin L, Zanon R, Park JM, Foster K, Opalenik H, Demonte M. Quantification of low levels ($<10\%$) of amorphous content in micronised active batches using dynamic vapour sorption and isothermal microcalorimetry. Int J Pharm 2002; 231:227–236.
29. Nyqvist H. Saturated salt solutions for maintaining specified relative humidities. Int J Pharm Tech Prod Mfr 1993; 4:47,48.
30. Buckton G. Isothermal microcalorimetry water sorption experiments: calibration issues. Thermochimica Acta 2000; 347:63–71.
31. Kawakami K, Numa T, Ida Y. Assessment of amorphous content by microcalorimetry. J Pharm Sci 2002; 91:417–423.
32. Vemuri NM, Chrzan Z, Cavatur R. Use of isothermal microcalorimetry in pharmaceutical preformulation studies, part II. Amorphous phase quantification in a predominantly crystalline phase. J Therm Anal Cal 2004; 78:55–62.
33. Al-Hadithi D, Buckton G, Brocchini S. Quantification of amorphous content in mixed systems: amorphous trehalose with lactose. Thermochimica Acta 2004; 417: 193–199.
34. Buckton G, Darcy P, Greenleaf D, Holbrook P. The use of isothermal microcalorimetry in the study of changes in the crystallinity of spray dried salbutamol sulphate. Int J Pharm 1995; 116:113–118.
35. Aso Y, Yoshioka S, Otsuka T, Kojima S. The physical stability of amorphous nifedipine determined by isothermal microcalorimetry. Chem Pharm Bull 1995; 43:300–303.
36. Sebhatu T, Angberg M, Ahlneck C. Assessment of the degree of disorder in crystalline solids by isothermal microcalorimetry. Int J Pharm 1994; 104:135–144.
37. Angberg M. Lactose and thermal analysis with special emphasis on microcalorimetry. Thermochimica Acta 1995; 248:161–176.
38. Buckton G, Darcy P, Mackellar AJ. The use of isothermal microcalorimetry in the study of small degrees of amorphous content of powders. Int J Pharm 1995; 117:253–256.

39. Chidavaenzi OC, Buckton G, Koosha F, Pathak R. The use of thermal techniques to assess the impact of feed concentration on the amorphous content and polymorphic forms present in spray dried lactose. Int J Pharm 1997; 159:67–74.
40. Darcy P, Buckton G. Quantitative assessments of powder crystallinity: estimates of heat and mass transfer to interpret isothermal microcalorimetry data. Thermochimica Acta 1998; 316:29–36.
41. Dilworth SE, Buckton G, Gaisford S, Ramos R. Approaches to determine the enthalpy of crystallization, and amorphous content, of lactose from isothermal calorimetric data. Int J Pharm 2004; 284:83–94.
42. Angberg M, Nyström C, Castensson S. Evaluation of heat-conduction microcalorimetry in pharmaceutical stability studies. V. A new approach for continuous measurements in abundant water vapour. Int J Pharm 1992; 81:153–167.
43. Angberg M, Nyström C, Castensson S. Evaluation of heat-conduction microcalorimetry in pharmaceutical stability studies. VI. Continuous monitoring of the interaction of water vapour with powders and powder mixtures at various relative humidities. Int J Pharm 1992; 83:11–23.
44. Hogan SE. Investigation of the amorphous and crystalline properties of lactose and raffinose. Ph.D. thesis, University of London, 2002.
45. Makower B, Dye WB. Equilibrium moisture content and crystallisation of amorphous sucrose and glucose. J Agric Food Chem 1956; 4:72–77.
46. Buckton G, Darcy P. Water mobility in amorphous lactose below and close to the glass transition temperature. Int J Pharm 1996; 136:141–146.
47. Angberg M, Nyström C, Castensson S. Evaluation of heat-conduction microcalorimetry in pharmaceutical stability studies. III. Crystallographic changes due to water vapour uptake in anhydrous lactose powder. Int J Pharm 1991; 73:209–220.
48. Price R, Young PM. Visualisation of the crystallisation of lactose from the amorphous state. J Pharm Sci 2004; 93:155–164.
49. Phillips EM. An approach to estimate the amorphous content of pharmaceutical powders using calorimetry with no calibration standards. Int J Pharm 1997; 149:267–271.
50. Hancock BC. Comments on "An approach to estimate the amorphous content of pharmaceutical powders using calorimetry with no calibration standards." Int J Pharm 1998; 160:131–133.
51. Phillips EM. Reply to: Comments on "An approach to estimate the amorphous content of pharmaceutical powders using calorimetry with no calibration standards." Int J Pharm 1988; 160:135.
52. Ramos R, Gaisford S, Buckton G. Calorimetric determination of the amorphous content in lactose: A note on the preparation of calibration curves. Int J Pharm 2005; 300:13–21.
53. Yff BTS, Royall PG, Brown MB, Martin GP. An investigation of calibration methods for solution calorimetry. Int J Pharm 2004; 269:361–372.
54. Conti S. Unpublished data, 2006.

7

Stability Assessment

INTRODUCTION

Formulated pharmaceuticals are complex and often heterogeneous systems that are difficult to analyze using standard analytical tools; this may be because the object of the study is the active pharmaceutical ingredient (API), and any analysis first requires its isolation from the rest of the medicine, or because the product contains materials that are progressing through both chemical and physical changes. In either case, isothermal calorimetry (IC) offers a potentially useful alternative to conventional stability assessments because it simply monitors a sample over time.

As mentioned in Chapter 1, for manufactured materials, it is generally the case that the initial state of a formulation is the one with the desired properties, and over time changes in its components will lead to deterioration in its properties and a consequential loss of quality. Some materials degrade because they are inherently unstable (such as the decay of a radioactive isotope), whereas others degrade because of the effects of external forces (such as hydrolysis or oxidation, photodegradation, mechanical degradations such as that resulting from shearing or compression, and thermal degradation). Although for some materials the effects of these forces may be negligible, for others they may be catastrophic and a formulation should be developed such that these effects are tempered and, hence, shelf life is maximized.

The ultimate aim of any stability assessment is the assignment of a shelf life or "use by" date. By using the product before the manufacturer's specified "use by" date, a consumer is assured that the product's performance conforms to the original manufacturing specification, assuming it has been stored correctly. Evidently, the assignment of an accurate "use by" date is of more importance for some products than others; it is likely that an emulsion paint, which had phase-separated after being left standing for a long period of time, could be restored to its original specification by stirring, but the same is unlikely to be true for a pharmaceutical cream. Similarly, if the batteries in a radio expire before their "use by" date, then some inconvenience may be caused. The situation could be fatal were the batteries in a pacemaker [indeed, calorimetry has been used to determine the shelf life of pacemaker batteries (1)].

The pharmaceutical industry in particular, then, has a strict obligation to ensure the safety and efficacy of its products. This requires knowledge of the stability of each of the components in a medicine both individually (characterized during preformulation) and in combination (characterized during formulation). Given the fact that even a simple immediate release tablet may contain more than 10 excipients and, thus, may undergo many simultaneous degradation processes, it is clear that stability assessment of pharmaceuticals is not a trivial task.

Confidence in the ability to predict stability and define accurate "use by" dates comes both from understanding the processes that lead to change and by relating these to any stress conditions the material may be placed under (during manufacture or storage, for instance). Stability data are recorded either in "real-time" (i.e., during a conventional long-term stability trial conducted under storage conditions) or during an accelerated stability trial [i.e., where the sample is placed under stress conditions, usually an elevated temperature and/ or relative humidity (RH)]. Comparison of the degradation rates for a series of related products allow on-going improvement programs to be followed, showing how different modifications may affect the stability of the final product.

Two issues present themselves as being problematic with these approaches. First, knowledge of the kinetics of a particular reaction gives no insight into the mechanism by which that reaction occurs, other than the order of reaction, and the cause of the (presumably) instability often remains unknown. Knowing the cause of a degradation reaction allows greater control over the development of a product, as the product can be specifically redesigned to make it more stable. Secondly, the experimental measurement times for both approaches are long (at least in the order of months, if not years), which means that stability trials are time and labor consuming and expensive.

IC offers the possibility to conduct rapid stability assays directly under storage conditions and in a noninvasive manner. Indeed, the principal benefit of the technique is the ability to predict long-term stability from short-term experimental measurements. This chapter reviews some of the approaches that may be adopted when using IC for stability assessment, and is illustrated with a number of examples from the recent literature.

CURRENT METHODS FOR STABILITY ASSESSMENT

Of the many analytical techniques used in materials characterization, perhaps those based on spectroscopic or chromatographic principles are the most widely used for pharmaceutical stability assessment. High-performance liquid chromatography (HPLC) is the principal technique used in the pharmaceutical industry to assess drug stability by quantifying the concentrations of the parent compound and any degradation products as a function of storage time. However, the use of HPLC is limited by two important factors; its relative insensitivity to small changes in concentration and the requirement that samples are dissolved in a suitable solvent prior to analysis. The former is a problem for all samples that exhibit slow degradation rates whereas the latter can affect samples not formulated in solution, as accelerated decomposition rates are often observed when compounds are solvated.

An additional concern, when using HPLC for the analysis of an API is that it cannot, of course, detect if a solid drug has changed polymorph, because dissolution of the sample before analysis removes any solid-state history, nor will it be of use if it is a change in the properties of an excipient, or an interaction between excipients, that causes the product to fail to meet its specification.

Because of the poor sensitivity of HPLC, it is often the case that stability assays are conducted under some condition of elevated stress, the applied stress usually being an increase in temperature and/or RH. The resulting increase in degradation rate allows a rate constant (k) to be determined at each temperature. The data are then plotted in accordance with the Arrhenius relationship [Equation (1)]:

$$k = Ae^{(-E_a)/RT} \tag{1}$$

where A is the preexponential constant, R the gas constant, T the absolute temperature, and E_a the activation energy.

This allows the rate constant under any desired (usually storage) condition to be obtained though data extrapolation. Necessarily, it is assumed that the analysis results in a linear relationship, and that any of the reaction processes occurring under stress conditions are the same as those that would occur under storage conditions.

There are two major drawbacks of using the Arrhenius relationship as the basis of an accelerated stability study. The first, as mentioned earlier, is the assumption that the reaction mechanism is independent of temperature, such that any extrapolation of data is valid. Over the narrow temperature ranges usually used for Arrhenius analyses, typically 35°C to 70°C, it is often difficult to observe non-Arrhenius behavior in experimental data, even though nonlinearity might be expected from the reaction mechanism (2). Indeed, even some complex biological processes may show apparent Arrhenius behavior over certain temperature ranges.[a] In general, this means that the use of the Arrhenius relationship is commonplace and, relatively, justified.

However, the complex nature of pharmaceuticals often means that non-Arrhenius behavior is observed; there are many potential events that may

---

[a]It has been shown, for example, that natural phenomena such as the frequency of flashing of fireflies and the rate of a terrapin's heartbeat will often exhibit Arrhenius behavior, when studied over a narrow temperature range (3).

cause such nonconformance to the model, including solvent evaporation, multiple reaction pathways, or a change in physical form. An illustrative example of non-Arrhenius behavior is provided by the decomposition of ampicillin. It is observed that, upon freezing, the rate of decomposition of ampicillin is increased (2). Such a decrease in stability at low temperatures is often observed for reactions that obey second- or higher-order reaction kinetics. Two possible explanations have been proposed to explain such behavior. The first hypothesis, for reactions obeying second- or higher-order kinetics, is that an increase in rate may be brought about by an increase in the concentration of the solute as freezing occurs, the solute molecules being excluded from the ice lattice. The second hypothesis suggests that an increase in rate may be caused by a change in pH upon freezing. It has been shown that citrate–sodium hydroxide and citrate–potassium phosphate buffers do not change pH upon freezing, but a citrate–sodium phosphate buffer decreases from pH 8 to 3.5 and a sodium hydrogen phosphate buffer decreases from pH 9 to 5.5 upon freezing (2). Such effects should be considered when studying the stability of freeze-dried products, for instance.

The second is that the order of reaction for the degradation needs to be known, and this may often be a time-consuming procedure. It has been suggested that, from classically based studies, at $<10\%$ degradation, and within the limits of error associated with stability studies, it is not possible to distinguish between zero-, first-, and second-order kinetics using curve-fitting techniques (4). Consequently, these authors suggest that, if first-order kinetics is assumed for a degradation process, errors are kept to an absolute minimum. In order to remove the requirement that the reaction order is known, Amirjahed (5) showed that there is a linear relationship between $t_{0.9}$ (the time taken for the concentration of a reactant to decrease to 90% of its initial value) and the reciprocal temperature, which is independent of the order of reaction for a series of drugs. It is suggested that the use of such linear plots to estimate $t_{0.9}$ at any desired temperature would provide a rapid, and sufficiently accurate, means of studying decomposition rates during drug development phases.

Moisture may also affect degradation rates. Storage of a product in an atmosphere of high RH will accelerate degradation reactions that arise as a result of hydrolysis. It has been shown that added moisture decreases the lag time and increases the zero-order rate constant for the decomposition of aminosalicylic acid (6), for example. Such tests may give mechanistic information on any degradation process occurring, and often give an indication of the level of RH that is tolerable by the sample. This information allows packaging to be designed that affords sufficient protection to the sample over the course of its shelf life. Tingstad et al. (7,8) quote an example of tablets containing a water-labile component. At 50°C, the component was considerably more stable in a water-permeable blister package than in a sealed glass bottle, but at room temperature and 70% RH, the situation was reversed. The behavior was attributed to the loss of considerable quantities of water through the film at 50°C, improving stability. At room temperature, the diffusion was reversed, and considerable quantities of water entered the container through the film, decreasing the stability.

It should also be noted that the effects caused by high temperatures may sometimes be confused with those that arise from the effects of low RH. Unless controlled, the RH in a high-temperature storage cabinet will be lower than the RH in the surrounding ambient conditions. This may cause the sample to lose some water content, causing an apparent increase in reactant concentration. If these changes in concentration are not accounted for in any subsequent degradation calculations, any decomposition that has occurred may not be observed. If there is any doubt about the extrapolation, then long-term storage studies (holding the sample directly under expected storage conditions) are conducted to confirm the findings.

When the moisture content is very high, the decomposition of solid materials may often be described by solution-phase kinetics. In such cases, the decomposition is generally zero order.

Although elevated temperature methods are less time consuming than a stability study conducted at room temperature, a complete study may still take a considerable period of time. It has been shown that for a compound with a degradation rate of 1%/yr at 25°C and an activation energy of 75 kJ/mol, the degradation rate would increase to 1.52%/wk at 75°C (9). It would therefore take about two weeks to obtain any useful data from such an experiment. Repeat experiments over the range of temperatures required to construct an Arrhenius plot would lead to a study time approaching 10 weeks.

The Arrhenius relationship is derived from solution-phase kinetics. Although degradations that occur in solid-state systems usually follow different kinetic mechanisms from those that occur in solution, it is still generally possible to treat rate constants that have been determined from solid-state mechanisms with the Arrhenius relationship, and hence to predict stability at any desired temperature. There are, however, exceptions to this rule; for example, those solids in which decomposition exhibits an approach to equilibrium. Examples of such systems include vitamin A in gelatine beads and vitamin E in lactose-base tablets (10). In such cases, the effect of temperature should be described by the van't Hoff equation [Equation (2)], rather than by the Arrhenius relationship:

$$\ln K = -\frac{\Delta H}{RT} + c \tag{2}$$

where K is the equilibrium constant; ΔH, the enthalpy of the process under study; and c, a constant.

Other issues concerning the formulation of a drug may also affect stability. For example, the pH of a drug solution may dramatically influence the rate of degradation. Depending on the reaction mechanism, a rate constant may be increased 10-fold with a change of just one pH unit. It is, therefore, essential to construct a pH versus rate profile for a formulated product, such that the best pH value for stability may be determined. Almost all pH versus rate profiles can be rationalized using an approach in which the reaction of each molecular species of the drug with the hydrogen ion, water, and the hydroxide ion is analyzed as a function of pH (2).

Photochemical degradation may also be induced by exposing a sample to high intensities of light. Usually, fluorescent tubes are used to simulate natural light, although care must be taken to avoid excessively heating a sample. Many pharmaceutical compounds have been shown to undergo photochemical degradation, including phenothiazine tranquillizers, hydrocortisone, prednisolone, riboflavin, ascorbic acid, and folic acid although, in most cases, the mechanisms of degradation are complex. Photosensitive compounds may be protected by storage in colored containers. Amber containers are often chosen as this color excludes light of wavelength <470 nm and so affords considerable protection to compounds sensitive to ultraviolet (UV) light. Similar information on stability may be obtained from these studies as can be obtained from studies conducted at higher humidities.

For many pharmaceuticals, it may be necessary to incorporate water-miscible solvents to solubilize the active component. Typical solvents are ethanol, propylene glycol, or polyethylene glycols (PEGs), and it can be difficult to predict any solvent effects that may arise. As well as altering the activity coefficients of the reactant molecules and the transition state, a change in solvent may also induce changes in physicochemical parameters such as pK_a, surface tension, or viscosity. For example, in the presence of ethanol, aspirin degrades via an extra pathway, forming an ethyl ester (11). Conversely, the stability of a drug may be enhanced by the selective use of a solvent. The degradation of aspirin in PEG arises, in part, from a transesterification process between the drug and the glycol moiety (12). By blocking the free hydroxy group on the PEG molecule, either by methylation or acetylation, the stability of aspirin is greatly enhanced (13).

Although not useful for accelerated stability studies, the effects of excipients present in formulations must also be considered. For example, it has been shown that the stability of aspirin in tablet formulations is markedly affected by different excipients (14). An increased degradation rate is observed using excipients in the order, hydrogenated vegetable oil < stearic acid < talc < aluminum stearate, and it has been suggested that stearate salts should be avoided as tablet lubricants if the active component is subject to hydroxyl ion-catalyzed degradation. The base used in the formulation of some suppositories may also affect the degradation rate of an active component. Jun (12) studied the decomposition of aspirin in several PEGs that are often incorporated into suppository bases. It was shown that degradation arises partly because of transesterification. The rate of decomposition was shown to follow pseudo first-order kinetics and increased considerably when fatty bases such as cocoa butter were used.

Once the method of stability testing has been decided, a protocol must be designed to ensure that results from different formulations are comparable.

Protocols normally define such variables as the temperature and RH of storage, the time elapsed before sampling, the number of batches sampled, the number of replicates within each batch, and details of any chemical assays used. Although pharmaceutical preparations must conform to strict governmental regulations within any one particular country, there are presently no uniformly defined conditions for storage and, as a result, stability-testing conditions may vary from company to company, even within the same company. Storage conditions may be tailored to individual formulations and formulations are tested in their intended final packaging. If the final packaging is unknown, the testing should be conducted using a range of packs and pack materials. For example, tablets may be stored in glass bottles, high-density polyethylene containers, aluminum foil, and polyvinyl chloride (PVC) or polyvinylidene chloride blisters. Batches of liquids may be stored in an inverted position to check for interactions between the active components and the liner of the lid. In order to develop improved stability assays, which do not depend on exposing samples to extreme conditions, it is necessary to understand the basic requirements that must be met if a reaction is to proceed, and to appreciate the principles of reaction kinetics.

CALORIMETRY FOR STABILITY ASSESSMENT

The primary aim of any stability assessment is to ascertain whether a compound, either alone or formulated with other actives and/or excipients, will degrade significantly over a defined period of time under specified environmental conditions. If no degradation is observed, then the compound is assumed to be stable and no further assessment is necessary. If degradation is observed, then either the system is abandoned or further experimentation is required so as to identify the cause (which may not necessarily arise from a chemical change in the sample; physical changes are likely in amorphous or polymorphic drugs or in heterogeneous drug delivery systems). In this sense, screens can therefore be qualitative (i.e., designed to give a yes/no, stable/unstable answer) or quantitative (i.e., designed to return a rate constant); careful experimental design allows IC to conduct both types of assessment.

Primary Screening

A primary screen is taken here to mean a rapid assessment of the stability of a compound or mixture during preformulation or early formulation; the output is generally qualitative, the formulator simply requiring a stable/unstable answer. The degradation of an API or excipient will almost certainly lead to a loss of potency of a product; as the lowest acceptable level of potency is usually 90% of the label claim (15), IC must be able to detect reactions on this order to be of practical value. It has been shown that by analysis of 50 hours of power–time data recorded at 25°C using an IC, for a reaction following first-order kinetics and occurring with a reasonable ΔH of -50 kJ/mol, it is possible to distinguish between rate constants of 1×10^{-11} and 2×10^{-11}/s (16). A reaction progressing with a rate constant of 1×10^{-11}/s has a half-life of approximately 2200 years (note also that this result is obtainable by recording data over two days under storage conditions).

Thus, an initial screen of a sample should detect rapidly any degradation process occurring, and the absence of any heat changes gives confidence that the system under investigation is stable (although it should be noted that exothermic and endothermic powers occurring simultaneously will reduce the observed net power signal).

The experimental protocol for the case of a single component is very simple and is represented by the flow diagram in Figure 1. Such an approach was first described by Pikal (17), who showed there was a correlation between the exothermic heat output of some pharmaceutical systems and their known degradation rates (previously determined using other analytical methods). The data also showed that degradation rates of the order of 2%/yr were easily quantified.

A similar approach can be adopted for testing binary mixtures of an API with an excipient; a typical process is represented by the flow diagram in Figure 2. In this case, however, a number of control experiments must be performed first. The thermal response of the active and the excipient are recorded separately and then summed, to give the "predicted" power response. This trace is then compared with that determined for the actual drug-excipient mixture. Any unexpected powers recorded in the drug-excipient mixture indicate a possible interaction.

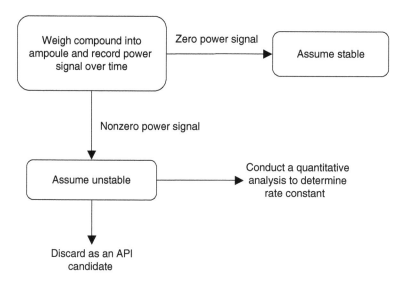

Figure 1 A flow diagram for a qualitative stability screen for an active pharmaceutical ingredient (API) alone.

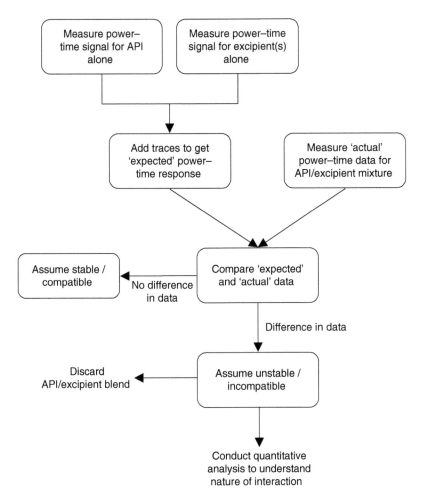

Figure 2 A flow diagram for a qualitative stability screen for a binary mixture of active pharmaceutical ingredient (API) and excipient.

This approach has been discussed in the literature (18,19) and provides an extremely versatile screening methodology. However, small variations between the experimental and the predicted data sets are often observed in these experiments. These can arise from various experimental factors such as small differences in sample masses or variations in the position of the baseline caused by different materials. One approach that has been suggested, therefore, is to set arbitrary upper and lower limits (which can be a power value such as $\pm 10 \ \mu$W, or a percentage value such as $\pm 10\%$); if the predicted and experimental data differ by more than the limits, the mixture is discarded. An example data set conducted under this regimen is shown in Figure 3.

Heat Flow (µW/g)

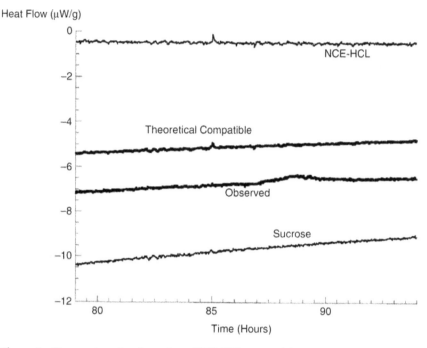

Time (Hours)

Figure 3   Power–time data for a drug (NCE-HCl), an excipient (sucrose), and a binary mixture of the two (observed). The observed trace is different from the theoretical compatible trace (which is the sum of the NCE-HCl and sucrose data), indicating a possible incompatibility. *Source*: From Ref. 17.

Calorimetric measurements have also been used to screen batch-to-batch variations in excipients (20). In this work, the authors used solution calorimetry to measure the heat of solution of a number of excipients [cetostearyl alcohol, microcrystalline cellulose (MCC), and a number of nonionic surfactants], and showed that batch variations could be quantified.

A further adaptation of this approach is to use water slurries instead of humidified samples (21). For instance, Schmitt et al. (22) developed a procedure that allows rapid assessment of API-excipient compatibility by studying two developmental drugs formulated with excipients that could undergo a Maillard reaction. Their recommended methodology is to add water (20% w/w) to a binary mixture of API and excipient (100 mg of each) and monitor the power–time signal at 50°C for three days. They note that comparison of the calorimetric results with actual formulation stability showed it was possible to predict relative stability within functional classes, but advised caution because the apparent reaction enthalpies varied three-fold among excipients within the same functional class.

Note here that primary screens, such as those described earlier, are usually designed to *maximize* the chance of seeing an interaction, if there is one, rather than simulate the conditions to be found in the intended formulation; thus, binary samples are usually mixed in a 1:1 mass ratio, ensuring the two materials have equivalent particle sizes, and a high RH may be used to facilitate reaction. Further screens under more representative conditions can then be conducted, if necessary, to see whether any instability observed in the primary screen is relevant to the intended formulation. These steps are summarized in the flow diagram presented in Figure 4.

Although not the focus of this text, it is worth briefly mentioning the role of differential scanning calorimetry (DSC) in primary compatibility screens [for example (23–25)], because its use is more widespread than IC. The basic principle is the same as for IC; the thermal responses for the API and excipient alone are recorded and summed to give the "predicted" response. These data are then compared with those recorded for physical blends of the API with a

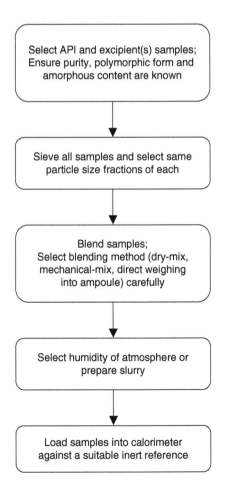

Figure 4 A flow diagram showing the steps involved in designing an active pharmaceutical ingredient (API)–excipient compatibility screen.

range of excipients. Any changes in the expected peaks, or the appearance of new thermal events, indicate a likely incompatibility.

For instance, ibuproxam has been shown to be compatible with corn starch, avicel, and sodium carboxymethylcellulose but incompatible with PEG 4000, stearic acid, and magnesium stearate using DSC data (26). The major benefit of using DSC for primary screens is one of time; an initial judgment on likely incompatibilities can be made in a few minutes. However, the same caveat applies to the interpretation of DSC stability data as for all elevated temperature stability assessments; it must be assumed that the reaction mechanism does not change as a function of temperature. It should also be noted that if one component dissolves in the other, then this would also give rise to an unexpected power that may be misinterpreted as an interaction.

The approach adopted in the studies discussed earlier (both IC and DSC) does not allow the quantification of the amount of degradation (unless complete degradation occurs within the time frame of the experiment, in which case the use of fractional areas allows the calculation of percent degradation with time) nor, without further analysis, does it indicate the exact nature of any degradation processes. It does, however, allow an initial judgment to be made on the likely stability of a compound or mixture, giving the formulator a valuable insight into the stability of the API to be formulated or which excipients are likely to result in a stable product. As such, the use of IC during initial product formulation offers the potential greatly to reduce the number of potential formulations undergoing stability assays, although IC data alone are not currently accepted by regulatory bodies to prove product stability.

One area where IC data can currently be used in regulatory submissions is in the continuing development and product line extension of existing products, whose stability has been assessed and accepted using classical methods. Comparison of the calorimetric data of the new formulation with the old, and the demonstration of equality between them, confirms that the stability profiles of both formulations are the same.

In a similar manner, the effectiveness of new packaging designs can be assessed using IC. Assuming the packaging is designed to isolate the formulation from a specific environmental condition, be it (commonly) oxygen or humidity, the thermal response of the formulation enclosed within the packaging is recorded under that environmental condition; the lack of a detectable heat signal provides good evidence that the packaging will ensure stability. An example of the use of packaging was given in Chapter 4 (27).

The lack of quantitative studies is a direct result of the difficulties that arise in the analysis of complex power–time signals; for instance, a low-enthalpy, fast-rate process may easily appear the same as a high-enthalpy, slow-rate process (28), or competing endothermic and exothermic processes may result in an apparently small thermal response (29). These problems were noted in the study of API-excipient stability using water slurries by Schmitt et al. (22), and made direct stability assessment comparisons between samples difficult. While a slow-rate process may have no consequence on product stability, a fast-rate process may be of some considerable significance and it is clear that an analysis that results in the derivation of a rate constant, reaction enthalpy, or both is necessary for true comparisons to be drawn. Indeed, it has been stated that although IC offers considerable benefits in determining product stability, ultimately it will never replace the need for chemical analysis (30); this area forms a considerable challenge that must be overcome if IC is to become more widely used for pharmaceutical stability assessment.

Stability Assessment of Individual Compounds

If the primary screen conducted in the calorimeter indicates some incompatibility or instability in the sample, and there is a need to understand the process on a molecular level or define its reaction kinetics and/or thermodynamics, then the traditional approach is to conduct further analyzes with complementary analytical techniques. However, it is usually the case that the power–time data so obtained already contain sufficient information that, if a suitable analysis can be undertaken, the process(es) occurring can be quantified with no further experimentation.

For instance, Koenigbauer et al. (31) determined the activation energies for the degradation of several drugs, including phenytoin, triamterene, digoxin, tetracycline, theophylline, and diltiazem, using the initial power rates measured using IC at several elevated temperatures. The results were compared with HPLC data recorded at a single temperature and it was shown that the IC data were more precise. Similarly, Hansen et al. (32) showed that the shelf life of a product, degrading via an autocatalytic reaction was inversely proportional to the rate of heat production during the induction period, using lovastatin as an example.

If there is only one reacting component, then the reaction order will be integral and it is relatively easy to determine the rate constant by replotting the data.

1. If the power signal is constant as a function of time, then the process is zero order. The rate constant can be determined from the power value if the number of moles of material and reaction enthalpy are known [the deflection is equal to $k\,\Delta H\,V$, where V is the volume of sample (Fig. 5)].
2. If a plot of ln(power) versus time is linear, then the process is first order and the rate constant is given by the gradient of the line (Fig. 6).
3. If a plot of power$^{-0.5}$ versus time is linear, then the process is second order and the rate constant is given by the gradient of the line (Fig. 7).

If the data do not appear to be zero, first, or second order, there may be multiple processes occurring or (unusually) the process may have a nonintegral order. In this case, a more complex analysis using some of the methods described in Chapter 3 (such as kinetic modeling, direct calculation, or chemometrics) must be used to discern the reaction order and/or number of reaction steps.

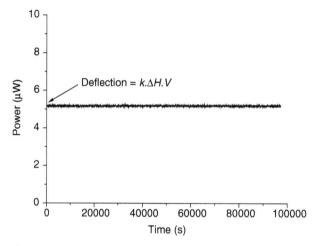

Figure 5   A linear power versus time plot indicates a zero-order process. *Abbreviations*: k, rate constant; ΔH, enthalpy; V, volume of sample.

The following sections discuss applications to real data on the basis of the most common degradation pathways: hydrolysis, oxidation, elevated RH, and photodegradation.

Hydrolysis Reactions

Many pharmaceuticals are susceptible to hydrolysis and, as water is difficult to remove entirely from a formulation, hydrolysis is a common cause of chemical

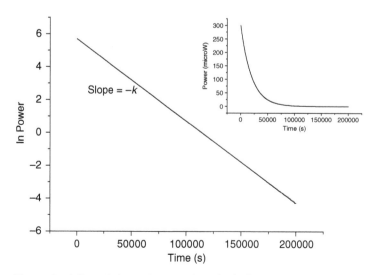

Figure 6   A linear ln(power) versus time plot indicates a first-order process (the raw data are shown on the inset graph). *Abbreviation*: k, rate constant.

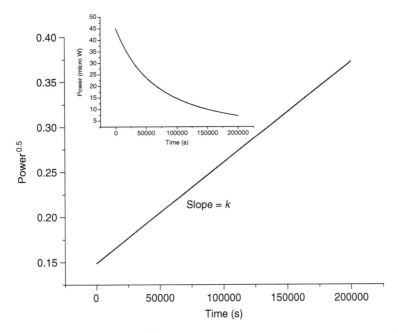

Figure 7 A linear power$^{-0.5}$ versus time plot indicates a second-order process (the raw data are shown on the inset graph) *Abbreviation*: k, rate constant.

instability. There are numerous examples in the literature where hydrolysis reactions have been studied calorimetrically. For instance, the degradation rate of meclofenoxate hydrochloride (MF), which hydrolyses in aqueous solution, has been determined using IC (33). By plotting ln(power) versus time, the degradation rate constants for MF at pH 6.4 and 2.9 were determined as $1.14 \times 10^{-4}/s$ and $9.7 \times 10^{-7}/s$, respectively. Comparison of these data with rate constant values determined using HPLC ($1.29 \times 10^{-4}/s$ and $9.0 \times 10^{-7}/s$) revealed the utility of the calorimetric technique. A similar approach has been used to determine the rate constants for ampicillin degradation in aqueous buffers (34) from pH 2 to 8 and a number of cephalosporins (35).

Kinetic models, such as those described in Chapter 3, have also been applied to systems degrading via hydrolyses, and allow the analysis of reactions progressing via a single step, parallel steps, or consecutive steps. Examples of the analysis of each type of pathway are discussed in the following case studies.

Case study: Single-step Degradation—Aspirin: Aspirin (acetylsalicylic acid) degrades to salicylic acid and acetic acid in aqueous solution, following pseudo first-order kinetics because water is held in excess (36–40). The rate of hydrolysis exhibits pH dependence, being fastest <pH 1.0 (acid catalysis predominates) and >pH 10 (base catalysis predominates); at pH 7.0 and 25°C, the reaction has a rate constant of $3.7 \times 10^{-6}/s$, corresponding to a half-life of 52 hours (41).

The first use of IC to study aspirin hydrolysis in aqueous solution as a function of pH was by Angberg and Nyström (38). In these studies, rate constants were derived from the gradient of ln(power) versus time plots at a series of temperatures (30–50°C). In 0.1 M HCl at 40°C, the rate constant was $9.0 \times 10^{-6}/\text{s}$, increasing to $22.5 \times 10^{-6}/\text{s}$ at 50°C (38), whereas in pH 4.8 acetic acid buffer rate constants of $14 \times 10^{-6}/\text{s}$ and $34.1 \times 10^{-6}/\text{s}$ were determined at 40 and 50°C, respectively (39).

A later study of aspirin hydrolysis by IC was conducted by Skaria and Gaisford (42), who used a kinetic model to fit the data. As the reaction progressed in a single, first-order step, Equation (3) was employed:

$$\frac{dq}{dt} = \Delta H\, v\, k\, [A_o]\, e^{-kt} \tag{3}$$

Typical power–time traces for aspirin hydrolysis in 0.1 M HCl at 25, 40, and 50°C are shown in Figure 8. The fit of each data set to Equation (3) is represented by the open circles. At 25°C, the power signal was very small (the noise being of the order of ± 0.025 μW), yet the data were sufficiently resolved to allow a complete analysis. Similar data sets for aspirin degrading in a pH 5.0 citrate buffer were obtained (data not shown). The average values determined for the rate constants of aspirin hydrolysis from this study, as a function of temperature and pH, are given in Table 1. The values were slightly lower than those observed in the earlier work, but nevertheless the data were compared well.

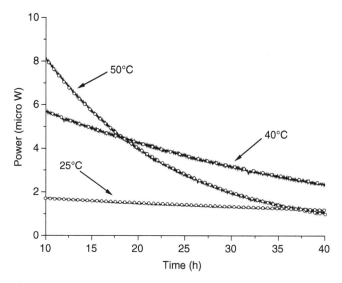

Figure 8   Power–time data for aspirin (0.01 M) in HCl (Hydrogen Chloride) aqueous solution (0.1 M) at 25°, 40°, and 50°C and the fit lines (○) generated by application of Equation (3). *Source*: From Ref. 42.

Table 1   Average Values for the Rate Constants and Reaction Enthalpies for the Hydrolysis of Aspirin in Various Media, Determined by Fitting Power–Time Data to Equation (3)

Temperature (°C)	0.1 M HCl		pH 5.0	
	k $(10^{-6}\,s^{-1})$ ($\pm$SD,n)	ΔH (kJ mol^{-1}) ($\pm$SD,n)	k $(10^{-6}\,s^{-1})$ ($\pm$SD,n)	ΔH (kJ mol^{-1}) ($\pm$SD,n)
25	2.8 $(\pm 0.2, n=6)$	-23.6 $(\pm 2.4, n=6)$	—	—
40	7.4 $(\pm 0.7, n=5)$	-34.4 $(\pm 2.9, n=5)$	12.1 $(\pm 1.0, n=8)$	-24.0 $(\pm 2.2, n=8)$
50	19.9 $(\pm 3.1, n=9)$	-30.3 $(\pm 3.9, n=9)$	29.3 $(\pm 5.4, n=8)$	-25.5 $(\pm 1.9, n=8)$

Source: From Ref. 42.

Angberg and Nyström (38), using the Arrhenius relationship, predicted a rate constant for aspirin degradation at 25°C in 0.1 M HCl of 2.3×10^{-6}/s. Advances in amplifier technology have led to improvements in the detection sensitivity of IC, and Skaria and Gaisford (42) were able to record data directly at 25°C, recording a rate constant of 2.8×10^{-6}/s. Although these values are similar, construction of an Arrhenius plot from the data in Table 1 did not result in a linear fit (Fig. 9), suggesting some changes in reaction mechanism as a function of temperature. This is perhaps to be expected, because aspirin can degrade through one of the four pathways, the predominant pathway being dependent upon pH. Extrapolation from the (admittedly limited number of) high-temperature values gave a predicted rate constant at 25°C of 1.5×10^{-6}/s.

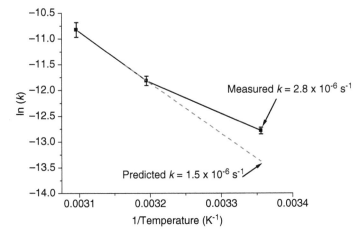

Figure 9   An Arrhenius plot for aspirin rate constants determined in HCl (Hydrogen Chloride) aqueous solution (0.1 M) at 25, 40, and 50°C (*solid points*) and the extrapolation of the high temperature data to 25°C (*dotted line*). *Source*: From Ref. 42.

An important conclusion from this study is that extrapolation of data recorded at elevated temperatures, even over a small range, can give a significantly different predicted rate constant from the experimentally determined rate constant, and such data should be treated with caution.

A further benefit of fitting experimental data to Equation (3) is that the reaction enthalpy can be determined directly (which obviates the need for knowing the initial power signal or letting the reaction run to completion). The enthalpy values determined for aspirin hydrolysis as a function of temperature and pH from Equation (3) are given in Table 1, and agree well with the value stated by Angberg (40) of approximately -29 kJ/mol. However, the enthalpy values determined by Skaria and Gaisford (42). in the higher pH medium are lower than those for degradation in 0.1 M HCl. As the start and end points of the degradation reaction are the same in each case, and the change in enthalpy is independent of the reaction pathway, it would be expected that these values would be identical. However, the calorimetric signal reflects the *net* heat-change from all the processes occurring in the sample cell. As the pK_as of both acids are below five (salicylic acid $\sim$3.0 and acetic acid $\sim$4.8), both will deprotonate in the higher pH medium; this event will be accompanied by an endothermic heat response, lowering the observed net enthalpy. Significant deprotonation will not occur in the lower pH medium and it is notable that these values are much closer to that determined by Angberg (40).

As noted earlier, data fitting also removes the need for reactions to run to completion, and it is interesting to quantify the minimum recording period necessary to return accurate results. Skaria and Gaisford (42) found that as the intensity of the power signal increased, the minimum number of data needed to recover the correct reaction parameters (i.e., within the standard errors specified in Table 1) reduced. At 25°C, it was found that a minimum of five hours of data (following equilibration) were required; this reduced to one hour at 40°C and to 30 minutes at 50°C. Given that the first hour of IC data are usually lost, this means that, even at 25°C, reaction parameters could be determined accurately with just six hours of data.

Case study: Parallel Degradation—Binary Paraben Mixtures: A typical formulated pharmaceutical may well have several independently degrading components and, although the degradation kinetics of the individual components of a medicine may be known, their behavior in combination may be significantly different. It is, therefore, not sufficient simply to know the stability profiles of the individual materials but to understand any synergism between them. The overall degradation profile for a mixture of compounds can be described by summation of the relevant kinetic mechanisms. In terms of a calorimetric analysis, this means data can be modeled by summation of the relevant kinetic expressions (such as those described in Chap. 3).

An example of this approach to analyze compounds degrading in parallel has been reported by Skaria et al. (43) who studied aqueous solutions of binary

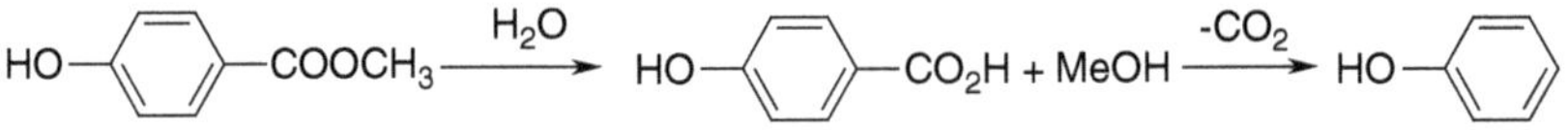

Figure 10 The reaction pathway for the degradation of methyl paraben.

mixtures of selected parabens. The parabens were selected for study because their degradation kinetics are known [and, as discussed in Chaps. 1 and 2, the base-catalyzed hydrolysis of methyl paraben (MP) has been suggested as a test reaction for flow microcalorimeters (44)]; they can degrade through consecutive steps dependent upon solution pH; they provide a model example of parallel degradation; and they have found widespread application in pharmaceuticals, foods, and cosmetics as preservatives (where it is usually the case that at least two parabens are in any particular formulation).

The degradation of MP is represented schematically in Figure 10 (the processes for ethyl paraben (EP) and *n*-propyl paraben (PP) are analogous, the only difference being the alcohol formed). The initial hydrolysis step follows pseudo first-order kinetics and is pH dependent (45), although no literature data are available at pH > 10.59. Depending upon solution pH, *p*-hydroxybenzoic acid (*P*-HBA) can decarboxylate to form phenol, a reaction first reported by Cazeneuve (46), whose data showed the acid to be stable in alkaline but unstable in acidic conditions. Quantitative kinetic data are only available for this reaction between pH 1.26 and 10.59, where it has been shown that the rate falls significantly at the higher pHs and the reaction follows first-order kinetics over four to five half-lives (45).

In their study, Skaria et al. (43) studied paraben degradation in NaOH solution (pH 12.3) to ensure that degradation stopped once *p*-HBA had formed. The power–time traces obtained for the degradation of the three individual parabens are shown in Figure 12. The data were fitted to Equation (3); the fit of each data set to the model is represented by the open circles in Figure 11, and the reaction parameters obtained are shown in Table 2. It is apparent that the degradation rate decreases as the hydrocarbon moiety increases in length, which would be expected on steric grounds, while the reaction enthalpies are roughly equivalent.

Initial experiments on binary systems were conducted on mixtures of MP and PP, as these components had the largest difference in rate constants. Power–time data obtained for MP–PP mixtures could not, as expected, be fitted by Equation (3) (this observation, in the absence of any prior knowledge of the system, would have immediately indicated the likelihood of there being more than one event occurring in the sample); the data were, however, described very well by Equation (4) (which describes two simultaneous first-order decay processes):

$$\frac{dq}{dt} = \Delta H_1 \, v \, k_1 [A_o] e^{-k_1 t} + \Delta H_2 \, v \, k_2 [B_o] e^{-k_2 t} \tag{4}$$

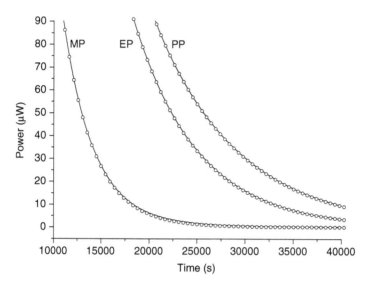

Figure 11 Power–time data for methyl paraben (0.05 M), ethyl paraben (0.05 M), and propyl paraben (0.05 M) in NaOH aqueous solution (0.5 M) at 25°C and the fit lines (○) generated by application of Equation (3). *Abbreviations*: EP, ethyl paraben; MP, methyl paraben; PP, propyl paraben. *Source*: From Ref. 43.

where the subscripts 1 and 2 refer to the individual reaction pathways and $[A_o]$ and $[B_o]$ refer to the initial concentrations of reactants A and B, respectively.

The fit to the data is shown in Figure 12 and the reaction parameters returned are given in Table 3. Interestingly, although the enthalpy values determined from the mixed system data were the same (within error) as those determined when the species were studied individually, both rate constants were lower than expected. The most likely explanation for this observation was that the two reactants present degraded to a common product, p-HBA. The rate of degradation of an individual paraben (MP in this case) is given by:

$$-\frac{d[MP]}{dt} = \frac{d[pHBA]}{dt} \tag{5}$$

Table 2 Average Values for the Rate Constants and Reaction Enthalpies for the Individual Parabens Determined by Fitting Experimental Data to Equation (3)

Ester	$k \ (s^{-1})_{(\pm SD,n)}$	$\Delta H \ (kJ \ mol^{-1})_{(\pm SD,n)}$
Methyl	3.1×10^{-4} $_{(\pm 0.01, n=3)}$	-59.2 $_{(\pm 0.4, n=3)}$
Ethyl	1.5×10^{-4} $_{(\pm 0.01, n=3)}$	-64.4 $_{(\pm 1.3, n=3)}$
n-Propyl	1.2×10^{-4} $_{(\pm 0.01, n=3)}$	-60.1 $_{(\pm 0.3, n=3)}$

Source: From Ref. 43.

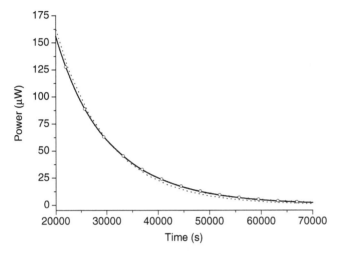

Figure 12 Power-time data for a binary mixture of methyl paraben (0.05 M), and propyl paraben (0.05 M) in NaOH (Sodium Hydroxide) aqueous solution (0.5 M) at 25°C and the fit lines generated by application of Equations (3) (*dotted line*) and (4) [represented by open circles—the fit line to Equation (4) overlays the raw data]. *Source*: From Ref. 43.

where [MP] is the concentration of MP and [pHBA] the concentration of *p*-HBA as a function of time (the rate laws are analogous for the other parabens). In the case of a binary mixture it is clear that two sources contribute to [pHBA]; this must inevitably cause the rates of disappearance of the two reactants to change.

When the authors measured MP degradation in a solution of base containing 0.05 M *p*-HBA (i.e., a solution that already contained one of the degradation products), the measured rate constant was 2.2×10^{-4}/s, a value that was identical to that recorded for MP in the mixed MP–PP system. Furthermore, the total amount of heat released from the binary mixture (16.7 ± 0.6 J) was the same, within error, of the sum of the heats released by the individual components

Table 3 Average Values for the Rate Constants and Reaction Enthalpies for Binary Mixtures of the Parabens Determined by Fitting Experimental Data to Equation (4)

Ester mix	k_1 (s^{-1}) $_{(\pm SD,n)}$	k_2 (s^{-1}) $_{(\pm SD,n)}$	ΔH_1 (kJ mol^{-1}) $_{(\pm SD,n)}$	ΔH_2 (kJ mol^{-1}) $_{(\pm SD,n)}$
MP/EP[a]	2.3×10^{-4} $_{(\pm 0.1, n=8)}$	1.1×10^{-4} $_{(\pm 0.1, n=8)}$	n/a	n/a
MP/PP	2.2×10^{-4} $_{(\pm 0.08, n=9)}$	8.0×10^{-5} $_{(\pm 0.01, n=9)}$	-58.2 $_{(\pm 2.1, n=9)}$	-54.4 $_{(\pm 1.9, n=9)}$
EP/PP[a]	1.2×10^{-4} $_{(\pm 0.1, n=8)}$	8.0×10^{-5} $_{(\pm 0.01, n=8)}$	n/a	n/a

[a]The rate constants for these systems were obtained by fixing the enthalpy values constant.
Abbreviations: EP, ethyl paraben; MP, methyl paraben; PP, *n*-propyl paraben.

(17.9 ± 0.6 J), indicating the same extent of reaction taking place in all cases. When the power–time data recorded for individual samples of MP and PP were summed [producing a data set that differed significantly from that recorded experimentally for the actual binary system (Fig. 13)] and fitted to Equation (4), rate constant values of 2.9×10^{-4} (± 0.1) and 1.1×10^{-4}/s (± 0.1) and enthalpies of -60.9 (± 2.0) and -53.7 (± 2.0) for the methyl and *n*-propyl esters were obtained, respectively, values that are in much better agreement with those presented in Table 2.

The observation that degradation rate constants may differ significantly from those expected when materials are formulated in combination is important; in this case, a shelf life could have been predicted on the basis of the stability data obtained for the individual materials, but the actual shelf life of the product would have been longer. It is clear that the alternative scenario could also have resulted. Other properties of the parabens have been observed to alter when formulated in combination. For instance, mixtures of parabens are more effective as preservatives than the individual parabens (47,48), and this highlights the importance of developing analytical techniques that allow the direct study of heterogeneous samples.

The data for the other two binary mixtures (MP:EP and EP:PP) were found not to be fitted by either Equation (3) or (4), when all the parameters were allowed to vary, but were successfully fitted to Equation (4) when the

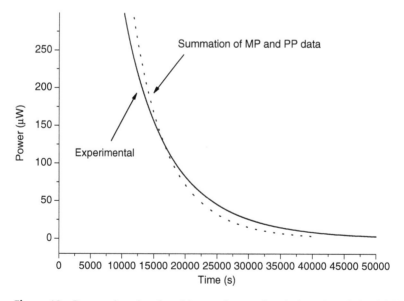

Figure 13 Power–time data for a binary mixture of methyl paraben (MP) (0.05 M) and propyl paraben (PP) (0.05 M) in NaOH (Sodium Hydroxide) aqueous solution (0.5 M) at 25°C (*solid line*) and the power–time trace obtained by summation of the data recorded for MP alone and PP alone (*dotted line*). *Abbreviations*: MP, methyl paraben; PP, propyl paraben. *Source*: From Ref. 43.

enthalpy values (as determined for the individual components) were fixed. As for the MP:PP binary system, the fitting again returned lower than the expected rate constants (Table 3), but again these systems have a common degradation product.

These data serve to show the practical limits of the resolution of the model-fitting technique to real data, and suggest that one rate constant needs to be at least twice the magnitude of the other to enable a successful analysis, assuming approximately equal enthalpies. If "ideal" data for either the MP:EP or EP:PP system are generated (for instance, using MathCad®), then using the data in Table 2 the model fitting successfully recovers the rate constants using Equation (4). This suggests that it is the inherent noise in the data that prevents successful analysis, rather than the parameters being too similar. Analysis using simulated data also showed that if one rate constant is more than three orders of magnitude larger than the other, the model fitting is similarly unable to recover both values, because one process dominates the observed signal.

As noted earlier, under certain conditions, paraben degradation can proceed via two consecutive steps. If, as expected, p-HBA is stable under alkaline conditions then paraben degradation should be described by Equation (3). However, if p-HBA subsequently degrades to phenol, then the data would be better described by a two-step consecutive model. As noted in Chapter 3, such data would be described by:

$$\frac{dq}{dt} = \Delta H_1 \, v \, k_1 \, [A_o] \, e^{-k_1 t} + k_1 \, k_2 \, H_2 \, v [A_o] \left(\frac{e^{-k_1 t} - e^{-k_2 t}}{k_2 - k_1} \right) \qquad (6)$$

where k_1 and k_2 are the rate constants, respectively, and ΔH_1 and ΔH_2 the enthalpies for the two reaction steps, respectively. Power–time data for the degradation of MP were fitted to Equation (6) and showed a better statistical measure of fit than that obtained when the data were fitted to Equation (3) [chi^2 = 0.008 for fit to Equation (6) compared with $\chi^2 = 0.125$ for fit to Equation (3)].

IC provides no direct molecular information and, in the absence of any other supporting data, it has been stated that the best approach to determine reaction mechanisms from calorimetric data is to fit the data to a range of models and select the one which gives the best fit with the fewest variables (49). As such, these results suggest that p-HBA is itself degrading and is not, as suggested by Cazeneuve (46), stable under alkaline conditions. However, the enthalpy value returned by the fitting process is very small ($\sim$0.3 kJ/mol); if the reaction parameters are used to construct the power–time traces for the two steps, it is clear that the degradation to phenol contributes very little to the observed heat-flow (Fig. 14). An alkaline solution of p-HBA gave no detectable heat-flow in the calorimeter over a period of four days; the same solution also showed no detectable change when analyzed by UV spectroscopy over the same time period (p-HBA λ_{max} 280, phenol λ_{max} 287). It may therefore be the case that Equation (6), having more variables, simply gives a better fit by generating artifacts that are not related to the reaction mechanism and, as stated earlier, should be disregarded.

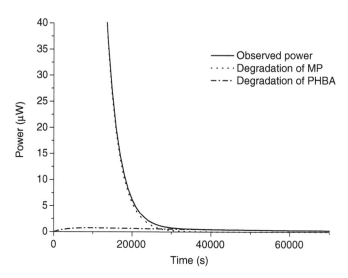

Figure 14 Power–time data for methyl paraben (MP) (0.05 M) in NaOH (Sodium Hydroxide) aqueous solution (0.5 M) at 25°C (*solid line*) and the theoretical contributions to the observed signal from the individual steps as determined using Equation (5); degradation of MP ($\cdots$) and degradation of p-hydroxybenzoic acid (—·—). *Abbreviations*: MP, methyl paraben; p-HBA, p-hydroxybenzoic acid. *Source*: From Ref. 43.

Case study: Consecutive Degradation—Potassium Hydroxylamine Trisulfonate: Gaisford et al. (49) studied the acid-catalyzed hydrolysis of the three-step reaction mechanism:

$$(SO_3)_2N \cdot O \cdot SO_3^{3-} + H_2O \xrightarrow{k_1, \Delta H_1} SO_3 \cdot NH \cdot O \cdot SO_3^{2-} + HSO_4^-$$

$$SO_3 \cdot NH \cdot O \cdot SO_3^{2-} + H_2O \xrightarrow{k_2, \Delta H_2} NH_2 \cdot O \cdot SO_3^- + HSO_4^-$$

$$NH_2 \cdot O \cdot SO_3^- + H_2O \xrightarrow{k_3, \Delta H_3} NH_2 \cdot OH + HSO_4^-$$

Each step proceeds via first-order kinetics and, hence, the power–time data for the overall reaction is described by Equation (7) (derived in Chap. 3):

$$
\begin{aligned}
\frac{dq_{obs}}{dt} = {} & k_1 \Delta H_1 A_o \, e^{-k_1 t} + k_1 k_2 \Delta H_2 A_o \frac{e^{-k_1 t} - e^{-k_2 t}}{k_2 - k_1} \\
& + \Delta H_3 \Bigg[A k_1 k_2 k_3 \Bigg(\frac{e^{-k_1 t}}{(k_2 - k_1)(k_3 - k_1)} + \frac{e^{-k_2 t}}{(k_1 - k_2)(k_3 - k_2)} \\
& + \frac{e^{-k_3 t}}{(k_1 - k_3)(k_2 - k_3)} \Bigg) \Bigg]
\end{aligned}
\tag{7}
$$

Table 4 Kinetic Data for the Acid Catalyzed Hydrolysis of Potassium Hydroxylamine Trisulfonate

Reaction	Temperature (°C)	[HClO$_4$] (M)	$10^4 k$ (s^{-1})	I (M)
Step 1[b]	25.0	0.0005	0.089	0.0011[a]
Step 2[b]	75.0	0.025	1.80	0.025
Step 3[c]	45.0	0.0708	0.042	1.00
Step 3[c]	25.0	1.01	0.085	1.05

[a]Ionic strength adjusted using NaClO$_4$.
[b][trisulfonate] = 0.01 M. *Source*: From Ref. 50.
[c][monosulfonate] = 0.04 M. *Source*: From Ref. 51.

A number of kinetic data were available for the reaction, as it had been the subject of earlier investigation (50,51). The hydrolysis of the trisulfonate ion is the fastest of the three steps, and produces the relatively stable hydroxylamine–NO–disulfonate ion. Potassium hydroxylamine–NO–disulfonate is stable enough to be recrystallized from dilute acid solution. The disulfonate is hydrolyzed slowly in dilute acid solution forming the hydroxylamine–O–sulfonate ion and, eventually, hydroxylamine and hydrogen sulfate. The published kinetic studies were based on titration, using ^{35}S-labeled compounds. The first hydrolytic step was observed to proceed at a much faster rate than the second and third hydrolyses, the final two hydrolyses proceeding at similar slow rates. The kinetic data for the three steps are represented in Table 4.

It is notable that the authors of the earlier work were not able to conduct the assay titrations at 25°C for the second hydrolytic step, as the rate of reaction was too slow, and that a very high concentration of acid was required to enable the study of the third hydrolytic step at 25°C. Using IC, it was possible to study the reaction directly at 25°, 30°, and 35°C.

The calorimetric data were fitted to Equation (7) in order to recover the values of the reaction parameters. Typical fit data are shown in Figure 15. Values for the rate constants returned from the fitting process for each data set are represented in Tables 5 and 6.

Although not recorded under the same conditions, some comparisons of the calorimetric data with the literature data could be made. At 25°C, the published value for k_1 is 8.9×10^{-6}/s, compared with the value determined from the fitting procedure of 4.2×10^{-4}/s, a 100-fold difference in reaction rate. The only other value determined at 25°C is k_3, determined to be 8.5×10^{-6}/s by Candlin and Wilkins (50,51) and 6.7×10^{-6}/s by calorimetry. However, the published value was determined with an added acid concentration of 1.01 M, far in excess of the 0.0005 M acid used in the calorimetric experiment.

It was noted earlier that, when choosing an equation with which to fit calorimetric data, the equation with the fewer or fewest variables but which gives an acceptable fit line should be selected. This applies to reactions where

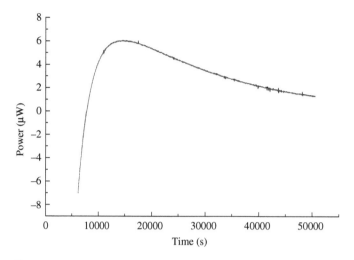

Figure 15 Power–time data for the acid catalyzed hydrolysis of potassium hydroxylamine trisulfonate at 35°C, with 0.0005 M acid, and the fit (*dotted*; overlays experimental data) line obtained after fitting the data to Equation (7). *Source*: From Ref. 52.

the exact mechanism of reaction is unknown. In the case of the hydrolysis of potassium hydroxylamine trisulfonate, it is known that the reaction follows a three-step consecutive reaction scheme and, hence, that was the model chosen for fitting. However, it might seem reasonable to assume that, at such a low added concentration of acid, the third hydrolytic step might not occur during the lifetime of the experiment. If that were the case, then the reaction would follow a two-step consecutive reaction scheme and, following the advice given earlier, should be fitted to a model with fewer variables. Thus, Gaisford (52) fitted the data to Equation (6); a typical fit line is shown in Figure 16. It can be

Table 5 Rate Constant Values for the Acid Catalyzed Hydrolysis of Potassium Hydroxylamine Trisulfonate, Determined by Fitting Power–Time Data to Equation (7)

Temperature (°C)	k_1 (s^{-1})	k_2 (s^{-1})	k_3 (s^{-1})
35	5.58×10^{-4}	4.39×10^{-5}	2.59×10^{-4}
35	4.71×10^{-4}	5.02×10^{-5}	2.05×10^{-4}
35 (average)	5.13×10^{-4}	4.75×10^{-5}	2.32×10^{-4}
30	4.38×10^{-4}	2.93×10^{-5}	5.80×10^{-5}
30	4.39×10^{-4}	2.60×10^{-5}	6.65×10^{-5}
30 (average)	4.38×10^{-4}	2.76×10^{-5}	6.23×10^{-5}
25	4.25×10^{-4}	4.56×10^{-6}	6.71×10^{-6}

Reactions were performed using an acid concentration of 0.0005 M.
Source: From Ref. 52.

Table 6 Rate Constant Values for the Acid Catalyzed Hydrolysis of Potassium Hydroxylamine Trisulfonate, Determined by Fitting Power–Time Data to Equation (7)

Temperature (°C)	k_1 (s^{-1})	k_2 (s^{-1})	k_3 (s^{-1})
35	5.11×10^{-4}	5.43×10^{-5}	5.43×10^{-5}
35	5.05×10^{-4}	5.26×10^{-5}	5.26×10^{-5}
35 (average)	5.08×10^{-4}	5.35×10^{-5}	5.35×10^{-5}
30	4.30×10^{-4}	1.91×10^{-5}	1.83×10^{-5}
30	4.51×10^{-4}	1.79×10^{-5}	1.79×10^{-5}
30 (average)	4.41×10^{-4}	1.85×10^{-5}	1.81×10^{-5}

Reactions were performed using an acid concentration of 0.0002 M.
Source: From Ref. 52.

seen that the two-step consecutive model did not give a good fit to the calorimetric data. The fit is good over the initial section of data, which might be expected because, over this region, the first two reaction steps predominate and the reaction is essentially a two-step process. As the reaction progresses, however, the third step begins to contribute to the observed signal, and the model no longer describes the overall reaction. This is good evidence that the data observed derive from a three-step process.

Using the data returned from the fitting, it was possible to deconvolute the power–time curves for the individual reactions, resulting in the power–time curves for the three individual hydrolytic steps (Fig. 17).

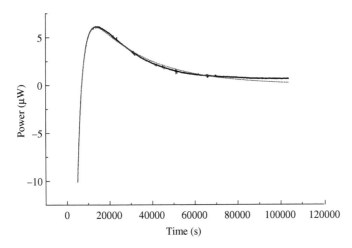

Figure 16 Power–time data for the acid-catalyzed hydrolysis of potassium hydroxylamine trisulfonate at 35°C, with 0.0005 M acid, and the fit (*dotted*) line obtained after fitting the data to Equation (6). *Source*: From Ref. 52.

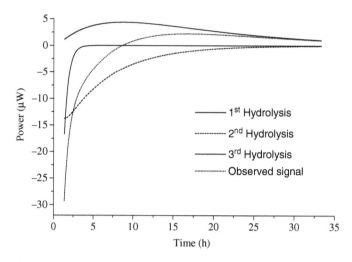

Figure 17 Deconvolution of the observed power–time data obtained for the hydrolysis of potassium hydroxylamine trisulfonate (0.001 M) at 35°C. *Source*: From Ref. 52.

Oxidation

As most calorimeters employ closed ampoules, it is possible to control the local atmosphere in the sample cell and, hence, to study oxidation reactions. Two simple methodologies to determine whether a sample is decomposing via oxidative degradation are

1. to run samples in closed ampoules under either air or nitrogen; differing power–time profiles indicates that oxidation is occurring,
2. for drugs in solution, to add a metal chelating agent (such as ethylenediaminetetraacetic acid [EDTA]), which will remove any free metal ions that may catalyze oxidation; differing power–time profiles in the presence and absence of chelating agent indicates that oxidation is occurring.

For instance, Otsuka et al. (33) investigated the oxidation of DL-α-tocopherol using IC. Samples of the drug (a slightly viscous liquid) were placed in glass ampoules that were left open to the atmosphere in ovens at 50°, 40°, 30°, and 23°C for varying lengths of time. Each sample was then capped before being placed in the calorimeter. Equivalent samples were analyzed using HPLC. First-order rate constants for the samples were determined at each temperature. An Arrhenius plot of the data revealed a linear correlation and an excellent agreement between the HPLC and calorimetric data.

   Another example of the use of calorimetry in this way to study oxidative degradation is provided by Tan et al. (53), who studied tretinoin (all-*trans*-retinoic acid). Tretinoin degrades upon exposure to heat, light, and/or oxygen

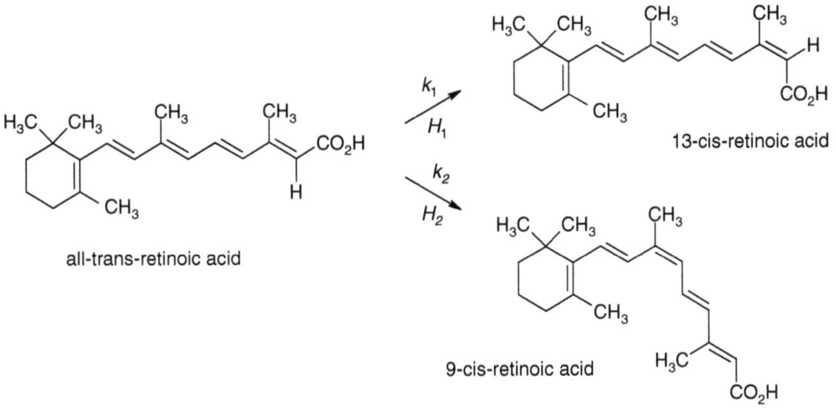

Figure 18   The principal degradation pathways of all-*trans*-retinoic acid.

principally to form one of two geometrical isomers, 13-*cis*-retinoic acid (isotretinoin) or 9-*cis*-retinoic acid (Fig. 18). While isotretinoin exhibits almost the same clinical activity as tretinoin, and indeed is formulated for systemic administration, 9-*cis*-retinoic acid is clinically inactive and, hence, the precise quantification of the rates of formation of the breakdown products is of some considerable importance. It was demonstrated that, under air, the decomposition of isotretinoin was autocatalytic and the decomposition of tretinoin followed zero-order kinetics. In both cases, HPLC analysis showed the appearance of degradation products, although the mechanisms were complex. Under a nitrogen atmosphere, both compounds showed a first-order kinetic event, but simultaneous HPLC analysis showed no evidence of chemical degradation, indicating a physical change was occurring.

A similar approach has shown that solid-state lovastatin degrades in the presence of oxygen (54). In this case, the degradation mechanisms were shown to change between 50°C and 60°C, an important observation in the context of elevated temperature studies, as discussed earlier.

A more in-depth discussion of the analysis of oxidative degradation by IC is provided using ascorbic acid as a case study.

Case study: Ascorbic Acid: Ascorbic acid oxidizes reversibly in aqueous solution to form dehydroascorbic acid, which is subsequently irreversibly hydrolyzed to give diketogluonic acid. Angberg et al. (55) noted that the powers measured for solutions of ascorbic acid in pH 4.9 and 3.9 buffers were greater for those samples that were prepared with an air space in the ampoule, compared with those that were not. Furthermore, if the solution was purged with nitrogen prior to loading, the power dropped nearly to zero. Willson et al. (56) noted, by observing a linear ln(power) versus time plot, that the oxidation was first-order with respect to oxygen concentration. Both studies suggested that the oxygen in the ampoule was exhausted after three to four hours by virtue of the fact that the measured power after this time reduced to zero.

The rate of ascorbic acid degradation is affected by a number of other factors including pH, ascorbic acid concentration, the presence of metal ions, and the presence of antioxidants. Angberg et al. (55) observed that the measured power increased with increasing ascorbic acid concentration up to a certain concentration, whereupon further increases did not increase the power response. It was presumed that the reaction became limited by the oxygen concentration at higher acid concentrations. Willson et al. (56) calculated the rate constants for the oxidation at varying ascorbic acid concentrations and found they were identical, concluding that the oxidation rate is independent of acid concentration. The presence of metal ions (e.g., copper or iron) is known to affect the oxidation rate of compounds in solution and is difficult to control, because only trace amounts are required to catalyze the reaction. Both authors conducted experiments in the presence of EDTA, a metal-chelating ligand, and observed that the measured powers fell substantially, compared with samples run in the absence of the metal binder.

Oxidation can also be investigated for solid samples although, as discussed in Chapter 3, these systems are often difficult to describe in kinetic terms. Using the calorimetric form of the Ng equation [Equation (8), the derivation of which was given in Chap. 3], Willson et al. (57) analyzed the solid-state degradation of ascorbic acid and samples of ascorbic acid with small quantities of added water:

$$\frac{dq}{dt} = A_0 \, k \Delta H \left(\frac{q}{A_0 \Delta H} \right)^m \left(1 - \frac{q}{A_0 \Delta H} \right)^n \tag{8}$$

Dry ascorbic acid degraded with a rate constant of $1.15 \times 10^{-6}/\text{s}$ and an enthalpy of $-199 \, \text{kJ/mol}$. Addition of small aliquots of water (from 20 to 200 μL to 0.5 g of acid) did not significantly change the reaction parameters. At added water amounts of 500 μL and greater, the reaction kinetics were best described by a solution phase model. Between 200 and 500 μL of added water, the data were not described by either model, suggesting a complex, mixed-phase system had been formed.

Elevated Relative Humidity Studies

Proper control of the water content of a formulation, or control over the supply of water to a formulation during storage, is vital if stability is to be assured. This is because water acts in many ways to degrade pharmaceuticals (which include both chemical and physical mechanisms). Water may, for instance, induce degradation via hydrolysis, cause an amorphous sample to recrystallize by lowering its glass transition (T_g) temperature, cause deliquescence of crystals, result in the collapse of a freeze-dried "cake," result in the formation of hydrates, or act as an intermediary between two solid components. It is, therefore, essential to know how a sample will behave in the presence of water and, if necessary, reformulate or repackage the product to ensure there is no loss of potency upon storage.

The use of IC to investigate the stability of pharmaceuticals in the presence of water is possible because experiments can be constructed where the RH in the sample ampoule can be accurately controlled. Thus, an IC approach allows measurement of stability under such conditions *directly* (for most other analytical techniques, the sample must be stressed under an elevated RH prior to measurement). As should also be familiar by now, calorimetric data also permit the quantification of both chemical and physical change, a facility perhaps most important to stability assessment under humid conditions.

The two principal methods by which the RH may be controlled in a calorimetric ampoule have already been discussed at length in Chapter 6, which dealt with amorphous content detection and quantification, and will only be briefly considered here. The reader is referred to Chapter 6 for further details. The simplest method is to place a small glass tube (known as a Durham tube, or hydrostat) holding water or a saturated salt solution directly within an air-tight ampoule containing the sample, the saturated salt solution acting to maintain a specific RH (listed in Table 2 of Chap. 1) (58). The alternative is to use gas perfusion, wherein the RH of a carrier gas, flowing at a constant rate, is controlled as it passes through the sample ampoule (in a typical system, two gas lines are routed into the sample ampoule; one contains dry gas and one contains gas saturated with water vapor—proportional mixing of the two gas lines using mass-flow controllers allows a specific RH in the cell to be maintained).

Before discussing applications, it should be noted that both approaches have advantages and disadvantages and, as always, care must be taken that unexpected or erroneous heat-flows have not arisen simply as a corollary of poor experimental design or execution. Most importantly, the effects of water evaporation and wetting of the ampoule need to be removed in order to analyze data quantitatively. For hydrostat experiments, this is most conveniently achieved by using an equivalent salt solution in a reference cell, connected in opposition to the sample cell. RH perfusion experiments are usually conducted using a stainless steel ampoule containing a suitable quantity of an inert material as a reference and, hence, this correction is not possible. Perhaps the most important consideration is to ensure that the supply of water vapor is sufficient such that the measured sample response is not rate limited. Rate limitation may occur if, for instance, the rate of evaporation of water from the salt reservoir is slower than the rate of water uptake by the sample (in the case of the hydrostat) or if the flow of carrier gas is too slow (in the case of RH perfusion). The hydrostat method gives rise to the generation and use of humidity within the measuring site of the calorimeter and, hence, much (but not all) of the wetting response for the sample is matched by the (almost) equal and opposite response for the generation of the humid air. Consequently, the hydrostat method is very useful for studying samples that change following exposure to humidity, whereas RH perfusion is well suited to the measurement of wetting (and possibly subsequent changes if they are large enough to be visible in the presence of a large wetting response).

Arguably, the principal application of elevated RH calorimetry in the pharmaceutical sciences is the determination of amorphous content in processed powders. However, this topic is the subject of Chapter 6 and shall not be further discussed here.

Gas perfusion calorimetry has been used to measure the interactions between water vapor and a number of amorphous pharmaceutical solids (sucrose, lactose, raffinose, and sodium indomethacin) (59). The power–time data exhibited general trends that aided an explanation of the effect of moisture content on the physical stability on the amorphous form at given storage temperatures. At some RH threshold (RH_m) the data showed a large increase in the energy of interaction between the water vapor and the sample that could not be explained by a phase or morphology change. Below RH_m, water sorption/desorption was reversible; above RH_m hysteresis was noted and water–water interactions dominated the thermal response. Samples stored in an atmosphere below RH_m showed no evidence of instability after several months.

Photostability

There are a number of reasons why an assessment of photostability may be desirable: to quantify the stability of a drug or product under "whole light" conditions, to determine stability as a function of wavelength (which may give mechanistic information on the degradation reaction), to develop packaging materials that can ameliorate the effects of photodegradation, and to estimate the in vivo photosensitizing potential of a drug from its in vitro photostability (photolabile drugs are often used for chemotherapy for instance). While many drugs decompose to some degree during exposure to light, the practical consequences can be serious; nifedipine, an extremely photolabile drug, has a photochemical half-life of only a few minutes (60).

Until recently, there were no established guidelines for photostability testing of drugs. As a consequence, there was no consensus on factors such as the presentation of samples, types of radiation source, spectral exposure levels, exposure time, and dosage-monitoring devices. Such variations made it difficult to correlate photostability results among different research groups. In October 1993 the "Stability Testing of New Drug substances and Products" guidelines were recommended for adoption by the International Conference on Harmonization (ICH). The Europe, Japan and the USA agreed to the harmonization tripartite guideline, which described the procedures for investigating the effects of temperature and humidity without light during stability studies. Subsequently, the ICH guidelines "Photostability Testing of New Drug Substances and Products" have been implemented in Europe (1996) and in the USA and Japan (1997) with a concomitant increasing requirement by regulatory bodies to include photostability data in regulatory submissions. Several attempts have been made to provide an overview of the practical interpretation of the guidelines and offer important insights into satisfying the test requirements (61–64). The most recent guidelines "Photostability testing of new active substances and medicinal products," issued by ICH in 2002, state that light testing should be an integral part of stress testing. However, at present, there is no requirement for the quantification of photostability; a simple pass/fail (stable/unstable) decision is required. This means that photocalorimetry already offers an excellent method to record photostability data and one which, in principle, has the scope also to produce quantitative kinetic data.

The first recorded use of a photocalorimeter was reported by Magee et al. (65), who studied the quantum efficiency of photosynthesis in algae (*Chlorella pyrenoidosa* or *Chlorella vulgaris*). The instrument consisted of a thermopile heat-conduction calorimeter with a small, thin-walled quartz cell mounted in an aluminum container within a double thermostat bath. A multijunction thermocouple was used; one thermopile was used as a sensor for heat flow, the other for the measurement of light transmittance. A 500 W projection lamp was used to introduce light through the front wall of the quartz cylinder. The calculation of the thermal efficiency of photosynthesizing algae was based on the rate of heat absorption by the processes of respiration and photosynthesis. The net amount of radiation dissipated as heat by the algae per unit time was then calculated by measuring the difference between the deflections caused by the two processes. The same apparatus was later used to study the quantum yields and the influence of oxygen in the kinetics of the photobromination of hydrocarbons (66). However, the design of the instrument was convoluted and it lacked long-term stability.

In the late 1960s, Seybold et al. (67) investigated the quantum yield of fluorescence dyes using a semi-adiabatic double calorimeter. One of the calorimetric vessels was charged with fluorescent solution and the other vessel was charged with a black solution to allow the measurement of the transmitted light. This design was later developed resulting in a Dewar calorimetric vessel equipped with quartz windows to allow passage of light from a 75 W Hg lamp source (68,69). A similar instrument was designed for photon flux measurements of solutions, consisting of a Dewar vessel equipped with quartz optical windows, a magnetic stirrer which assured thermal uniformity of the solution, a calibration heater, and a thermistor used as a temperature sensor (70,71). Similarly, Adamson et al. (72) reported a photocalorimeter for the determination of enthalpies of photolysis of *trans*-azobenzene, ferrioxalate and cobaltioxalate ions, hexacarbonylchromium, and decarbonyldihenium.

The determination of the enthalpy change for photoreaction for all of these instruments is simple and has been discussed by Teixiera and Wadsö (73). For an inert reference material, or assuming no photochemical reaction occurs in the sample, all the light entering the calorimeter cell, E_r (J/s), is either absorbed by the sample or by the opaque walls of the calorimetric vessel and is converted into heat:

$$Q_r = E_r t \tag{9}$$

where Q_r is the heat produced by light input and t is the time of irradiation. If, however, the light induces a photochemical reaction in the sample, an additional heat of reaction is observed, over and above of the incident light energy itself:

$$Q_{tot} = Q_r + Q_p \qquad (10)$$

where Q_p is the heat produced by photochemical reaction and Q_{tot} the observed heat measured by the calorimeter. Thus:

$$\Delta H = \frac{(Q_{tot} - Q_r)}{n} \qquad (11)$$

where n is the number of moles of material photodegraded in time t.

In practice, photocalorimetric systems are not so simple because some of the incident radiation is reflected by components of the calorimetric vessel such as stirrers, light guides, and so on and, in the case of transparent vessels, the incident light is transmitted to the surroundings. Similarly, if luminescence occurs, the corresponding energy will be lost to the surroundings. However, the reflection terms are usually negligible and, if nontransparent vessels are used, Equation (11) becomes valid.

More modern instrument designs obviate the need for transparent windows by using optical light guides (either fiber-optic cables or liquid-filled light guides) to introduce light into the ampoules. These enable the calorimetric cell to be fully enclosed in a thermostat and to be made of a single material. Schaarschmidt and Lamprecht (74) first described the use of light guides in photocalorimetry in developing photocalorimeters for the study of living yeast cells. Cooper and Converse (75) transformed an LKB (now Thermometric AB) batch calorimeter into a photocalorimeter by fitting it with quartz optical fiber light guides for their study of the photochemistry of rhodopsin.

Teixeira and Wadsö (76) developed a differential photocalorimeter using two twin calorimeters (i.e., employing four vessels; two samples with two corresponding references). Fiber-optic bundles guided light from a 100 W tungsten lamp through a monochromator, before being split equally into the two sample vessels. One vessel, a stirred perfusion/titration vessel, was used for the measurement of the thermal power during a photochemical reaction. The other served as a photo-inert reference. The differential signal was recorded for each of the two twin calorimetric vessels.

Subsequent work by Dias et al. (77) utilized a similar design, but the two light guides were fed into the sample and reference channels of one twin calorimeter. The reference side is photo-inert, quantitatively transforming the entire incident light. The position of the optical fibers is adjusted to maintain a stable baseline with and without illumination. In such conditions, the same heat output is measured in both vessels and thus the net output is zero.

The first use of photocalorimetry specifically for the assessment of pharmaceutical photostability was by Lehto et al. (78). Their apparatus consisted of a 75 W Xe-arc lamp, which was used to introduce light through a grating monochromator via an assembly of focusing mirrors and a shutter. The light entry was split into two parts through two identical 1 mm optical light cables and introduced into the calorimetric vessels. As in the design discussed earlier, one of the vessels was used as a photocalorimeteric vessel that recorded both the thermal activity of the photosensitive reaction and absorption of the light, whereas the other vessel served as a reference, recording solely the light absorption. The instrument was used for photodegradation studies of photoliable compounds, nifedipine and L-ascorbic acid, at different wavelengths in solution and in the solid-state.

Most recently, the development of a photocalorimeter for pharmaceutical photostability assessment has been pursued by Morris (79) and, subsequently, Dhuna et al. (80) using an instrument based on the design of Lehto et al. (78). This instrument uses a 300 W Xe-arc lamp and liquid light guides to direct light into sample and reference ampoules. The beam is split and focused using an arrangement of mirrors and lenses. This arrangement allows a fine adjustment of the light flux entering both cells, ensuring a zero power signal is obtained with the light on (and no sample loaded).

Stability Assessment of Active Pharmaceutical Ingredient–Excipient Mixtures

As discussed earlier, the principal methodology used to detect API–excipient interactions using IC, is to compare the thermal responses recorded for the active and excipient alone, with that obtained for an active–excipient blend. An unexpected heat-flow in the blend indicates some interaction is occurring. This approach was first suggested by Elmqvist et al. (81) for use in stability testing of explosive components, and a typical pharmaceutical application was shown earlier in Figure 3.

The method can be used with a wide range of drugs and excipients and there is, therefore, no associated method development required for each test subject as there is with, for instance, HPLC analysis. Another advantage is that the experiment requires just 15 hours to complete.

Such a qualitative analysis does not usually indicate the nature of the interaction, but it does allow a rapid assessment of the likely stability of a number of potential formulations. There are few reported studies where IC data of active–excipient blends have been analyzed quantitatively, usually because of the complexity of the data, although this forms perhaps the most exciting potential growth area for the use of IC in pharmaceutical sciences.

The well-documented incompatibility between aspirin and magnesium stearate (82) provides an ideal system to demonstrate the use of IC to investigate API–excipient mixtures. Under an RH of 100%, a binary mixture of aspirin and lactose showed no thermal response whereas a binary mixture of aspirin and magnesium stearate showed a large exothermic signal (Fig. 19) (83). The heat output appeared to follow zero-order kinetics and lasted for several days. Analysis of the aspirin content in the ampoule after the power had returned to zero showed that

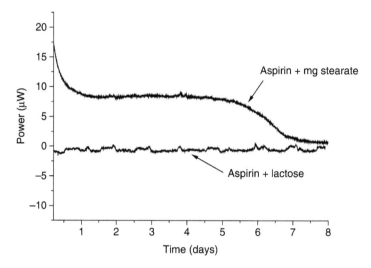

Figure 19 Power–time data for binary mixtures of aspirin/lactose and aspirin/magnesium stearate, both under an relative humidity of 100%. *Source*: From Ref. 83.

the entire drug sample had degraded. It has been suggested that the interaction between the drug and the metal carboxylate arises from a reduction in the melting point of aspirin, generating a liquid layer of drug on the surface of the magnesium stearate particles that accelerates degradation (82).

Interactions between a solid active and a range of excipients, including potato starch, α-lactose-monohydrate, MCC, and talc have been investigated using IC, albeit between elevated temperatures of 60°C to 80°C (84). Large exothermic heat responses were observed for mixtures of drug with MCC, potato starch, and lactose, indicating these systems were unstable. A similar study has looked at the interactions of an active compressed into a tablet (85).

Flow microcalorimetry has been used to study the interaction between heparin sodium and dopamine hydrochloride in two parenteral formulations (86). A significant interaction between the drugs was noted when dextrose was included in the parenteral solution, which was not seen in normal saline formulations.

The effect of menadione and prednisone on the physical stability of various microemulsions has been investigated by IC (87). It was shown that neither drug influenced the stability of the formulation. Gaisford et al. (88) used IC for studying the swelling of PEG-based hydrogels in water. In these experiments, a segment of dry hydrogel (xerogel) was immersed in water (1 mL) and the heat response from swelling recorded. A typical trace showed a two-phase process (Fig. 20), which was ascribed to the hydration of the polymer core and subsequent relaxation of the polymeric network. The break point time between the two processes was observed to reduce with an increase in storage temperature.

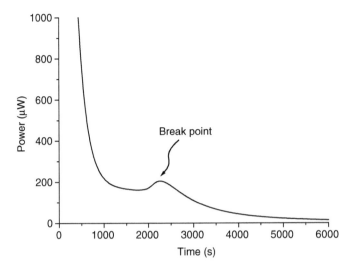

Figure 20 Power–time data for the swelling of a hydrogel segment, showing the break point between hydration of the gel and relaxation of the polymer network. *Source*: From Ref. 88.

Stability Assessment of Formulated Products

The benefits of using IC for stability assessment should be clear by this point, but the benefits for formulated products are perhaps greater than those for any other sample. This is because of the ability of the technique simply to monitor the sample over time. The only issue facing the operator is to select a fraction of the sample for investigation that is representative of the whole (assuming the formulation in its entirety does not fit wholly within the calorimetric ampoule). However, applications in this area are not widespread in the literature, for many of the reasons previously raised in this chapter.

A good example of the use of IC to quantify the stability of an active in a formulation is provided by Zaman et al. (89), who studied the degradation of benzoyl peroxide (BPO) with a number of excipients in an aqueous base. It was found that nearly all of the excipients commonly used to formulate BPO actually stabilized the system (data shown in Table 7); two of the excipients, Monawet and Pluronic 234, actually resulted in degradation rate constants two orders of magnitude below that seen for BPO alone. Only Sipernat was seen to decrease stability, which was ascribed to it being a Lewis acid.

One area where IC has the potential for stability assessment is in hospital pharmacy where, because of pressing clinical need, it is often the case that admixtures are prepared containing multiple actives, formulations are prepared in greater strength than available on prescription, or drugs that are very unstable are formulated immediately prior to administration. In these circumstances, there may be little stability data available to guide the pharmacist.

Table 7 Rate Constant Values for the Degradation of
Benzyol Peroxide in Combination with Various Excipients

Excipient	$k\ (\mathrm{s}^{-1})$
BPO alone	2.0×10^{-7}
2% polytrap	1.04×10^{-7}
1% sipernat	1.11×10^{-6}
0.15% monawet	3.40×10^{-8}
5% propylene glycol; 2.5% glycerine; 0.15% monawet	8.42×10^{-9}
5% propylene glycol; 2.5% glycerine; 2% polytrap	3.41×10^{-7}
Pluronic 234	7.24×10^{-9}

Abbreviation: BPO, benzoyl peroxide.
Source: From Ref. 89.

IC has been used to assess the likelihood of any incompatibilities in a three-component admixture used as a premed in pediatric medicine (90). The admixture and its components [benzyl penicillin (BP), metronidazole (M), and gentamicin (G)] were studied using IC at 37°C. Both M and G are formulated (and hence stable) in solution, and consequently their calorimetric outputs were negligible. BP, however, hydrolyzes rapidly in aqueous solution and is reconstituted prior to administration; following reconstitution, BP has a shelf life of 48 hours. The calorimetric data for BP alone, BP + M, BP + G, and the admixture are shown in Figure 21. The data for BP alone are complex, reflecting its three-step degradation in acidic media, and it was difficult to recover any reaction parameters. However, as the aim of the study was to assess the likelihood of any interactions between the components, the binary and tertiary data were of more importance. It is clear from the data in Figure 21 that the degradation of BP was altered both in the presence of G and M. Moreover, the tertiary mixture showed the greatest difference from the BP-alone data. Although no mechanistic information on the interactions occurring was determined, the data strongly suggested that there were incompatibility issues with that particular admixture and care should be exercised with its use.

A further example of the potential application of IC in clinical pharmacy is provided by a study of busulfan (a powerful agent used in antileukemic therapy) (91). In its commercial formulation (Busilvex®b), busulfan (60 mg) is dissolved in a nonaqueous solvent (10 mL), comprising anhydrous dimethylacetamide (33% v/v) and PEG 400 (PEG400, 67% v/v) (92). The solution is then diluted to 100 mL with 0.9% w/w sodium chloride prior to administration. The undiluted (nonaqueous) product is quite stable and has a shelf life of two years. However,

---

bBusilvex® is a registered trademark of Pierre-Fabre Ltd., Winchester, UK.

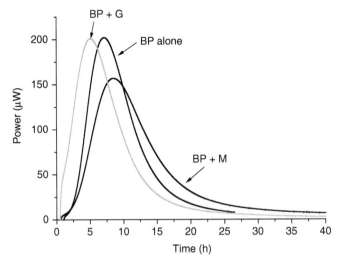

Figure 21 Power–time data for the degradation of benzyl penicillin (BP) alone and for binary mixtures of BP with mentronidazole and BP with gentamicin. *Abbreviations*: BP, benzyl penicillin; G, gentamicin; M, metronidazole. *Source*: From Ref. 90.

following dilution with saline, the stability of the active is dramatically reduced; stability has been demonstrated for 12 hours after reconstitution to 100 mL with 0.9% w/w saline stored at 5 ± 3°C, followed by 3 hours stored at 20 ± 5°C (93). In practice, a regimen of 12 hours at 6°C followed by 3 hours at 25°C is applied. Clinically, the short timescale of this regimen is problematic; often, staff need to work outside of normal working hours to prepare the formulation or the solution is destroyed because it has passed its short stability window. The high cost of the treatment means either scenario is undesirable. IC was used to determine the rate of degradation of the active at a number of elevated temperatures using two reconstitution regimens (dilution to 50 or 100 mL with saline); the data were then extrapolated to the in-use temperatures to determine whether the shelf life could be safely extended.

Extrapolation of the Arrhenius plots allowed the rate constants for busulfan degradation to be determined at 6°C; the values were determined as $3.9 \times 10^{-7}/s$ and $9.7 \times 10^{-7}/s$ for dilution to 50 and 100 mL, respectively. At 25°C, the rate constants were measured directly from the calorimetric data (2.58 and $5.08 \times 10^{-6}/s$ for dilution to 50 and 100 mL, respectively). The rate constant data were then used to calculate the extent of degradation that would be expected for either reconstitution and use regimen using the following relationship:

$$\%\text{Drug} = \left[\frac{A_0 \cdot e^{-k_1 t_1}) \cdot e^{-k_2 t_2}}{A_0} \right] \cdot 100 \qquad (12)$$

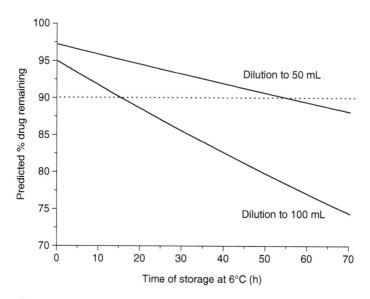

Figure 22   The predicted percentage drug remaining versus storage time at 6°C plots for reconstitution of Busilvex® to either 50 or 100 mL. Note that the calculations assume a three-hour period at 25°C (for infusion), which is why neither line intersects the y-axis at 100%. The dotted line represents the arbitrary 90% drug remaining limit. *Source*: From Ref. 91.

where the subscripts 1 and 2 denote the parameters that relate to storage at temperatures 1 and 2, respectively. The results are shown graphically in Figure 22. The authors note two important points for these data (91). First, for this calculation, a 3-hour period at 25°C is assumed in all cases (this allows time for infusion of the solution and represents the worst-case temperature); the data in Figure 22, therefore, represent the percent drug remaining as a function of storage time at 6°C (and is thus the reason the data do not intercept the y-axis at 100%). Secondly, the calculation assumes an instantaneous increase in the temperature of the sample from 6°C to 25°C; clearly, in practice, there will be a finite period of time over which this temperature rise will occur, which will result in slightly less degradation than predicted. The dotted line in Figure 22 represents the point at which 90% drug remains (which, in practice, is set as the arbitrary maximum amount of drug degradation permissible). This point is reached after 55 hours for dilution to 50 mL and 15 hours for dilution to 100 mL.

The aim of the study was to determine whether the shelf life of busulfan could be extended from its current period of 15 hours (12 hours of which must be at 6°C). The data showed that a three-hour extension would be acceptable, with the current reconstitution regimen, although it is doubtful whether this would offer any practical advantage. Conversely, diluting with saline to 50 mL, rather than 100 mL, had a dramatic effect on stability, increasing the acceptable shelf life to 58 hours (55 hours of which must be at 6°C). Assuming the slight increase in viscosity of the solution, and higher concentration, were not problematic this approach would appear to offer a significant practical benefit, in that a shelf life of 27 hours (24 hours at 6°C followed by 3 hours at 25°C) could be assigned.

SUMMARY

IC offers great potential for stability assessment of pharmaceuticals, but its use is not currently widespread. The methodology for characterizing the stability of individual pharmaceuticals is relatively simple, assuming any degradation pathway is not overly complex, and it is easily possible to determine degradation rate constants. The number of examples of the use of IC in this area is growing. For formulated materials, however, the technique is not so widely used, despite its considerable advantage of being able to study whole samples without any purification steps. Several potential reasons for this present themselves. First, the data recorded may be complex, containing contributions from both chemical and physical changes. Secondly (and fairly obviously), medicines are formulated to be stable; hence, little degradation would be expected in a formulated product. The challenge facing the user of any analytical technique is, therefore, to make a judgment as to whether a zero signal means there is no change in the sample material or the magnitude of any change is below the detection level of the instrument. Finally, IC data give no mechanistic insight into the process under study, and IC data in isolation are not currently accepted in regulatory documents. Developments in data analysis routines and experimental design can address the first issue, while improvements in sensitivity give the confidence to make sound judgments for the second issue. Acceptance of IC data in regulatory submissions will only come through publication of IC data and comparison of the results with classical analytical techniques; this then forms what is perhaps the most pressing area for current research efforts and academic–industrial collaboration.

REFERENCES

1. Hansen LD, Hart RM. The characterization of internal power losses in pacemaker batteries by calorimetry. J Electrochem Soc 1978; 125:842–845.
2. Fung H-O. Chemical kinetics and drug stability. In: Banker GS, ed. Modern Pharmaceutics. 2nd ed., Chapter 6. New York: Dekker, 1990.
3. Laidler KJ. Unconventional applications of the Arrhenius law. J Chem Ed 1972; 49:343–344.
4. Londi NG, Scott MW. Design and application to accelerated stability testing of pharmaceuticals. J Pharm Sci 1965; 54:531–537.
5. Amirjahed AK. Simplified method to study stability of pharmaceutical preparations. J Pharm Sci 1977; 66:785–789.

6. Kornblum SS, Bartley JS. Decarboxylation of *p*-amino-salicylic acid in the solid-state. J Pharm Sci 1964; 53:935–941.
7. Tingstad J, Dudzinski J. Preformulation studies II. Stability of drug substances in solid pharmaceutical systems. J Pharm Sci 1973; 62:1856–1860.
8. Tingstad J, Dudzinski J, Lachman L, Shami E. Simplified method for determining chemical stability of drug substances in pharmaceutical suspensions. J Pharm Sci 1973; 62:1361–1363.
9. Koenigbauer MJ, Brooks SH, Rullo G, Couch RA. Solid-state stability testing of drugs by isothermal calorimetry. Pharm Res 1992; 9:939–944.
10. Carstensen JT. Stability patterns of vitamin A in various pharmaceutical dosage forms. J Pharm Sci 1964; 53:839–840.
11. Garrett ER. The neutral solvolysis of the aspirin anion in aqueous and mixed solvents. J Org Chem 1961; 26:3660–3663.
12. Jun HW, Whitworth CW, Luzzi LA. Decomposition of aspirin in poly-ethylene glycols. J Pharm Sci 1972; 61:1160–1162.
13. Whitworth CW, Jun HW, Luzzi LA. Stability of aspirin in liquid and semi-solid bases I. Substituted and non-substituted polyethylene gylcols. J Pharm Sci 1973; 62:1184–1185.
14. Kornblum SS, Zoglio MA. Pharmaceutical heterogeneous systems I. Hydrolysis of aspirin in combination with tablet lubricants in an aqueous suspension. J Pharm Sci 1967; 56:1569–1575.
15. Kommanaboyina B, Rhodes CT. Trends in stability testing, with emphasis on stability during distribution and storage. Drug Dev Ind Pharm 1999; 25:857–868.
16. Willson RJ. Isothermal microcalarimetry: Theoretical development and experimental studies. Ph.D. thesis, University of Kent at Canterbury, UK, 1995.
17. Pikal MJ. Results of evaluation of the LKB2277 calorimeter for stability testing of pharmaceuticals, Application Note 335, Thermometric AB, Järfälla, Sweden, 1983.
18. Phipps MA, Winnike RA, Long ST, Viscomi F. Excipient compatibility as assessed by isothermal microcalorimetry. J Pharm Pharmacol 1998; 50(S):9.
19. Phipps MA, Mackin LA. Application of isothermal microcalorimetry in solid state drug development. PSTT 2000; 3:9–17.
20. Rowe RC, Parker MD, Bray D. Batch and source variations in excipients—quantification using microcalorimetry. Pharm Tech Eur Feb 1994; 26–30.
21. Schmitt EA, Peck K, Sun Y, Geoffroy JM. Rapid, practical and predictive excipient compatibility screening using isothermal microcalorimetry. Thermochim Acta 2001; 380:175–183.
22. Schmitt EA. Excipient compatibility screening by isothermal calorimetry. 53rd Calorimetry Conference, Midland, Michigan, U.S.A., Aug 9–14, 1998.
23. Bruni G, Amici L, Berbenni V, Marini A, Orlandi A. Drug-excipient compatibility studies. Search of interaction parameters. J Therm Anal Cal 2002; 68:561–573.
24. Mura P, Bettinetti GP, Faucci MT, Manderioli A, Parrini PL. Differential scanning calorimetry in compatibility testing of picotamide with pharmaceutical excipients. Thermochim Acta 1998; 321:59–65.
25. Mura P, Faucci MT, Manderioli A, Furlanetto S, Pinzauti S. Thermal analysis as a screening technique in preformulation studies of picotamide solid dosage forms. Drug Dev Ind Pharm 1998; 24:747–756.
26. Mura P, Faucci MT, Manderioli A, Bramanti G, Ceccarelli L. Compatibility study between ibuproxam and pharmaceutical excipients using differential scanning

calorimetry, hot-stage microscopy and scanning electron microscopy. J Pharmaceut Biomed Anal 1998; 18:151–163.

27. O'Neill MAA. 2006, unpublished data.
28. Buckton G. Applications of isothermal microcalorimetry in the pharmaceutical sciences. Thermochim Acta 1995; 248:117–129.
29. Gaisford S, Buckton G. Potential applications of microcalorimetry for the study of physical processes in pharmaceuticals. Thermochim Acta 2001; 380:185–198.
30. The Pharmaceutical Codex. 12th ed. London: Pharmaceutical Press, 1994.
31. Koenigbauer MJ, Brooks SH, Rullo G, Couch RA. Solid-state stability testing of drugs by isothermal calorimetry. Pharm Res 1992; 9:939–944.
32. Hansen LD, Eatough DJ, Lewis EA, Bergstrom RG, Degraft-Johnson D, Cassidy-Thompson K. Shelf-life prediction from induction period calorimetric measurements on materials undergoing autocatalytic decomposition. Can J Chem 1990; 68:2111–2114.
33. Otsuka T, Yoshioka S, Aso Y, Terao T. Application of microcalorimetry to stability testing of meclofenoxate hydrochloride and *dl-α*-tocopherol. Chem Pharm Bull 1994; 42:130–132.
34. Oliyai R, Lindenbaum S. Stability testing of pharmaceuticals by isothermal heat conduction calorimetry: ampicillin in aqueous solution. Int J Pharm 1991; 73:33–36.
35. Pikal MJ, Dellerman KM. Stability testing of pharmaceuticals by high-sensitivity isothermal calorimetry at 25°C: cephalosporins in the solid and aqueous states. Int J Pharm 1989; 50:233–252.
36. Kelly CA. Determination of the decomposition of aspirin. J Pharm Sci 1970; 59:1053–1079.
37. Connors KA, Amidon GL, Kennon L. Chemical Stability of Pharmaceuticals; A Handbook for Pharmacists. 1st ed. New York: John Wiley and Sons, 1979.
38. Angberg M, Nyström C. Evaluation of heat-conduction microcalorimetry in pharmaceutical stability studies. I. Precision and accuracy for static experiments in glass vials. Acta Pharm Suec 1988; 25:307–320.
39. Angberg M, Nyström C, Castensson S. Evaluation of heat-conduction microcalorimetry in pharmaceutical stability studies. II. Methods to evaluate the microcalorimetric response. Int J Pharm 1990; 61:67–77.
40. Angberg M. Evaluation of isothermal heat-conduction microcalorimetry in pharmaceutical stability studies. Ph.D. dissertation, Uppsala University, Sweden, 1992.
41. Garrett ER. The kinetics of solvolysis of acyl esters of salicylic acid. J Am Chem Soc 1957; 79:3401–3408.
42. Skaria CV, Gaisford S. 2006; unpublished data.
43. Skaria CV, Gaisford S, O'Neill MAA, Buckton G, Beezer AE. Stability assessment of pharmaceuticals by isothermal calorimetry: two component systems. Int J Pharm 2005; 292:127–135.
44. O'Neill MAA, Beezer AE, Labetoulle C, et al. The base catalysed hydrolysis of methyl paraben: a test reaction for microcalorimeters used for determination of both kinetic and thermodynamic parameters. Thermochim Acta 2003; 399:63–71.
45. Sunderland VB, Watts D. Kinetics of the degradation of methyl, ethyl, and n-propyl 4-hydroxybenzoate esters in aqueous solution. Int J Pharm 1984; 19:1–15.
46. Cazeneuve MP. Recherches sur la decomposition des acides-phenols derives du benzene et du naphthalene. Bull Soc Chim Fr 1896; 15:73–82.

47. Littlejohn OM, Husa WJ. The potentizing effect of antimolding agents in syrups. J Am Pharm Assoc Sci Ed 1955; 44:305–308.

48. Schimmel J, Husa WJ. The effect of various preservatives on microorganisms isolated from deteriorated syrups. J Am Pharm Assoc Sci Ed 1956; 45:204–208.

49. Gaisford S, Hills AK, Beezer AE, Mitchell JC. Modelling and fitting of isothermal microcalorimetric data; applications to consecutive reaction schemes. Thermochim Acta 1999; 328:39–45.

50. Candlin JP, Wilkins RG. Sulphur-nitrogen compounds. Part I. The hydrolysis of sulphamate ion in perchloric acid. J Chem Soc 1960; 4236–4241.

51. Candlin JP, Wilkins RG. Sulphur-nitrogen compounds. Part II. The hydrolysis of hydroxylamine trisulphonate and hydroxylamine-NO-disulphonate ions in perchloric acid. J Chem Soc 1961; 3625–3633.

52. Gaisford S. Kinetic and thermodynamic investigations of a series of pharmaceutical excipients. Ph.D. thesis, University of Kent, Canterbury, UK, 1997.

53. Tan X, Meltzer N, Lindenbaum S. Solid-state stability studies of 13-cis-retinoic acid and all-trans-retinoic acid using microcalorimetry and HPLC analysis. Pharm Res 1992; 9:1203–1208.

54. Hansen LD, Lewis EA, Eatough DJ, Bergstrom RG, Degraft-Johnson D. Kinetics of drug decomposition by heat conduction calorimetry. Pharm Res 1989; 6:20–27.

55. Angberg M, Nyström C, Castensson S. Evaluation of heat-conduction microcalorimetry in pharmaceutical stability studies. VII. Oxidation of ascorbic acid in aqueous solution. Int J Pharm 1993; 90:19–33.

56. Willson RJ, Beezer AE, Mitchell JC. A kinetic study of the oxidation of L-ascorbic acid (vitamin C) in solution using an isothermal microcalorimeter. Thermochim Acta 1995; 264:27–40.

57. Willson RJ, Beezer AE, Mitchell JC. Solid-state reactions studied by isothermal microcalorimetry; the solid-state oxidation of ascorbic acid. Int J Pharm 1996; 132:45–51.

58. Nyqvist H. Saturated salt solutions for maintaining specified relative humidities. Int J Pharm Tech Prod Mfr 1993; 4:47–48.

59. Lechuga-Ballesteros D, Bakri A, Miller DP. Microcalorimetric measurement of the interactions between water vapour and amorphous pharmaceutical solids. Pharm Res 2003; 20:308–318.

60. Thoma K, Klimek R. Photostabilization of drugs in dosage forms without protection from packaging and materials. Int J Pharm 1991; 67:169–175.

61. Piechocki JT. Use of actinometry in light-stability studies. Pharm Tech 1993; 17:46–52.

62. Piechocki JT. Light stability studies: a misnomer. Pharm Tech 1994; 18:60–65.

63. Thatcher SR, Mansfield RK, Miller RB, Davis CW, Baertschi SW. "Pharmaceutical photostability" a technical guide and practical interpretation of the ICH guideline and its application to pharmaceutical stability—Part I. Pharm Tech 2001; 25(3):98–110.

64. Thatcher SR, Mansfield RK, Miller RB, Davis CW, Baertschi SW. "Pharmaceutical photostability" A technical guide and practical interpretation of the ICH guideline and its application to pharmaceutical stability—Part II. Pharm Tech 2001; 25(4):58–64.

65. Magee JL, DeWitt TW, Smith EC, Daniels F. A photocalorimeter: The quantum efficiency of photosynthesis in algae. J Am Chem Soc 1939; 61:3529–3533.

66. Magee JL, Daniels F. The heat of photobromination of the phenyl methanes and cinnamic acid, and the influence of oxygen. J Am Chem Soc 1940; 62:2825–2833.

67. Seybold PG, Gouterman M, Callis J. Calorimetric, photometric and lifetime determinations of fluorescence yields of fluorescein dyes. Photochem Photobiol 1969; 9:229–242.

68. Madelli M, Olmsted III J. Calorimetric determination of the 9,10–diphenyl-anthracene fluorescence quantum yield. J Photochem 1977; 7:277–285.

69. Magde D, Brannon JH, Cremers TL, Olmsted III J. Absolute luminescence yield of cresyl violet. A standard for the red. J Phys Chem 1979; 83:696–699.

70. Olmsted III J. Photon flux measurements using calorimetry. Rev Sci Instr 1979; 50:1256–1259.

71. Olmsted III J. Photocalorimetric studies of singlet oxygen reactions. J Am Chem Soc 1980; 102:66–71.

72. Adamson AW, Vogler A, Kunkely H, Wachter R. Photocalorimetry. Enthalpies of photolysis of *trans*-azobenzene, ferrioxalate and cobaltioxalate ions, chromium hexacarbonyl, and dirhenium decarbonyl. J Am Chem Soc 1978; 100:1298–1300.

73. Teixeira C, Wadsö I. Solution photocalorimeters. Netsu Sokutei 1993; 21:29–39.

74. Schaarschmidt B, Lamprecht I. UV-irradiation and measuring of the optical density of microorganisms in a microcalorimeter. Experientia 1973; 29:505–506.

75. Cooper A, Converse CA. Energetics of primary process in visual excitation: Photocalorimetry of rhodopsin in rod outer segment membranes. Biochem 1976; 15:2970–2978.

76. Teixeira C, Wadsö I. A microcalorimetric system for photochemical processes in solution. J Chem Thermodyn 1990; 22:703–713.

77. Dias PB, Teixeira C, Dias AR. Photocalorimetry: photosubstitution of carbonyl by phosphites in [Mn $(\eta^5\text{-}C_5H_4CH_3)(CO_3)$]. J Organomet Chem 1994; 482:111–118.

78. Lehto VP, Salonen J, Laine E. Real time detection of photoreactivity in pharmaceutical solids and solutions with isothermal microcalorimetry. Pharm Res 1999; 16:368–373.

79. Morris AC. Photocalorimetry: Design, development and test considerations. Ph.D. dissertation, University of Greenwich, UK, 2004.

80. Dhuna M, Morris AC, Beezer AE, Clapham D, Connor JC, Gaisford S, O'Neill MAA. Photostability of pharmaceutical materials determined by calorimetry. 60th US Calorimetry Conference, Gaithersburg, U.S.A., 26th June–1st July, 2005.

81. Elmqvist CJ, Lagerkvist PE, Svensson LG. Stability and compatibility testing using a microcalorimetric method. J Hazard Mater 1983; 7:281–290.

82. Mroso PV, Li Wan Po A, Irwin WJ. Solid-state stability of aspirin in the presence of excipients: Kinetic interpretation, modeling and prediction. J Pharm Sci 1982; 71:1096–1101.

83. Potluri K. Use of isothermal microcalorimetry for drug-excipient incompatibility testing. M.Sc. dissertation, University of London, 2003.

84. Selzer T, Radau M, Kreuter J. Use of isothermal heat-conduction microcalorimetry to evaluate stability and excipient compatibility of a solid drug. Int J Pharm 1998; 171:227–241.

85. Selzer T, Radau M, Kreuter J. The use of isothermal heat conduction microcalorimetry to evaluate drug stability in tablets. Int J Pharm 1999; 184:199–206.

86. Pereira-Rosario R, Utamura T, Perrin JH. Interaction of heparin sodium and dopamine hydrochloride in admixtures studied by microcalorimetry. Am J Hosp Pharm 1988; 45:1350–1352.

87. Fubini B, Gasco MR, Gallarate M. Microcalorimetric study of microemulsions as potential drug delivery systems. 2. Evaluation of enthalpy in the presence of drugs. Int J Pharm 1989; 50:213–217.

88. Gaisford S, Buckton G, Forsyth W, Monteith D. Isothermal microcalorimetry as a tool to investigate the swelling and drug release of a PEG-based hydrogel. J Pharm Pharmacol 2000; 52(suppl):304.

89. Zaman F, Beezer AE, Mitchell JC, Clarkson Q, Elliot J, Nisbet M, Davis AF. The stability of benzoyl peroxide formulations determined from isothermal microcalorimetric studies. Int J Pharm 2001; 225:135–143.

90. Gaisford S, O'Neill MAA, Garrett S, Chan K-L. 2006, unpublished data.

91. Gaisford S, O'Neill MAA, Thompson L, Chan K-L. Shelf-life prediction of busulfan I.V. (Busilvex^{TM}) by isothermal calorimetry. Hospital Pharmacist 2006; Accepted for publication, 2006.

92. www.rxlist.com.

93. Busilvex$^®$ data sheet, Pierre Fabre Ltd., Winchester, UK.

8

Future Developments and Areas of Application

INTRODUCTION

The preceding seven chapters have described the main principles of isothermal calorimetry (IC) in the context of pharmaceuticals. Necessarily, such a discussion can be predicated solely on currently available technology, and it should be clear that calorimetric instrumentation lends itself to a wide variety of application areas. However, instrumentation is constantly improving and often the development of new technology opens new areas of application. It is the purpose of this discourse to examine some of the potential areas of application for current calorimetric technology and to consider some of the future technological developments that might underpin the next generation of calorimetric instrumentation. In particular, attention will be paid to measurements made on whole systems, applications to processes scale up and process analytical technology (PAT), and the development of high-throughput screening (HTS) methodologies.

WHOLE SYSTEM MEASUREMENTS

The drive to make measurements directly on "whole" systems stems from a desire to minimize the number of assumptions made when performing analytical measurements. Here, a whole system is taken to mean either the entire sample (if it fits within the calorimetric ampoule) or a fraction of the sample that is entirely representative of the whole. Examples of such measurements were provided in Chapter 5, when approaches to quantify the efficacy of antibiotics against microorganisms were discussed. The drugs were designed to target specific biological targets (and their structures were probably optimized in specific assays against those targets), but the calorimetric measurements were performed on the whole organism. Thus, in those cases, the measurements of efficacy included all the factors involved in the mechanism of the drug, including membrane transport and diffusion, as well as binding to a target. Such measurements are much more representative of the in vivo performance of the drug and afford greater confidence in the ranking of drug efficacies. Indeed, the ability to produce better in vitro–in vivo correlations is one main benefit of this type of measurement.

Although, in principle, the use of IC to make such measurements is straightforward, in practice, the number of applications is limited. In part, this is a result from the (sometimes) complex nature of the data produced; the lack of suitable equipment and experimental methodologies may also be a factor. However, with care, virtually any system can be studied and meaningful conclusions drawn. The subsections subsequently highlight some potential application areas for direct calorimetric measurements and, while they are limited in number, should be illustrative of the type of information that can be obtained.

Measurement of Enthalpies of Transfer

An important event in the sequence of steps that occurs when a drug reaches its target is the partitioning of a drug from one phase to another (hydrophilic to hydrophobic or vice versa). In particular, partitioning into (and out of) membranes is an important parameter to quantify and is relatively easy to measure; either estimates of the partition coefficient are made using solvent models (often octanol and water), or partitioning into micelles or cells is measured directly (the micellar/cellular phase is removed by centrifugation). In either method, drug concentrations in the aqueous phase (usually) are determined spectroscopically, which enables calculation of the partition coefficient.

Calorimetric methods permit the direct study of partitioning, an event that is reported by the enthalpy change as the solute partitions from one phase to the other (the enthalpy of transfer, $\Delta_{trans}H$). It was noted in Chapter 5 that interpretation of events (binding affinities in that case) through equilibrium constants can be aided considerably with complementary thermodynamic data; the same is true here, as the partition coefficient is an equilibrium constant. Hence, knowledge of $\Delta_{trans}H$ can be a valuable asset in understanding the mechanism of partitioning of drugs.

Arnot et al. (1) demonstrated the potential of solution calorimetry for making measurements of $\Delta_{trans}H$ by studying the dissolution of two model compounds, propranolol HCl and mannitol, into buffered media and simulated intestinal fluids (SIFs). The SIFs were selected to represent the fed (FeSSIF) and fasted (FaSSIF) states; both contained bile salt and lipid components that formed mixed micelles. The buffered medium [Hanks' balanced salt solution (HBSS)], on the other hand, did not contain any species capable of forming micelles. The dissolution of the compounds into the three media was always observed to be an endothermic event. However, the enthalpy of solution for propranolol in the SIFs was lower than that observed in HBSS; in addition, the enthalpy of solution was lower in FeSSIF than FaSSIF. No significant changes were observed for the mannitol data. Subtraction of the enthalpy of solution in HBSS from the enthalpy of solution in the SIFs yielded the enthalpy of transfer of propranolol from solvent to the micelles. The $\Delta_{trans}H$ values for FeSSIF and FaSSIF were calculated to be -10.3 and -2.1 kJ/mol, respectively.

In a similar study, Patel et al. (2) used isothermal titration calorimetry (ITC) to measure $\Delta_{trans}H$ for simvastatin into a range (SDS, HTAB, SDCH, and Brij 35) of surfactant micelles. The $\Delta_{trans}H$ data were then compared with the solubility enhancements determined for each surfactant using high performance liquid chromatography assays. It was found that there was a correlation between the free energy of transfer for the drug to each surfactant and the solubility enhancement of that surfactant, so long as there was a favorable free energy of interaction. Although the data set is limited, the results suggested that ITC screening of a range of surfactants against a poorly water soluble drug might allow the selection of the best potential solubilizing surfactants.

Measurement of the Swelling of Polymers, Compacts, and Hydrogels

Many drug delivery systems (DDS) are polymeric, the polymer network forming an ideal matrix for the encapsulation, and subsequent controlled or sustained release, of a drug. For an erodable matrix, following addition to a dissolution medium the DDS will swell (a process involving multiple steps, including hydration, swelling, gelation, and dissolution); as swelling proceeds the drug is released, the rate of release being governed by the rate of swelling. Clearly knowledge of, and control over, the swelling characteristics of a DDS is, thus, essential in order to tailor drug release to a specific pharmacological profile.

The measurement of swelling is not, however, straightforward. Two common approaches are (*i*) to remove the DDS from the dissolution medium periodically and assess the extent of swelling visually (by light microscopy for instance) or (*ii*) to measure the rate of drug release from the DDS (dissolution testing). Neither approach is ideal. In the former, the polymer must be removed from the dissolution medium and excess liquid must be blotted off prior to measurement. This can be hampered both by the shape of the system and if any erosion has occurred. It is also invasive and not practicable for microparticulate systems. In the latter, it must be assumed that drug release from the polymer matrix occurs instantaneously following swelling and is not a rate-limiting step. This being so (often model drugs with high solubilities are used), the method is not a direct measurement of swelling, and so all conclusions drawn about the behavior of the polymer are inferred.

The fact that swelling will be accompanied by a change in heat means that the event is amenable to study by calorimetric methods. Conti et al. (3) have demonstrated how heat conduction solution calorimetry can be used to quantify polymer swelling, using various grades of hydroxypropyl methyl cellulose (HPMC) and sodium carboxymethylcellulose (NaCMC) as model polymers.

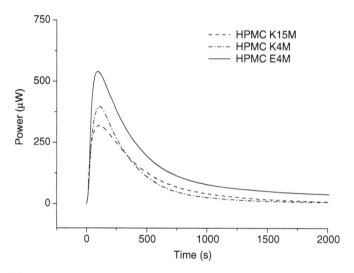

Figure 1 Power–time data for the dissolution of various grades of hydroxypropyl methyl cellulose into water. *Abbreviation*: HPMC, hydroxypropyl methyl cellulose. *Source*: From Ref. 3.

Typical data sets, for the dispersion of the individual polymers, are shown in Figure 1.

The swelling responses of the HPMC grades were different, indicating that solution calorimetry offers a level of sensitivity to swelling, sufficient to discriminate between polymer grades. In order to compare the calorimetric data with other more conventional approaches, Conti et al. (3) converted them to a swelling ratio by plotting q_t/Q versus time (where Q is the total heat output during the experiment and q_t is the heat output to any time t); typical plots of this type are shown in Figure 2.

The data were subsequently interpreted using a derivative of the model first derived by Ritger and Peppas (4):

$$q_t/Q = kt^n \tag{1}$$

where k is a constant and n a parameter, the value of which indicates the mechanism of swelling. Hence, a plot of log q_t/Q versus log t gives a straight line, the slope of which gives the value of n directly. These plots are given in Figure 3. It was noted that the values of n were all greater than 1, which the authors ascribed to the fact that powdered samples were used.

Having established that calorimetry could monitor swelling, Conti et al. (3) then determined the sensitivity of the technique to changes in swelling caused by differences in experimental parameters. Two principal factors were selected: the particle size distribution of the polymer and the pH of the dissolution medium. In the former case, two particle size fractions were obtained by sieving the samples

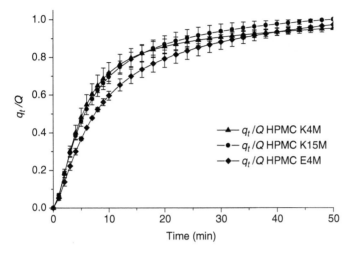

Figure 2 The swelling profiles of various grades of hydroxypropyl methyl cellulose calculated from the power–time data shown in Figure 1. *Abbreviation*: HPMC, hydroxypropyl methyl cellulose. *Source*: From Ref. 3.

(90–125 μm and 45–53 μm). The calorimetric responses for both particle size fractions of E4M (represented in Fig. 4) showed significantly different responses (the values of n were 1.2 and 1.36 for the 90–125 μm and 45–53 μm fractions, respectively).

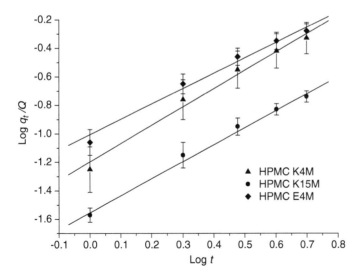

Figure 3 Plots of log q_t/Q versus log t for the three grades of hydroxypropyl methyl cellulose swelling in water. *Abbreviation*: HPMC, hydroxypropyl methyl cellulose. *Source*: From Ref. 3.

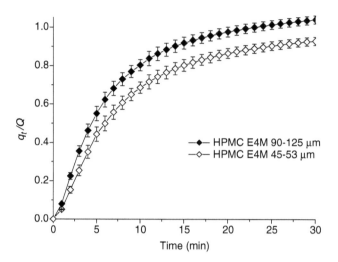

Figure 4   The swelling response for different particle size fractions of hydroxypropyl methyl cellulose E4M in water. *Abbreviation*: HPMC, hydroxypropyl methyl cellulose. *Source*: From Ref. 3.

To examine the effect of pH, swelling experiments were conducted in two buffered systems (pH 2.2 and pH 6.8 McIlvaine citrate buffers). HMPC polymers have no ionizable functional groups and, hence, should give the same swelling responses regardless of pH. This was indeed observed to be the case (Fig. 5). NaCMC, on the

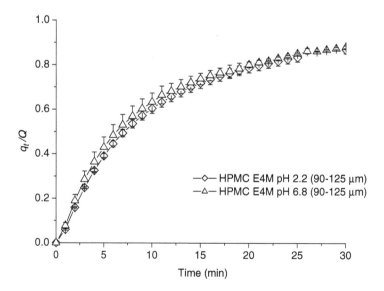

Figure 5   The effect of particle size distribution on the swelling of hydroxypropyl methyl cellulose E4M as a function of buffer pH. *Abbreviation*: HPMC, hydroxypropyl methyl cellulose. *Source*: From Ref. 3.

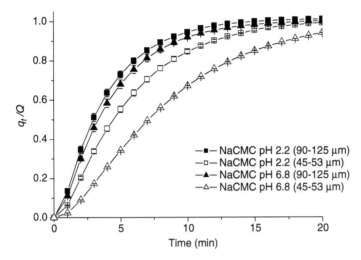

Figure 6 The effect of particle size distribution on the swelling of sodium carboxymethylcellulose as a function of buffer pH. *Abbreviation*: NaCMC, sodium carboxymethylcellulose. *Source*: From Ref. 3.

other hand, has carboxyl functional groups and is affected by pH and, accordingly, its swelling response was observed to alter as a function of pH (Fig. 6).

The authors then studied the swelling response of a binary mixture of two polymers, HPMC E4M and NaCMC, as previous dissolution studies using matrix tablets of the two polymers had indicated some synergistic effects as a function of pH. The calorimetric swelling responses of HPMC, NaCMC, and their 1:1 mixture at pH 2.2 are represented in Figure 7. NaCMC showed the fastest swelling,

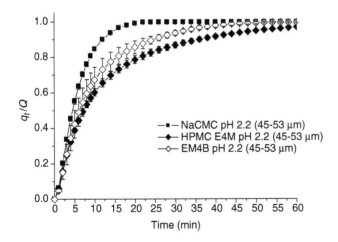

Figure 7 The swelling response for hydroxypropyl methyl cellulose E4M and sodium carboxymethylcellulose, both alone and in combination at pH 2.2. *Abbreviations*: HPMC, hydroxypropyl methyl cellulose; NaCMC, sodium carboxymethylcellulose. *Source*: From Ref. 3.

completion being reached in about 20 minutes. Conversely, the swelling of HPMC was slower, total heat release occurring in about 60 minutes. The mixture of the two polymers behaved in an intermediate way, the thermal profile lying in between those of the individual polymers. The values of n did not show a significant difference, all being higher than 1, again indicating a swelling mechanism mainly associated with the dissolution of the polymer from the surface exposed to the fluid.

To highlight and quantify the differences between the samples, the initial swelling rates were determined by taking the slope of the first part of the curve. It was clear that the hydration and swelling processes of NaCMC were faster compared with the other samples. HPMC E4M showed the lowest rate, and the rate of the sample containing the polymer blend lay in between the rate of each individual material.

Dissolution studies, on drug-loaded tablets prepared by direct compression of the two polymers and a model drug, showed that when formulated in combination, these polymers exhibited a synergistic effect at pH 6.8, releasing loaded drug more slowly than was observed when either polymer was used alone. The time to release 90% of the loaded drug (t_{90}) was 16.4 hour for a HMPC E4M matrix and 10.8 hour for an NaCMC matrix. This value increased to 19.10 hour when the two polymers were formulated in combination. No synergistic effects were noted at lower pH values.

The calorimetric responses of the two polymers both alone and in combination at pH 6.8 (Fig. 8) showed trends, which were similar to the dissolution

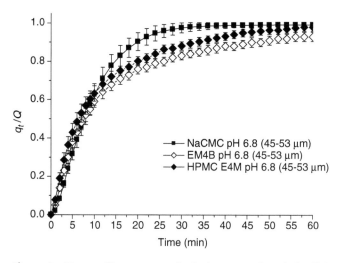

Figure 8 The swelling response for hydroxypropyl methyl cellulose E4M and sodium carboxymethylcellulose, both alone and in combination at pH 6.8. *Abbreviations*: HPMC, hydroxypropyl methyl cellulose; NaCMC, sodium carboxymethylcellulose. *Source*: From Ref. 3.

profiles already noted for matrix tablets containing the same polymers and a model drug. In fact, the time required for the total release of heat by NaCMC was faster, compared with both HPMC and the mixture. This meant that the swelling process, including hydration, swelling, gelling, and dissolution, occurred fastest for this polymer. The presence of carboxylic acid groups on the main chain of NaCMC appeared to confer this polymer with a higher affinity to water compared with HPMC. The higher hydrophilicity of the NaCMC leads to a higher hydration rate and a faster swelling and dissolution process and, when the polymer is formulated in a drug-loaded matrix tablet, to a faster drug release profile. The data for the HPMC E4M sample suggested a slower swelling process compared with NaCMC. This is in agreement with the t_{90} value calculated from dissolution profiles of matrices containing the same polymer noted earlier.

The polymer mixture showed the slowest swelling process whereas the t_{90} values for tablets containing this polymer blend were the longest. This could attest to an interaction between the two polymers that leads to a slower hydration, swelling, gelling, and dissolution process. The n values calculated for these samples were 1.61 (NaCMC), 1.09 (HPMC E4M), and 1.32 (HPMC E4M:NaCMC 1:1 mixture). The swelling mechanism for NaCMC seemed to be highly determined by polymer dissolution from the surface exposed to the fluid. This phenomenon was less marked for HPMC E4M whereas the sample containing the two polymers showed an intermediate n value. No significant differences were noticed in the initial rate of NaCMC and HPMC E4M, but a slower rate was clearly shown by the mixture.

Measurement of Drug Dissolution

Typically, drug dissolution is monitored using a dissolution test conducted to a specific pharmacopoeial specification. However, a major limitation of this approach is that the assessment of drug concentration is invariably conducted by spectroscopic measurement; this then requires filtration of solid particles from the dissolution medium prior to spectroscopic measurement as well as optical clarity of the dissolution medium itself. Though filtration is easy, the requirement for optical clarity means that it is often not possible to conduct dissolution experiments using biologically relevant media (such as SIF).

Noting this as a limitation, Ashby et al. (5) proposed the use of IC as an alternative method for measuring drug dissolution, using a flow-calorimetric arrangement. The tablet (half a tablet for the initial studies) was sited in the calorimetric chamber; dissolution medium flowed into the cell via an inlet in the ampoule's base and exited via an outlet at its top. The calorimeter was operated at 37°C while the temperature of the dissolution medium was maintained at 37°C using an external water bath. A fine mesh over the outlet tube prevented egress of any undissolved material.

Experiments were conducted using placebo tablets and tablets containing phyllocontin in a variety of media (buffer, intralipid/buffer, ensure/buffer). It was noted that the calorimetric power for the dissolution of drug-loaded tablets was ca. three times greater than that for the placebo, demonstrating that the majority of the response arose from drug dissolution. The dissolution profiles were observed to exhibit two distinct regions, both occurring with first-order kinetics. These were ascribed to two depot release phases from the tablet. Plotting ln(power) versus time enabled the derivation of rate constants for these phases. It was noted that the rate constants for the first phase were more affected by changes in the dissolution medium than the second. However, the authors also note that because of the different hydrodynamics of the calorimetric cell compared with conventional dissolution apparatus, the kinetic data obtained from each method were not analogous. Nevertheless, the potential of the calorimetric approach was demonstrated.

In a later study, Buckton et al. (6) extended the use of IC for measuring dissolution by investigating the effect of foodstuffs in the dissolution media. Again, this is an extremely difficult effect to quantify using conventional apparatus because of the requirement for optical clarity. Tetracycline was selected as a model drug because it is a well-known example of a drug whose bioavailability reduces in the presence of milk and milk products and calcium (the drug is removed from solution by chelation). Tetracycline was dissolved in a molten Gelucire base and prepared in a size of 00 gelatin capsule. As in the earlier study, the sample was sited in the calorimetric ampoule (now large enough to accommodate the entire dosage form), and various dissolution media were circulated from an external reservoir.

More complex (than that observed for phyllocontin) release profiles were seen with the dissolution of the Gelucire/gelatin base accounting for a considerable heat. However, it was still possible to derive apparent rate functions (distinct from rate constants because of the complex nature of the release mechanism). Two distinct rate behaviors were seen: for dissolution in HCl solution and Intralipid and for dissolution in calcium solution, milk, and Ensure. These corresponded to fasted and fed states, respectively. The rate functions for release in calcium solution, milk, and Ensure were indistinguishable, suggesting that the ion concentrations were all in excess and that the rate function was controlled by the rate of tetracycline release from the capsule.

Assessment of Photostability

A brief overview of the development of photocalorimetry was given in Chapter 7, and it was seen that the technique was first developed many years ago. However, the technique is still not at the point where commercial instruments are widely available. Dhuna et al. (7) have recently described the design of a compact instrument that is capable of irradiating a sample with either a full spectrum or a specific wavelength of light while simultaneously recording the power output from that sample.

An important feature of their work was the development of the instrument to the point where input of light to the calorimetric ampoule did not result in significant deflection of the baseline from zero. They achieved this by using liquid light guides and a system of shutters to introduce light into both the sample and reference ampoules; in practice, the light intensity in the sample side is set to a specific value whereas the light intensity in the reference side is adjusted until a zero baseline is seen. This methodology, in effect, means that the calorimeter is used as a null adjust balance. Note that this approach does not mean that equal light intensities are introduced to the sample and reference sides. It does mean, however, that light intensities can be specified for the sample side, which should allow comparisons to be made between instruments.

Dhuna et al. (8) have studied a number of solution and solid phase systems using the photocalorimeter. In an attempt to demonstrate the utility of the photo-degradation of 2-nitrobenzaldehyde (2-NB) as a potential test and reference reaction, they showed that what should be a simple zero-order degradation event was, in fact, a composite of multiple processes, including an oxidation event followed by a hydrolysis. A typical data set is shown in Figure 9. A subsequent investigation of the system using pH titration showed that the oxidation event occurred first, causing the nonzero-order behavior seen in the calorimeter. Addition of a metal chelator (EDTA) stopped this oxidation process.

A similar study on solid-state systems demonstrated that it is possible to observe the photodegradation of drugs (8). Here, the degradation of retinoic acid was shown to give a zero-order deflection upon irradiation (Fig. 10).

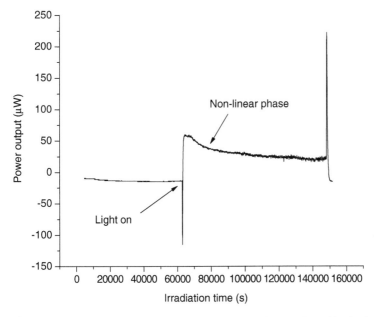

Figure 9 A typical power–time curve for the photodegradation of 2-nitrobenzaldehyde, showing the initial nonlinear phase. *Source*: From Ref. 8.

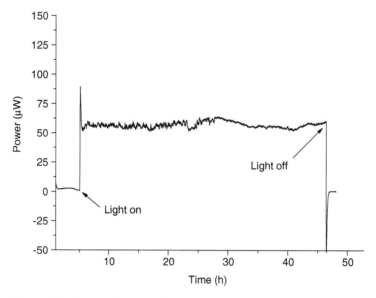

Figure 10 A typical power–time curve for the photodegradation of retinoic acid. *Source*: From Ref. 8.

Although only preliminary, these studies demonstrate that photocalorimetry has much potential for pharmaceutical stability assessment, now that the technical challenges of irradiating samples, without causing an instrumental heat effect, have been overcome. Future applications will see the assessment of the photo-stability of the whole systems, such as tablets and compacts, where the ability of film coats and packaging materials to resist photodegradation will be able to determined quantitatively.

APPLICATIONS TO PROCESS SCALE-UP AND PROCESS ANALYTICAL TECHNOLOGY

Nearly, all of the applications thus far discussed have centered on stages in the preformulation and formulation steps of a pharmaceutical. However, an equally important aspect encompassed by pharmaceutics is the manufacture of medicines. It is not straightforward to assume that a process that produces a specific product (be that a polymorph, an amorphous form, a powder blend, or a specific yield of a biological for instance) on a laboratory scale will do so on an industrial scale. Scale-up of processes is thus a critical aspect if a product is to be made a commercial success.

Recently, there has been a drive by regulatory authorities to introduce PAT. In effect, PAT aims to understand and control the manufacturing process itself, partly by using on-line analytical measurements, rather than (as is currently the case) retrospectively to assess the process by measuring final product quality (note that although the term analytical is used, this is taken in a broad sense to mean the integration of chemical, physical, microbiological, mathematical, and risk-analysis data). In so doing, the manufacturing of pharmaceuticals will move toward more automation, with feedback systems incorporated constantly to adjust the experimental conditions, removing product variability through human error and increasing product yields. There is thus a need to understand the impact of physicochemical parameters on the individual stages in a manufacturing process and also to introduce technologies that are capable of monitoring (and altering) process conditions in real time to maximize product yield.

Many technologies are capable of being adapted to on-line monitoring; however, many are invasive and/or involve sampling. Calorimetric techniques are not, and are hence well suited to PAT applications; reaction calorimetry, for instance, is very commonly used during processing. A review of reaction calorimeters has been published by Marison et al. (9). More recently, heat-balance instruments capable of both monitoring and controlling reaction conditions have been developed.

Process Monitoring

An often critical parameter during manufacturing is temperature; ensuring accurate control of the temperature of a vessel is central to optimize product quality and yield. A process may require a constant temperature or may necessitate one or more temperature steps. One method of temperature control is to use a heat-exchange fluid. The fluid circulates around the reaction vessel and passes through a heat-exchanger. Conventionally, the temperature of the fluid is altered, which then regulates the temperature of the vessel (and its contents). The amount of heat added or removed (Q) is then given by:

$$Q = UAT_d \qquad (2)$$

where U is the heat transfer coefficient, A the area covered by the heat-exchange fluid, and T_d the temperature difference between the vessel and the heat-exchange fluid. In effect, T_d is altered to maintain control.

A recently patented technology (Coflux[®a]) acts in a different way; the circulating heat-exchange fluid is kept at a constant temperature while the heat-exchange area is altered (hence, constant flux). Thus, A is altered to maintain control. This simple idea requires novel instrument design. With Coflux, the reactor vessel is encircled by a series of copper tubes, each distinct from one another. The inlets and outlets of each tube are arranged vertically in two manifolds. The manifolds, carrying the circulating heat-exchange fluid, contain

---

[a]Coflux is a registered trademark of AsheMorris Ltd.

pistons that can rise or fall. This allows the number of copper tubes carrying fluid to be increased or decreased, as necessary, to maintain proper temperature control. This simple system is both more sensitive than conventional systems (which are relatively insensitive because a change in heat content of the system leads to a small change in temperature of the circulating fluid) and allows a faster instrumental response time. Thus, the system can be used as an on-line calorimeter, allowing information on the heat output (or input) of a system to be garnered. Relating the heat change to product quality or yield means that this form of calorimetry offers much for potential for PAT applications.

HIGH-THROUGHPUT SCREENING METHODOLOGIES

There is little doubt that because of the relentless drive by industry to screen an ever increasing number of samples in as fast a time as possible that those technologies that are, or can be made to be, operable in high-throughput mode are likely to become predominant, even if they may not be most apt for the measurement required. While it is patent that calorimetric methods offer many advantages for pharmaceutical systems, it is equally obvious that a major factor limiting their application is the time (and sample quantities) required for measurement. In many instances, it takes 30 minutes or more for a calorimeter to reach equilibrium following loading of a sample; this considerable period of time is the price that must be paid to have sensitive instruments.

Many current technological platforms are, thus, simply not amenable to high-throughput analysis, although some instruments are starting to be produced that ameliorate this issue, usually by increasing the number of sample ampoules that can be accommodated. A good example is provided by the TAM III (Thermometric AB), which houses 48 differential calorimeters in a single thermostatted bath. In the case of the TAM III, reference ampoules are contained below the sample ampoules in each calorimeter; thus, only a single channel is required for each calorimeter, compared with the two needed in a TAM. This increases the number of calorimeters that can be housed in the thermostatting bath. A further example is provided by the development of the micro-Reaction Calorimeter[TM] (Thermal Hazard Technology Company) that positions seven sample chambers around a central common reference.

However, the area with perhaps the most exciting potential for developing calorimetric screening into a technology with true HTS potential are solid-state (integrated circuit) calorimeters.

Integrated Circuit Calorimeters

As implied by the name, integrated circuit calorimeters are constructed directly into a silicon wafer. A typical design will have a thermocouple (or a series of thermocouples) and a heater arranged in a convenient pattern in a silicon wafer with a supporting membrane on the surface, both to provide structural support and to prevent chemical attack of the silicon by the sample to be studied (10).

Commercial calorimeters of this type are available (for instance, from Xensor Integration BV) and can be found in industrial instruments (such as the Setsys range of calorimeters, Setaram).

There are some important design elements to be considered with this type of instrument. First, it is imperative that all heat-flows pass down through the thermocouple(s); this is usually facilitated by the fact that samples placed on the membrane are surrounded by air (which has a very low thermal conductivity). In addition, where liquid samples are used evaporation becomes an important factor (which can easily be of a sufficient magnitude to mask any study interaction); one approach to mitigate this effect is to ensure the air in the calorimetric chamber is saturated with solvent, often by inclusion of a hydrating reservoir.

The integrated circuit calorimeter is sited within a much larger carrier. The carrier, usually made of aluminium, acts as a heat-sink, facilitating the set-up of the calorimeter as a heat conduction unit and increasing the sensitivity of the measurement by preventing temperature and other environmental fluctuations. Utilizing this basic design, it is possible to construct a variety of instrumentation. Lerchner et al. (10) show four possibilities.

1. Liquid batch calorimeter—The sample studied is in solution, the drop (as little as 4 μL) being placed directly on the integrated circuit calorimeter. If so desired, a second solution can be added via a syringe.
2. Temperature scanning calorimeter—The aluminum heat-shield contains electrical heaters, thus turning it into a furnace. Again, the sample is placed directly on the membrane.
3. Flow-through calorimeter—A sample is immobilized on the surface on the calorimeter (such as an enzyme) and a substrate solution flows over the surface. An alternative arrangement would be to investigate gas–solid interactions, whereby a solid sample is placed on the surface of the calorimeter and a gas is routed through the chamber.
4. Mixing reaction calorimeter—Two inlet channels introduce sample to the surface of the calorimeter, where mixing occurs. The mixture is then routed to waste.

Integrated circuit calorimeters have been shown to have many applications, including the measurement of the heats of absorption of enantiomers onto chiral coatings (11), the detection of volatile organic compounds (12), and the heat of absorption of gases onto thin coatings (13). Their use for studying biochemical measurements, using immobilized enzymes on the surface of the calorimeter, has been discussed by van Herwaarden et al. (14). Heat-pulse techniques have been shown to allow the determination of heat capacities for small samples (15). The use of integrated circuit calorimeters operating in DSC mode has also been investigated, with heating rates of up to 10,000 K/s being possible (16).

Calorimetric Arrays

In terms of applicability to HTS, perhaps the most important aspect to integrated circuit designs is that they can be incorporated into an array, allowing true HTS capacity. The use of calorimeter arrays has recently been discussed by Torres et al. (17).

An example of a calorimeter array based on integrated circuit technology nearing production is provided by the MiDiCal® (Vivactis BV, Belgium, Europe). The MiDiCal uses an array of 96 integrated circuit calorimeters arranged in an 8×12 grid. The size of the layout is designed to match that of a 96-well plate (used in conventional HS assays). The array is housed in a humidity-controlled chamber that also includes a liquid-handling system. Thus, biological solutions can be prepared and then titrated directly onto the array. This technology has the sensitivity and capacity to determine extents of reaction for quantities as low as 2.5×10^{-10} mole.

SUMMARY

The development of new calorimetric apparatus, some of which were discussed already, has opened new areas of application and, doubtless, further applications remain to be exploited. It is notable that the two main developments highlighted earlier (heat-balance calorimetry for PAT applications and integrated circuit calorimetry) address divergent needs, one catering for large sample batches and the other for microlitre droplets. Thus, perhaps uniquely among its contemporary analytical peers, calorimetry can be applied to virtually any area of pharmaceutical research and development. As has been noted several times already, the only limitation really is that of ensuring a sample that is representative of the whole (if the whole sample does not fit within the ampoule).

With such advanced technologies already available, maybe the most promising area for the future development of calorimetric techniques lies in methods for data interpretation (such as those discussed in Chap. 3). If successful, the development of such methodologies will allow the increasing study of whole systems; as well as exploiting a major advantage of calorimetric technology, the study of whole systems is preferable because the number of assumptions necessary to interpret the data, and relate them to real systems, is minimized.

There is a certain sense of frustration that the data resulting from the study of the (living) systems investigated by Laviosier and Laplace when they constructed the first "modern" calorimeter remain as qualitative today as they were then. It is ironic that the first systems studied by calorimetry were, in fact, the most complex possible. In many respects, the role of calorimetry has progressed nearly full circle, from these early studies on living animals, through very precise thermodynamic measurements on pure systems, back to in vivo measurements. It is the functionality of calorimetry that allows it to act as a tool for the determination of precise thermodynamic data through to acting as a process monitor that has, and will continue to ensure, its universal application.

REFERENCES

1. Arnot LF, Minet A, Patel N, Royall PG, Forbes B. Solution calorimetry as a tool for investigating drug interaction with intestinal fluid. Thermochim Acta 2004; 419:259–266.
2. Patel R, Buckton G, Gaisford S. Isothermal titration calorimetry to assess the solubility enhancement of simvastatin with a range of surfactants. Thermochim Acta 2006. Submitted.
3. Conti S, Gaisford S, Buckton G, Conte U. The direct measurement of swelling by solution calorimetry. I. Polymers and polymer blends. Thermochim Acta 2006. Submitted.
4. Ritger PL, Peppas NA. A simple equation for description of solute release. II. Fickian and anomalous release from swellable devices. J Control Release 1987; 5:37–42.
5. Ashby L, Beezer AE, Buckton G. In vitro dissolution testing of oral controlled release preparations in the presence of artificial foodstuffs. I. Exploration of alternative methodology: microcalorimetry. Int J Pharm 1989; 51:245–251.
6. Buckton G, Beezer AE, Chatham SM, Patel KK. In vitro dissolution testing of oral controlled release preparations in the presence of artificial foodstuffs. II. Probing drug/food interactions using microcalorimetry. Int J Pharm 1989; 56:151–157.
7. Dhuna M, Morris AC, Beezer AE, et al. Photostability of pharmaceutical materials determined by calorimetry. 60th US Calorimetry Conference, Gaithersburg, USA, June 26 to July 1, 2005.
8. Dhuna M, Gaisford S, O'Neill MAA, Beezer AE, Connor JA, Clapham D. The potential of 2-nitrobenzaldehyde as a chemical test and reference reaction for photocalorimetry. 61st US Calorimetry Conference, Colorado, USA, July 31 to Aug 4, 2006.
9. Marison I, Liu J-S, Ampuero S, Von Stockar U, Schenker B. Biological reaction calorimetry: development of high sensitivity bio-calorimeters. Thermochim Acta 1998; 309:157–173.
10. Lerchner J, Wolf A, Wolf G. Recent developments in integrated circuit calorimetry. J Therm Anal Cal 1999; 57:241–251.
11. Lerchner J, Kirchner R, Seidel J, Waehlisch D, Wolf G. Determination of molar heats of absorption of enantiomers into thin chiral coatings by combined IC calorimetric and microgravimetric (QMB) measurements: 1. IC calorimetric measurement of heats of absorption. Thermochim Acta 2004; 415:27–34.
12. Lerchner J, Caspary D, Wolf G. Calorimetric detection of volatile organic compounds. Sensor and Actuator B 2000; 70:57–66.
13. Caspary D, Schröpfer M, Lerchner J, Wolf G. A high resolution IC-calorimeter for the determination of heats of absorption onto thin coatings. Thermochim Acta 1999; 337:19–26.
14. van Herwaarden AW, Sarro PM, Gardner JW, Bataillard P. Liquid and gas microcalorimeters for (bio)chemical measurements. Sensor and Actuator A 1994; 43: 24–30.
15. Winter W, Höhne GWH. Chip-calorimeter for small samples. Thermochim Acta 2003; 403:43–53.
16. Adamovsky S, Schick C. Ultra-fast isothermal calorimetry using thin film sensors. Thermochim Acta 2004; 415:1–7.
17. Torres FE, Kuhn P, De Bruyker D, Bell AG, Wolkin MV, Peeters E. Enthalpy arrays. Proc Nat Acad Sci USA 2004; 101:9517–9522.

Index